PLANT SECONDARY METABOLITES

Volume 1

Biological and Therapeutic Significance

PLANT SECONDARY METABOLITES

Volume 1

Biological and Therapeutic Significance

Edited by

Mohammed Wasim Siddiqui, PhD

Kamlesh Prasad, PhD

AAP | APPLE ACADEMIC PRESS

Apple Academic Press Inc.
3333 Mistwell Crescent
Oakville, ON L6L 0A2
Canada

Apple Academic Press Inc.
9 Spinnaker Way
Waretown, NJ 08758
USA

First issued in paperback 2021

Exclusive worldwide distribution by CRC Press, a member of Taylor & Francis Group

ISBN 13: 978-1-77463-107-2 (pbk)
ISBN 13: 978-1-77188-352-8 (hbk)

Library and Archives Canada Cataloguing in Publication

Plant secondary metabolites.

Includes bibliographical references and indexes.
Contents: Volume 1. Biological and therapeutic significance / edited by Mohammed Wasim Siddiqui, PhD, Kamlesh Prasad, PhD
Issued in print and electronic formats.
ISBN 978-1-77188-352-8 (v. 1 : hardcover).--ISBN 978-1-77188-353-5 (v. 1 : pdf)
1. Plant metabolites. 2. Plants, Edible--Metabolism. 3. Medicinal plants--Metabolism. 4. Metabolism, Secondary. I. Siddiqui, Mohammed Wasim, author, editor II. Prasad, Kamlesh, author, editor
QK881.P63 2016 572'.42 C2016-904969-8 C2016-904970-1

Library of Congress Cataloging-in-Publication Data

Names: Siddiqui, Mohammed Wasim, editor. | Prasad, Kamlesh, editor.
Title: Plant secondary metabolites / editors, Mohammed Wasim Siddiqui, Kamlesh Prasad.
Other titles: Plant secondary metabolites (Siddiqui)
Description: New Jersey : Apple Academic Press, Inc., [2017-] | Includes bibliographical references and index.
Identifiers: LCCN 2016030295 (print) | LCCN 2016031056 (ebook) | ISBN 9781771883528 (hardcover : alk. paper) | ISBN 9781771883535 (ebook) | ISBN 9781771883535 ()
Subjects: LCSH: Plant metabolites. | Plants, Edible--Metabolism. | Medicinal plants--Metabolism. | Metabolism, Secondary. | MESH: Plants, Edible--metabolism | Plant Extracts--chemistry | Phytochemicals | Plants, Medicinal Classification: LCC QK881 .P5526 2017 (print) | LCC QK881 (ebook) | NLM QK 98.5.A1 | DDC 572/.42--dc23
LC record available at https://lccn.loc.gov/2016030295

Apple Academic Press also publishes its books in a variety of electronic formats. Some content that appears in print may not be available in electronic format. For information about Apple Academic Press products, visit our website at **www.appleacademicpress.com** and the CRC Press website at **www.crcpress.com**

Plant Secondary Metabolites:

Volume 1: Biological and Therapeutic Significance

Editors: Mohammed Wasim Siddiqui, PhD, and Kamlesh Prasad, PhD

Plant Secondary Metabolites:

Volume 2: Stimulation, Extraction, and Utilization

Editors: Mohammed Wasim Siddiqui, PhD, Vasudha Bansal, PhD, and Kamlesh Prasad, PhD

Plant Secondary Metabolites:

Volume 3: Their Roles in Stress Ecophysiology

Editors: Mohammed Wasim Siddiqui, PhD, and Vasudha Bansal, PhD

ABOUT THE EDITORS

Mohammed Wasim Siddiqui, PhD

Dr. Mohammed Wasim Siddiqui is an Assistant Professor and Scientist in the Department of Food Science and Postharvest Technology, Bihar Agricultural University, Sabour, India, and the author or co-author of more than 33 peer-reviewed journal articles, 24 book chapters, and 18 conference papers. He has six edited and one authored books to his credit, published by Elsevier, USA; Springer, USA; CRC Press, USA; and Apple Academic Press, USA. Dr. Siddiqui has established an international peer-reviewed *Journal of Postharvest Technology*. He has been honored to be the Editor-in-Chief of two book series, Postharvest Biology and Technology and Innovations in Horticultural Science, both published by Apple Academic Press, New Jersey, USA, where he is a Senior Acquisitions Editor for Horticultural Science. He is also as an editorial board member of several journals.

Recently, Dr. Siddiqui received the Young Achiever Award 2014 for outstanding research work by the Society for Advancement of Human and Nature (SADHNA), Nauni, Himachal Pradesh, India, where he is also an Honorary Board Member. He has been an active member of organizing committees of several national and international seminars, conferences, and summits.

Dr. Siddiqui acquired BSc (Agriculture) degree from Jawaharlal Nehru Krishi Vishwa Vidyalaya, Jabalpur, India. He received MSc (Horticulture) and PhD (Horticulture) degrees from Bidhan Chandra Krishi Viswavidyalaya, Mohanpur, Nadia, India, with specialization in the Postharvest Technology. He was awarded a Maulana Azad National Fellowship Award from the University Grants Commission, New Delhi, India. He is a member of the Core Research Group at the Bihar Agricultural University (BAU), providing appropriate direction and assisting with sensiting priority of the research. He has received several grants from various funding agencies to carry out his research projects that are associated with postharvest technology and processing aspects of horticultural crops. Dr. Siddiqui is dynamically involved in teaching (graduate and doctorate students) and research, and he has proved himself as an active scientist in the area of Postharvest Technology.

Kamlesh Prasad, PhD

Prof. Kamlesh Prasad is a food technologist with a postgraduate degree in Food Technology from the Central Food Technological Research Institute, Mysore, India. He received a doctorate in Food Technology and Process and Food Engineering from GB Pant University of Agriculture and Technology, Pantnagar, India. His specialization is in fruits and vegetable processing. His research interest includes nondestructive testing, image analysis, nanotechnology, engineering properties of foods, and various related aspects of food processing.

Presently, he is a Professor in the Department of Food Engineering and Technology, Sant Longowal Institute of Engineering and Technology, Longowal, India. As an author or co-author, he has published more than 100 research papers in national and international journals of repute. He has five authored books and two edited books, and has contributed several chapters in books, published by leading presses of India and abroad. He has conducted a short-term course on "soft computing in process and product optimization" and chaired several conferences.

He is an active board member of the Agricultural and Food Management Institute (AFMI), Mysore, India. He is also the reviewer of many international and national journals and has edited many books and proceedings. He is associated with many reputed professional societies such as ASABE, MSI, ISNT, AFST (I), NSI, PAS, USI, ICC, and DTS (I). He is recognized as one of 2000 Outstanding Intellectuals of the 21st Century and one of the Top 100 Educators in 2010 by the International Biographical Centre, Cambridge, England. Also, his biography is published by Asia/Pacific International Biographical Centre, New Delhi and Who's Who in Science and Engineering by Marquis Who's Who, USA. As a co-coordinator, he has engaged himself in bringing the World Bank project under technical education quality improvement program (TEQIP) project to the institute for technical reform in education. He is involved in the process of social reform and in providing justice to deprived segments of society.

DEDICATION

This Book

Is

Affectionately Dedicated

to Our Beloved Parents

CONTENTS

LIST OF CONTRIBUTORS

Ankita Anu
Department of Horticulture, Bihar Agricultural College, Bhagalpur, Bihar, India.

Vasudha Bansal
Department of Civil & Environmental Engineering, Hanyang University, 22 Wangsimni-Ro, Seoul 133 791, South Korea.

S. K. Chaturvedi
Indian Institute of Pulse Research, Kanpur 208 024, India.

A. K. Chaudhary
ICAR RCER Regional Centre, Darbhanga 846 005, India.

P. S. Daris
Plant Cell Biotechnology, CSIR—Central Food Technological Research Institute, Mysore 570 020, Karnataka, India.

Idayu Muhamad Ida
Bioprocess Engineering Department, Faculty of Chemical Engineering, Universiti Teknologi Malaysia, 81310 Johor Bahru, Johor, Malaysia.

Azly Zahan Khairul
Section of Bioengineering Technology, Malaysian Institute of Chemical and Bioengineering Technology, Universiti Kuala Lumpur, Lot 1988, Bandar Vendor Taboh Naning, 78000 Alor Gajah, Melaka, Malaysia.

Pankaj Kumar
Department of Molecular Biology and Genetic Engineering, Bihar Agricultural College, Sabour, Bhagalpur, Bihar, India.

K. G. Mallikarjuna
Plant Cell Biotechnology, CSIR—Central Food Technological Research Institute, Mysore 570 020, Karnataka, India.

V. S. Neeha
Department of Food Engineering and Technology, SLIET, Longowal 148 106, Punjab, India.

Pa'e Norhayati
IJN-UTM Cardio Engineering Centre, V01 FBME, Universiti Teknologi Malaysia, 81310 Johor Bahru, Johor, Malaysia.

Bishun Deo Prasad
Department of Molecular Biology and Genetic Engineering, Bihar Agricultural College, Sabour, Bhagalpur, Bihar, India. E-mail: dev.bishnu@gmail.com.

K. K. Prasad
Department of Gastroenterology, Post Graduate Institute of Medical Education and Research, Chandigarh 160 012, India. E-mail: profkprasad@gmail.com.

Kamlesh Prasad
Department of Food Engineering and Technology, SLIET, Longowal 148 106, Punjab, India.

P. Ratnakumar
National Institute of Abiotic Stress Management (NIASM), Baramati, Pune 413 115, Maharashtra, India.

Sangita Sahni
Department of Plant Pathology, T.C.A., Dholi, RAU, Pusa, Samastipur, Bihar, India.

C. V. Sameer
International Crops Research Institute for the Semi-Arid Tropics (ICRISAT), Patancheru, Telangana, India.

R. Sarada
Plant Cell Biotechnology, CSIR—Central Food Technological Research Institute, Mysore 570 020, Karnataka, India. E-mail: sarada_ravi@yahoo.com.

A. B. Sharangi
Department of Spices and Plantation Crops, Faculty of Horticulture, Bidhan Chandra Krishi Viswavidyalaya, Agricultural University, Nadia, Mohanpur 741252, West Bengal, India. E-mail: dr_absharangi@yahoo.co.in.

Ashraf Mahdy Sharoba
Food Science Department, Faculty of Agriculture, Benha University, Khalifa, Egypt.

Mohammed Wasim Siddiqui
Department of Food Science and Postharvest Technology, Bihar Agricultural University, Sabour, Bhagalpur 813 210, Bihar, India.

R. Singh
Department of Gastroenterology, Post Graduate Institute of Medical Education and Research, Chandigarh 160 012, India.

Ravi S. Singh
Department of Plant Breeding and Genetics, Bihar Agricultural University, Sabour, Bhagalpur 813 210, Bihar, India

Rafat Sultana
Department of Plant Breeding and Genetics, Bihar Agricultural University, Sabour, Bhagalpur 813 210, Bihar, India. E-mail: rafat.hayat@gmail.com.

Nidhi Verma
National Bureau of Plant Genetic Resources, New Delhi, India.

Vidyashankar
Plant Cell Biotechnology, CSIR—Central Food Technological Research Institute, Mysore 570 020, Karnataka, India.

LIST OF ABBREVIATIONS

AA	arachidonic acid
ALA	alpha-linolenic acid
AM	*Aegle marmelos*
AMD	age-related macular degeneration
ANF	antinutritional factor
APC	allophycocyanin
BHT	butylated hydroxyl toluene
BLG	β-lacto globulin
BSG	basil seed gum
CAM	complementary and alternative medicine
CMC	carboxymethyl cellulose
COX2	cyclooxygenase 2
CSTR	continuous stirrer tank reactor
CVDs	cardiovascular disease
DHA	docosahexaenoic acid
DM	diabetes mellitus
DM	dry matter
DPPH	2,2-diphenyl-1-picryl-hydrazyl
EFSA	European Food Safety Authority
EOS	essential oils
EPA	eicosapentaenoic acid
FAO	Food and Agricultural Organization
FDA	Food and Drug Administration
G6PD	glucose-6-P-dehydrogenase
GC	gas chromatography
GLA	gamma-linolenic acid
GSH	glutathione
HDL	high-density lipoprotein
IBD	inflammatory bowel disease
IBS	inflammatory bowel syndrome
LA	linoleic acid
LDL	low-density lipoprotein
LPSS	low-pressure superheated steam

ME	methyl eugenol
MI	myocardial infarction
MS	mass spectroscopy
NAFLD	nonalcoholic fatty liver disease
NCD	non-communicable diseases
NFHS	National Family Health Survey
NO	nitric oxide
NPAA	nonprotein amino acid
NPs	natural products
NSCLC	non-small-cell lung cancer
PBPs	phycobiliproteins
PBRs	photobioreactors
PC	phycocyanin
PE	phycoerythrin
PUFAs	polyunsaturated fatty acids
RBC	red blood cell
RDA	Recommended Dietary Allowances
RDR	rotary discs reactor
ROS	reactive oxygen species
SFE	supercritical fluid extraction
SMs	secondary metabolites
STZ	streptozotocin
TAG	triacylglycerol
TIU	trypsin inhibitor unit
TNF	tumor necrosis factor
UC	ulcerative colitis
WBCs	white blood cells
WHO	World Health Organization

PREFACE

Plant-based formulations have been used since the medieval time of the human history against the ailments of living beings. Taking the evidence of natural functional components as secondary metabolites present in the plants, various medicines have been developed all over the world. Also, with the changing scenario of consumer perspective, there is more inclination toward natural healing compounds today. Owing to that, health supplements such as food additives, nutraceuticals, taste enhancers, and flavor retainers have engulfed the commercial food market by focusing and highlighting the natural ingredients of bioactive components present in them.

This book, *Plant Secondary Metabolites: Volume 1: Biological and Therapeutic Significance,* includes chapters based on the curative and/or therapeutic role of secondary metabolites present in different natural food groups. Efforts have been made not to just showcase the applications of herbal-based group but also to include the effective utility of other plant-based food categories as well. In addition to the clinical role of secondary metabolites, other natural sources like microalgae and bacterial cellulose are also presented as the efficacious source of functional components.

The chapters included in this book demonstrate the innovative and exploratory potential applications of secondary metabolites toward nutritional enrichment of food as well as their beneficial aspects. Chapter 1 illustrates microalgae as the source of nutritional and clinical metabolites; Chapter 2 covers the role of medicinal plants in the prevention of several ailments; Chapter 3 showcases the role of basil and the strength of its functional compounds; Chapter 4 explores the significance of spices as the medium of bioactive compounds; Chapter 5 discusses the production, processing, and medicinal prospective of bacterial cellulose; Chapter 6 describes the different therapeutic significance of horticultural crops; Chapter 7 is comprised of the details of pulses and their composition of metabolites for the sustenance of human and animal health; and Chapter 8 embraces the health benefits and functional components of Spirulina.

The editors and authors have contributed the chapters in order to delineate the natural components in a summarized manner. It is hoped that food scientists and chemists will find this book extremely useful. This book will also render an overview to the technical food industry for converting the functional compounds present in vivid nutritional sources to novel products.

ACKNOWLEDGMENTS

It was almost impossible to express the deepest sense of veneration to all without whose precious exhortation this book project could not be completed. At the onset of the acknowledgment, we ascribe all glory to the gracious "Almighty God" from whom all blessings come. We would like to thank for His blessing to write this book.

With a profound and unfading sense of gratitude, we sincerely thank the Bihar Agricultural University (BAU), India, CSIR—Central Scientific Instruments Organisation (CSIO), India, and Sant Longowal Institute of Engineering & Technology (SLIET), India, for providing us the opportunity and facilities to execute such an exciting project, and for supporting us toward our research and other intellectual activities around the globe. We convey special thanks to our colleagues and other research team members for their support and encouragement for helping us in every footstep to accomplish this venture. We would like to thank Mr. Ashish Kumar, Ms. Sandy Jones Sickels, and Mr. Rakesh Kumar of Apple Academic Press for their continuous support to complete the project.

Our vocabulary will remain insufficient in expressing our indebtedness to our beloved parents and family members for their infinitive love, cordial affection, incessant inspiration, and silent prayer to "God" for our well-being and confidence.

CHAPTER 1

MICROALGAE AS A SOURCE OF NUTRITIONAL AND THERAPEUTIC METABOLITES

SRIVATSAN VIDYASHANKAR, P. SIMON DARIS, K. G. MALLIKARJUNA, and RAVI SARADA*

Plant Cell Biotechnology, CSIR—Central Food Technological Research Institute, Mysore 570020, Karnataka, India

**Corresponding author, Tel.: +91 821 2516501; fax: + 91 821 2517233; E-mail: sarada_ravi@yahoo.com*

CONTENTS

ABSTRACT

Microalgae are polyphyletic group of photosynthetic organisms that convert solar energy and CO_2 to myriad molecules of high commercial value. Microalgae are the primary producers of the aquatic ecosystems and are ubiquitous in their distribution. They exist in different habitats such as freshwater, marine, brackish, polar regions, acidic/alkali waters, hot water springs, and nonpotable wastewaters. The ability of the microalgal forms to survive in these extreme environments could be attributed to their adaptive mechanisms, which involves production of secondary metabolites. These metabolites such as carotenoids, phycobiliproteins, and polyunsaturated fatty acids have high commercial importance as nutraceuticals. The present chapter discusses the various aspects of secondary metabolite production such as microalgal biodiversity, production systems, extraction of these secondary metabolites, and their applications.

1.1 INTRODUCTION

Microalgae are photosynthetic microorganisms that convert light energy and CO_2 into wide array of chemicals with potential industrial applications. They are the primary producers in the aquatic ecosystem and are ubiquitous in distribution including freshwater, brackish, and marine environments. Microalgae are polyphyletic in their origin with multiple parallel lineage comprising mainly of eukaryotic species. However, cyanobacteria a prokaryotic lineage is generally included with algae owing to the similarity in their photosynthetic and reproductive behavior (Barsanti & Gualiteri, 2006). Microalgae have been advocated as potential source of high value nutraceuticals such as pigments (Pangestuti and Kim, 2011), antioxidants, and polyunsaturated fatty acids (PUFAs) (Spolaore et al., 2006). The primary advantage in industrial exploitation of microalgae for production of these nutraceuticals is that these organisms are photosynthetic in nature and can be cultivated in nonarable land using nonpotable waters (Chisti, 2007). The photosynthetic efficiency and surface area productivity of microalgae are 3–4 times higher compared to terrestrial crop plants (Rittmann, 2008; Larkum, 2010). Further, the ease of scalability and continuous biomass production throughout a year gives microalgae an edge over other industrial crops.

Historically, many countries have been consuming algae as food or use them in folk medicine. Microalgae such as *Spirulina* have also had a long history in both human and animal nutrition. *Spirulina platensis* was consumed by the native population of the sub-Saharan region of Kanem, northeast of Lake Chad, and Belgium (Dangeard, 1940; Leonard & Compere, 1967). The nutritional benefits of microalgae are well exploited in aquaculture industry as source of primary live feed for feeding commercially important fish, crustaceans, and molluscs (Muller-Feuga, 2004). Microalgae supply the nutritional requirements of these marine organisms such as carotenoids and fatty acids and support their early growth (larvae and juvenile) stages mainly stress tolerance (Ibañez et al., 2012; Brett & Muller-Navarra, 1997). Further, certain microalgae such as *Spirulina* sp. and *Chlorella* sp. have been promoted as single cell proteins owing to their high biomass protein content (Becker, 2004). The list of commonly utilized algae for production of high value nutraceuticals are presented in Table 1.1.

TABLE 1.1 Algal Sources of Nutritionally Important Products.

Nutritionally important products	Algal source
Whole biomass—food supplements, SCPs, protein source	*Spirulina*, *Chlorella*, Seaweeds—*Palmaria palmata*, *Porphyra tenera*, *Ulva* sp., *Euchema*, etc.
Lipids/fatty acids, PUFAs	*Spirulina* (gamma-linolenic acid), *Porphyridium cruentum* (arachidonic acid), *Isochrysis* sp., *Pavlova* sp., *Nannochloropsis* sp. (eicosapentenoic acid), *Schizochytrium* sp., *Crypthecodinium* sp. (docosahexenoic acid)
Natural food colorants and carotenoids	*Dunaliella* (β-carotene), *Haematococcus pluvialis* (astaxanthin), Green microalgae (lutein), *Spirulina* (phycobilin–C-phycocyanin)
Dietary fibers	*Chlorella* sp., *Undaria pinnatifida* (wakame), *Ulva* sp. (sea lettuce), *P. tenera* (nori)
Minerals	*Spirulina* (iron), Brown seaweeds—*Fucus* sp. (iodine)
Vitamins	*Spirulina* (vitamin B_{12}), *Ascophyllum*, *Fucus* (vitamin E)
Hydrocolloids—carragenan, agar, alginates, etc.	*Kappaphycus alvarezii*, *Chondrus crispus*, *Euchema* sp., *Gelidium* sp., *Gracilaria* sp., *Laminaria* sp., *Ascophyllum* sp., *Macrocystis* sp.

Source: Spolaore et al. (2006), Ravishankar et al. (2008, 2012), Lee and Marino (2010).

Microalgae accumulate these nutraceuticals as an adaptive response to fluctuating environmental and cultivation conditions. Some of the common environmental fluctuations that microalgae face are high light intensity (photooxidation), nutrient fluxes, high salinity (salt stress), hydrodynamic fluctuations, wide temperature and pH variations, and others. These fluctuations induce stressful environmental conditions that include a phenomenon called "oxidative stress" which leads to accumulation of reactive oxygen species (ROS) that hamper the normal physiology of the organism. The ROS include wide range of oxygen free radicals such as superoxide radicals, hydroxyl radicals, hydrogen peroxide, peroxyl radicals, and others (Mallick & Mohn, 2000). Microalgae accumulate antioxidants as major class of secondary metabolites to prevent the inhibitory effects of stressful environmental conditions mainly to counter the oxidative stress. Some of the commonly observed antioxidant molecules in microalgae are listed in Table 1.2.

TABLE 1.2 Commonly Observed Antioxidants/Secondary Metabolites in Microalgae.

Enzymatic antioxidants	Superoxide dismutase
	Peroxidase
	Catalase
	Glutathione reductase
Nonenzymatic antioxidants	GSH (reduced glutathione)
	Ascorbic acid
	Tocopherols
	Carotenoids
	Flavonoids
	Hydroquinones
	Phycocyanin
	Proline
	Mannitol
	Myoinositol
	Phenolics
	Polyamines

Adapted from Mallick and Mohn (2000).

Among the various antioxidants observed in microalgae, carotenoids, phycobiliproteins (PBPs), and PUFAs have been shown to possess high commercial value. These metabolites have wide range of applications in food, pharmaceutical, and cosmetics industry. The surge in demand for natural food colorants and nutraceuticals has created a niche opportunity for microalgal biotechnology. However, a thorough understanding of the physiology and downstream processing techniques for production of these specialty chemicals is essential. Hence, the present chapter is focused toward stimulation, extraction methods, and applications of three important classes of compounds, namely, PBPs, carotenoids, and PUFAs obtained from microalgae.

1.2 ALGAL PIGMENTS: PHYCOBILIPROTEINS

Phycobiliproteins are brightly colored, fluorescent components present in the photosynthetic apparatus of prokaryotic cyanobacteria (blue-green algae) and few eukaryotic algal classes such as rhodophytes (red algae) and cryptomonads. Phycobiliproteins are protein–chromophore conjugate, which are formed by covalent cysteine thioether linkages between apoprotein complex made of two subunits (α and β) and chromophoric bilins (linear tetrapyrroles) which have light-harvesting function (shown in Fig. 1.1). These phycobilins are linear tetrapyrroles biosynthetically derived from heme via biliverdin (Glazer, 1994). The PBPs exhibit a wide spectrum of colors such as blue-colored phycocyanobilin, red-colored phycoerythrobilin, the yellow-colored phycourobilin, or the purple-colored phycobiliviolin, also named cryptoviolin (Sekar & Chandramohan, 2008). Phycobiliproteins are assembled into an organized cellular structure known as phycobilisomes that are attached in regular arrays to the external surface of the thylakoid membrane and act as major light-harvesting pigments in cyanobacteria and red algae. The term "phycobilisome" was first coined by Gantt and Conti (1966) on the basis of their size and shape, as visualized by an electron microscope. The electron micrographs showed a series of large granules aligned regularly on the thylakoid membranes of different cyanobacteria and red alga (De Marsac & Cohen-Bazire, 1977).

In cyanobacteria and red algae, four main classes of PBPs exist, which are allophycocyanin (APC, bluish green), phycocyanin (PC, blue), phycoerythrin (PE, purple), and phycoerythrocyanin (orange) having absorption maxima of 650–655 nm, 615–640 nm, 565–575 nm, and 575 nm, respectively, and emission at 660 nm, 637 nm, 577 nm, and 607 nm, respectively (Bryant et al., 1979). The major biological role of PBPs is photosynthetic light harvesting; however, under certain environmental stress conditions PBPs have a secondary role as intracellular nitrogen-storage compounds and are mobilized during nitrogen deplete conditions during the growth of the algae (Boussiba & Richmond, 1980).

In purified form, they may exist as monomers (α, β) or dimers or trimers or hexamers or equilibrium mixtures of two or more of these aggregates depending on pH, ionic strength, solvent composition, protein concentration, and temperature. The proteins are stable over a pH range of 5.0–9.0 (Sarada et al., 1999). Unlike other photosynthetic pigments, PBPs are highly soluble in water, intensely colored, and fluorescent compounds.

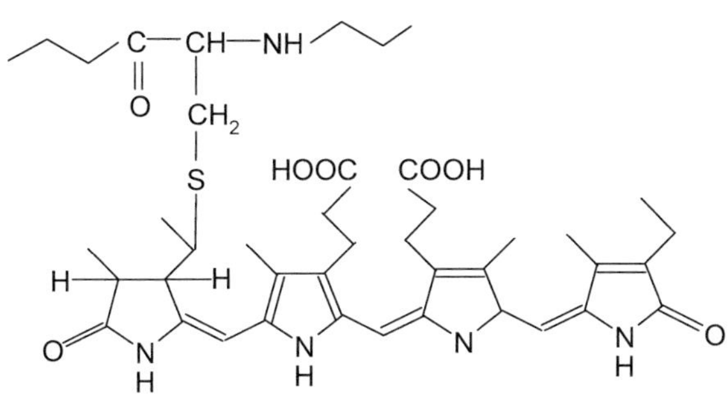

FIGURE 1.1 Structure of phycocyanobilin.

The PBPs have gained commercial importance owing to their water-soluble nature thus finding major applications in food and cosmetic industries. They are used as natural colorants in food products such as chewing gum, candies, soft drinks, dairy products, and cosmetics like lipstick and eyeliners. These pigments are highly sensitive fluorescent dyes. The high absorbance coefficients over a wide spectral range, high quantum yield (Q = 0.3–0.9) and their fluorescence stability over wide pH range and non-quenching of their emission by other biomolecules make them a highly suitable fluorescent markers (phycoflour probes). These markers can be easily linked to immunoglobulins, biotin, avidin, and lectins. The phycobiliprotein conjugates are excellent reagents for analysis of multiple cells surface markers by fluorescence-activated cell-sorting method. Since the emissions of PE, PC, and APC can readily be distinguished; multiple phycofluors can be simultaneously used to quantify the amounts of several markers on cell surfaces. The major organisms exploited for production are the cyanobacterium *Spirulina* for PC and the red alga *Porphyridium* for PE (Roman et al., 2002).

1.3 C-PC AS A NATURAL DYE

Most of the blue-green algae contain C-PC as a major PBP. The intense blue color in blue-green algae is due to the presence of PC, which emits red fluorescence on excitation. This pigment has a single visible absorbance maximum between 615 and 620 nm. It has maximum fluorescence emission at around 640 nm. The molecular weight of pigment is between 70 and 110 kDa and is composed of two subunits of α and β chain, which occur in equal numbers. However, the exact number of αβ pairs may vary among different species. In addition to absorbing light directly, this intensely blue pigment accepts quanta from PE by fluorescent energy transfer in organisms in which PE is present. Also, the C-PC pigment is widely used as a natural dye for various purposes due to its deep and intense blue color. They are well suited as a fluorescent reagent without any toxic effect for immunological analysis since they have a broad excitation spectrum and fluorescence with a high quantum yield. They can be used as a valuable fluorescent probes for analysis of cells and molecules (Kulkarni et al., 1996).

Phycobiliproteins are being used in the commercial sector as natural dyes and PC finds application as natural food colorant to replace the current synthetic pigments. The safety of C-PC as natural blue colorant is well established in several studies, for example, PC at high concentrations—0.25–5.0 g/kg body weight (w/w) did not induce any symptoms of toxicity nor mortality of the animals (Kuddus et al., 2013; Naidu et al., 1999). Native pigment prices of PBPs products are US$ 3–US$ 25/mg and they can reach US$ 1500/mg for certain cross-linked pigments. PC derived from *S. platensis* is used as a coloring agent in food items such as chewing gum, ice sherbets, popsicles, candies, soft drinks, dairy products, and jellies. In addition, it is being used as colorant agent in lipstick and eyeliners (Santiago-Santos et al., 2004). C-PC is used as natural protein dye in the food industry (Sekar & Chandramohan, 2008). PE derived from *Porphyridium aerugineum* and *S. platensis* is also used in color confectionary, gelatin deserts, fermented milk products, ice creams, sweet cake decoration, milk shakes, and cosmetics. Besides coloring properties, PE has yellow fluorescence properties and this fluorescent color is used to make transparent lollipops originating from sugar solution, dry sugar-drop candies for cake decoration, and soft drinks and alcoholic beverages (Dufossé et al., 2005).

1.4 FACTORS AFFECTING THE PHYCOBILIPROTEINS PRODUCTION

Cyanobacteria species, such as *Anabaena* sp., *Nostoc* sp., *Phormidium valderianum*, *Spirulina fusiformis*, *S. platensis*, contain C-PC in the range 8–20% dry weight (Sekar & Chandramohan, 2008). However, *Spirulina*, a filamentous fresh-water algae, is a preferred species for commercial production of C-PC. The optimal growth temperature for *Spirulina* is in the range of 35–38°C. *Spirulina* can tolerate low-to-moderate concentrations (in a range of tens to hundreds of millimole) of NaCl. However, exposure of cells to salt concentrations >0.75 M NaCl may significantly reduce the growth and biochemical composition (Vonshak & Tomaselli, 2000; Hu, 2004). In photoautotrophic culture, a light intensity above 50 W m^{-2} caused photoinhibition and reduced the growth and biomass productivity of *Spirulina* and the optimum light intensity was between 30 W m^{-2} and 50 W m^{-2} (Chojnacka & Noworyta, 2004). Commercial production of C-PC involves outdoor photoautotrophic cultivation in open raceway ponds (Lee, 1997; Pulz, 2001; Spolaore et al., 2006). Generally, *S. platensis* tolerate alkaline conditions and are grown at pH values up to pH 10.5 making them an obligatory alkalophile (Richmond & Grobbelaar, 1986). The tolerance to high alkalinity gives *S. platensis* a selective advantage over other competing species when cultivated under open conditions (Richmond & Grobbelaar, 1986; Richmond et al., 1990). The productivity of C-PC in microalgal and cyanobacterial cultures is determined by the growth properties and specific C-PC content in the biomass. In outdoor open raceways, the areal productivities of dry biomass and C-PC in cultures of *S. platensis* and *Anabaena* sp. have reached values of 14–23.5 g m^{-2} day^{-1} and 0.82–1.32 g m^{-2} day^{-1}, respectively (Jiménez et al., 2003; Pushparaj et al., 1997; Moreno et al., 2003). These values correspond to volumetric productivities of 0.05–0.32 g biomass L^{-1} day^{-1} and 3–24 mg C-PC L^{-1} day^{-1}, respectively (Eriksen, 2008).

The growth and biomass productivity of microalgae, specifically *S. platensis* is determined by the supply of light. Open raceway cultivation contains shallow liquid depths in the range of 10 cm to 30 cm for efficient absorption of incident light (Borowitzka, 1999). Light penetrates only the top layers of the culture, and each cell spends a large proportion of its time in darkness where the productivity is zero or even negative. Cells close to the culture surface experience high light intensities leading to photodamage and growth inhibition (Grobbelaar, 2007). Hence, for efficient light distribution, the cultures are continuously mixed using paddle wheels. Continuous mixing also helps in removal of oxygen accumulated during photosynthesis which otherwise may lead to oxidative stress (Eriksen, 2008).

The growth and biomass productivity of *S. platensis* could be improved by use of closed photobioreactors (PBRs). High cell density is easily achievable in closed PBRs (Posten, 2009). The closed PBRs are of three types, namely, tubular reactors, bubble-column reactors, and flat-plate reactors. The bioreactors are designed in such a way that rapid liquid circulations occur between light and dark zones leading to a uniform distribution of light. When the exposure to the high surface light intensities is short, the photosynthetic efficiency is increased and photoinhibition decreased (Eriksen, 2008). The volumetric productivity is generally 5–20 times higher compared to open raceway ponds. As the biomass productivity is increased, the C-PC productivity increases logarithmically. Enclosed PBRs offer better temperature control, which improves culture productivities of *S. platensis* in temperate climates, and very importantly, cultures can be maintained axenic. However, large-scale cultivation of *S. platensis* in closed PBRs is energy intensive and cost competitive.

1.5 MIXOTROPHIC CULTIVATION OF *S. PLATENSIS* FOR C-PC PRODUCTION

Mixotrophic cultivation of *S. platensis* involves growth of cells in the presence of light and a reduced carbon source such as sugars simultaneously. Mixotrophy increases the growth rate and biomass productivity of culture (Marquez et al., 1993). Mixotrophic cultivation can overcome respiratory loss occurring at night time. Glucose is the commonly used reduced carbon source during mixotrophic cultivation. According to Chen and Zhang (1997) the volumetric productivity of *S. platensis* increased up

TABLE 1.3 Biomass and Phycocyanin Productivity of *Spirulina platensis* in Different Culture Conditions.

Growth conditions	Culture volume (L)	Biomass yield (g L^{-1})	Biomass productivity (g m^{-2} day^{-1})	Specific phycocyanin content (mg g^{-1} biomass)	Phycocyanin productivity (g m^{-2} day^{-1})	References
Phototrophy—open raceway ponds	135,000	0.47	14	61	0.85	Jiménez et al. (2003)
Phototrophy—tubular reactor	11	5.0	47.7	70	3.33	Carlozzi (2003)
Phototrophy—alveolar flat panel	6	6.9	15.8	70	1.11	Tredici et al. (1991)
Mixotrophy (glucose supplementation)—batch*	0.1	2.52	1.66	131	0.22	Marquez et al. (1993)
Mixotrophy (glucose supplementation)—fed-batch*	2.5	10.2	0.82	107	0.087	Chen and Zhang (1997)
Mixotrophy (acetate supplementation)	0.1	1.65	NA	137	NA	Chen et al. (2006)

*Volumetric productivity g L^{-1} day^{-1}, NA—not available.

to 10 g L^{-1} during fed batch glucose supplementation. However, glucose does not increase specific C-PC concentration in *S. platensis*, but the significant increase in biomass productivity consequently results in higher C-PC productivity over autotrophic cultivation (Marquez et al., 1993, 1995; Chen et al., 1996). Dark heterotrophic cultivation of *S. platensis* did not result in C-PC production indicating requirement of light for C-PC accumulation (Chojnacka & Noworyta, 2004). Similarly, excess sugar (glucose) concentration reduced cell growth and biomass production and consequently PC yield. However, heterotrophic cultivation for production of PC was achieved using a red algae *Galdieria sulphuraria*. This species is an extremophile which can grow in highly acidic environment (between pH 1 and 3) and is heat resistant (40 °C) (Gross & Schnarrenberger, 1995). The PC produced by *G. sulphuraria* is similar to that of cyanobacteria. Therefore, *G. sulphuraria* could be exploited for production of PC.

Though light is an essential parameter for production of PC, nitrogen supplementation also plays an equally important role in biosynthesis of PC. Nitrogen limitation results in almost complete depletion of PBPs. PC acts as a storage protein and nitrogen reservoir and supplement elemental N for cell survival in nitrogen-depleted growth environments (Boussiba & Richmond, 1980; Sloth et al., 2006). Nitrate salts such as sodium nitrate and potassium nitrate are commonly preferred source of nitrogen; however, similar biomass growth and PC productivities were obtained with urea and ammonium carbonate as nitrogen sources. Use of urea as nitrogen source could reduce the production costs since they are cheaper. Further, *Spirulina* biomass could be produced using N:P:K fertilizers that are commonly used in agriculture (Venkataraman & Becker, 1985). The biomass and C-PC productivity of *Spirulina* cultivated in different conditions are presented in Table 1.3.

Spirulina biomass production has been demonstrated with use of different carbon sources such as complex sugars, acetate salts, and others. Among complex mixed sugars, molasses a by-product of sugar industry with sugar content >50% has been successfully utilized for biomass production. Andrade and Costa (2007) reported successful cultivation of *Spirulina* in a mixotrophic medium containing molasses up to 0.75 g L^{-1}. A maximum biomass yield of 2.94 g L^{-1} and a volumetric productivity of 0.32 g L^{-1} day^{-1} was achieved. Chen et al. (2006) reported mixotrophic cultivation of *Spirulina* with acetate as carbon source. Maximum biomass yield of 1.65 g L^{-1} and a PC content of 137 mg g^{-1} biomass were obtained with 4 g L^{-1} acetate supplementation.

TABLE 1.4 Different Methods of Extraction and Purification of Phycocyanin.

Method	Reference
Aqueous two phase system	Zhang et al. (2015), Antelo et al. (2015), Zhao et al. (2014), Rathnasamy et al. (2014), Patil and Raghavarao, 2007, Patil et al. (2008)
Ammonium sulphate precipitation, DEAE–sepharose fast-flow chromatography	Zhu et al. (2015)
Freezing and thawing, homogenization (using mortar and pestle), water extraction	Sarada et al. (1999)
Ion-exchange chromatography	Moraes et al. (2015)
Sea water-based media	Sandeep et al. (2015)
Aqueous extract	Jensen et al. (2015)
Freezing and thawing method	Boussiba and Richmond (1979), Setyaningsih et al. (2015)
Ammonium sulphate and DEAE cellulose	Kumar et al. (2014)
Freezing and thawing, ammonium sulphate, Sephacryl™ S-300 high-resolution column	Chen et al. (2014)
Phosphate buffer, 10 kDa (Amicon cell equipped with a YM10 cellulose membrane)	Martelli et al. (2014)
Ultra sonication, Sephadex G-150, DEAE–cellulose	Wang et al. (2014)
Solid–liquid extraction	Su et al. (2014)
Phosphate buffer saline and hydroxyapatite column	Scoglio et al. (2014)
Microwave assisted	Juin et al. (2014)
Phosphate buffer saline, Sephadex G-25, and hydroxyapatite column	Cai et al. (2014)
Ammonium sulphate fractionation	Sonani et al. (2014)
Treatments of alpha-ketoglutaric acid, succinic acid, and malic acid were extracted from the supernatants of algal culture after ultrasonic treatment	Bai et al. (2014)
Freezing and thawing and ammonium sulphate fractionation	Moon et al. (2014)

TABLE 1.4 *(Continued)*

Method	Reference
Repeated freeze–thaw, ammonium sulphate, Sephadex G-150, DEAE cellulose	Soni et al. (2006)
Cell disruption by homogenization, Sephadex G-150, DEAE A-50	Binder et al. (1972)
Repeated Freezing and thawing, ammonium sulphate fractionation, ion exchange and DEAE–sepharose CL-6B column	Patel et al. (2005)
Ammonium sulphate extraction, phenyl–sepharose column, and hydroxyapatite column chromatography	Jian-Feng et al. (2007)
Lysozyme and Q-sepaharose column chromatography	Santiago-Santos et al. (2004)
Ammonium sulphate precipitation, hydroxyapatite column, gel filtration supharose 12 HR 10/30 by FPLC method	Madhyastha et al. (2006)
Ammonium sulphate, Sephadex G-25 column	Minkova et al. (2003)
Freezing and thawing, ultra-filtration membrane (50 kDa from millipore) followed by ion exchange DEAE–cellulose column chromatography	Singh et al. (2009)
One step for phycoerythrin DEAE–sepharose fast flow chromatography	Liu et al. (2005)
Purification of recombinant allophycocyanin, ultrasonication, phenyl column (XK × 30)	Ge et al. (2005)
Ammonium sulphate fractionation, hydroxyapatite column (2.5 × 25 cm	Benedetti et al. (2006)
Different solvents and buffers	Silveira et al. (2007)
Comparative study of different extraction and chromatographic methods	Moraes and Kalil (2009)
Amicon centricon Y-100	Nield et al. (2003)
Ammonium sulphate fractionation, DEAE–sepharose ion exchange and size exclusion by sephacryl S-300	Chen et al. (2006)
Sucrose gradient, Sephadex G-25, DEAE–cellulose, ammonium sulphate fractionation, sepharose column by FPLC	Kupka and Scheer (2008)

1.6 EXTRACTION AND PURIFICATION OF PHYCOBILIPROTEINS (PHYCOCYANIN)

Phycobiliproteins are soluble in water; therefore, they can be easily isolated as protein–pigment complexes. Thus, the purification procedure of PBPs is facilitated, allowing one to separate these molecules from other lipid-soluble pigments. These molecules have been extracted from cyanobacteria by different methods but appropriate control of pH and ionic strength during the extraction procedure is crucial for complex stability. Several factors can influence the PC extraction. The most important are freezing and thawing at two different temperatures (Sarada et al., 1999), cellular-disruption method, type of solvent, biomass–solvent ratio, and extraction time (Abalde et al., 1998; Reis et al., 1998). Various methods have been reported for the extraction and purification of PBPs (Table 1.4). Conventional methods for purification of PBPs involve two steps: pretreatment of the sample to liberate the intracellular material and isolation of C-PC, which include ultrasonication, buffer, salts, freezing–thawing, and combination of several techniques includes centrifugation, ammonium sulphate precipitation, ion-exchange chromatography, gel-filtration chromatography, chromatography on hydroxyapatite expanded bed adsorption chromatography, and FPLC methods. Freezing–thawing is a mild, nondenaturing condition for the extraction of PBPs as compared to other methods. All extraction methods may need to be studied for each strain of cyanobacteria depending upon the cell wall rigidity and are standardized. During the process of extraction, cyanobacterial cells are lysed by freezing at −20°C and thawing at room temperature, which also allows APC and PE with other proteins and nucleic acids to be extracted along with C-PC. The absorption spectra of crude extract of *Spirulina*, *Phormidium*, and *Lyngbya* spp. demonstrate the presence of C-PC, APC, other proteins, and nucleic acids corresponding to their maximum absorption at 620, 652, 280, and 260 nm wavelengths, respectively.

1.7 ESTIMATION OF PHYCOCYANIN

The absorbances of PBP-containing supernatant are generally measured on a UV-visible spectrophotometer at wavelengths 620, 652, and 562 nm for calculating the concentrations of C-PC, APC, and PE, respectively, using the following equations (Siegelman & Kycia, 1978):

$$\text{C-PC}\ (\text{mg ml}^{-1}) = \frac{\left[\text{OD}_{615} - 0.474\left(\text{OD}_{652}\right)\right]}{5.34}, \tag{1.1}$$

$$\text{C-APC}\ (\text{mg ml}^{-1}) = \frac{\left[\text{OD}_{652} - 0.208\left(\text{OD}_{620}\right)\right]}{5.09}, \tag{1.2}$$

$$\text{C-PE}\ (\text{mg ml}^{-1}) = \frac{[(\text{OD}_{562} - 2.41[\text{C-PC}] - 0.849(\text{C-APC})]}{9.6}, \tag{1.3}$$

1.8 HEALTH AND NUTRACEUTIAL APPLICATIONS OF PHYCOCYANIN

Apart from previously mentioned food and cosmetic applications, C-PC has been attributed with many therapeutic applications of which the primary one is its high antioxidant activity (Yan et al., 2014). A very low concentration (10 μM) of C-PC has been shown to inhibit the steady state peroxyl radicals generation by 50% (Bhat & Madyastha, 2000). Further, C-PC has been shown to significantly inhibit lipid peroxidation at a concentration of 500 μM in rat microsomes and also CCl_4-induced lipid peroxidation *in vivo*. Romay et al. (1998) and Reddy et al. (2003) identified the anti-inflammatory activity of C-PC and demonstrated its cyclooxygenase-2 enzyme inhibitory mechanism in animal models. In addition to its high antioxidant property, C-PC has been shown to possess detoxifying and immune stimulatory activity in experimental animal models. Hayashi et al. (2006) reported that C-PC enhanced proliferation and differentiation of bone marrow hematopoietic cells thereby increasing the levels of various cytokines like IL-1β, IFN-g, GMCSF, and IL-3. Further, C-PC increases the expression of liver and kidney enzymes such as cytochrome P-450, super oxide dismutase, catalase, alanine transaminase, and aspartate transaminase, which are involved in detoxification mechanism. The various health benefits attributed to C-PC as reported by several researchers are listed in Table 1.5.

TABLE 1.5 Biological Activities of Phycocyanin *In Vitro* and *In Vivo* Models.

Biological activities	References
Ameliorative	Roy et al. (2008), Hussein et al. (2015)
Antiallergic	Liu et al. (2015)
Selenite-induced cataractogenic rat model	Kumari et al. (2015)
Monosodium glutamate induced oxidative stress	Bertolin et al. (2011)
Photodynamic	Cai et al. (2014)
Immunomodulatory	Chen et al. (2014)
Oxalate-mediated oxidative stress	Farooq et al. (2014)
Oxalate-mediated renal cell injury	Farooq et al. (2004)
Cisplatin-induced nephrotoxicity oxidative stress	Fernández-Rojas et al. (2014a,b)
Induced apoptosis	Gantar et al. (2012)
Antitumor	Gardeva et al. (2014), Li et al. (2015)
Platelet aggregation	Hsiao et al. (2005), Chiu et al. (2006)
Salicylate-induced tinnitus	Hwang et al. (2013)
Sodium selenite-mediated catractogenesis	Kumari and Anbarasu (2014)
Beta cell apoptosis	Li et al. (2014)
Higher efficacy of *in vitro* antioxidant activity when exposed to blue light	Madhyastha et al. (2009)
Thermal stability	Martelli et al. (2014)
Toxicity assessment	Naidu et al. (1999)
Human erythrocytes	Pleonsil et al. (2013)
Antibacterial and toxicity assessment	Sabarinathan and Ganesan (2008)
Colon carcinogenesis	Saini and Sanyal (2014a)
Cyclooxygenese-2 inhibitor	Reddy et al. (2000), Saini and Sanyal (2014b)
Phagocytic	Satyantini et al. (2014)

TABLE 1.5 *(Continued)*

Biological activities	References
Antiproliferative	Tantirapan and Suwanwong (2014), Thangam et al. (2013)
Inhibitory effect on growth of HeLa cells *in vitro*	Yang et al. (2014)
Anticancer	Zhang et al. (2011)
Antoxidative, antiproliferative, and nuroprotective	Romay et al. (1998)
Peroxyle radical scavenging activity	Bhat and Madhysthata (2000)
Hepatoprotective	Ou et al. (2010)

1.9 MICROALGAL CAROTENOIDS

Carotenoids are tetraterpenoids which are composed of eight condensed C5 isoprene (2-methyl-1,3-butadiene) precursors that generate a C40 linear backbone. The carotenoids are of two types, namely, carotenes and xanthophylls. Carotenes are conjugated hydrocarbons whereas xathophylls are oxygenated derivatives of carotenes. The biosynthesis of carotenoids is by the successive condensations of the two interconvertible forms of active isoprene; isopentenyl diphosphate and its double bond-containing isomer dimethyl allyl diphosphate (Domonkos et al., 2013). Xanthophylls are more polar than carotenes because they contain oxygen as hydroxyl groups and/or as pairs of hydrogen atoms that are substituted by oxygen atoms acting as a bridge (epoxide). Overall, carotenoids are relatively hydrophobic and are found to attach with membranes or specific proteins. In most of the photosynthetic organisms, they are attached with chlorophyll-binding polypeptides in the photosynthetic machinery. In nonphotosynthetic organism, they are associated with cytoplasmic or cell wall membranes and contribute to the membrane fluidity. Microalgal carotenoid biosynthetic pathway is similar to that of other eukaryotic organisms and higher plants with same intermediates (Fig. 1.2).

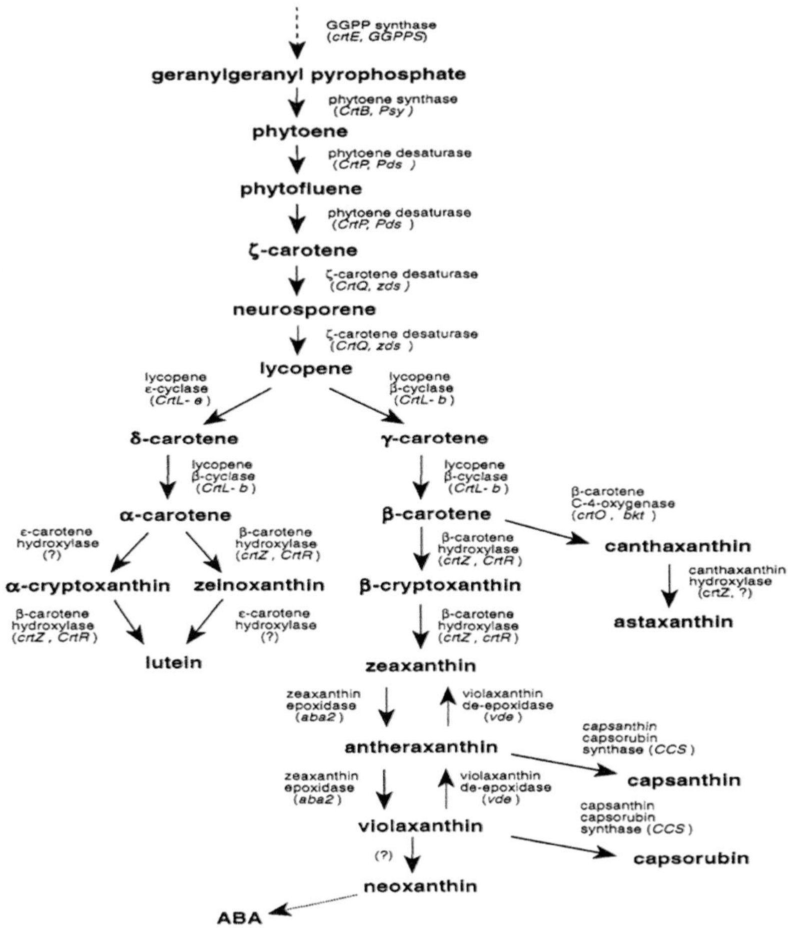

FIGURE 1.2 Carotenoid biosynthesis pathway.

Microalgae accumulate various types of carotenoid/xanthophyll pigments and they are class specific. Shahidi et al. (1998) listed about a total of 71 different types of carotenes and xanthophylls distributed in prokaryotic and eukaryotic microalgae. The list of few major carotenoids observed in microalgae is given in Table 1.6 and Figure 1.3. Among these, only few carotenoids such as β-carotene, astaxanthin, and lutein have been extensively studied owing to their high yields, extractability, stability, bioavailability, and wide nutritional/therapeutic applications (Guedes et al., 2011; Rangarao et al., 2013a,b).

TABLE 1.6 Major Carotenoids Observed in Microalgae.

Pigment	Microalgae	Yield	References
Astaxanthin	*Haematococcus pluvialis*	1.30 mg g^{-1} 30 mg g^{-1}	Park and Lee (2001), Lorenz and Cysewski (2000), Rangarao et al. (2014a,b)
	Chlorella zofingiensis	<1 mg g^{-1}	Ip and Chen (2005)
Canthaxanthin	*Chlorella emersonii*	0.6 μg ml^{-1}	Arad et al. (1993)
Lutein	*Muriellopsis* sp.	4–6 mg g^{-1}	Del Campo et al. (2001)
	Chlorella zofingiensis	3.4 mg g^{-1}	Shi et al. (2006)
	Chlorella protothecoides	4.6 mg g^{-1}	Shi et al. (2006)
	Scenedesmus almeriensis	4.5–5.5 mg g^{-1}	Sánchez et al. (2008)
	Botryococcus braunii	2.9 mg g^{-1}	Rangarao et al. (2013a,b)
Zeaxanthin	*Neospongiococcum excentricum*	0.35 mg g^{-1}	Liao et al. (1995)
	Dunaliella salina	6 mg g^{-1}	Jin et al. (2003)
β-Carotene	*Dunaliella salina*	100 mg g^{-1}	Garcia-Gonzalez et al. (2005)
	Dunaliella bardawil	20.1 pg $cell^{-1}$	Ben-Amotz (1996)

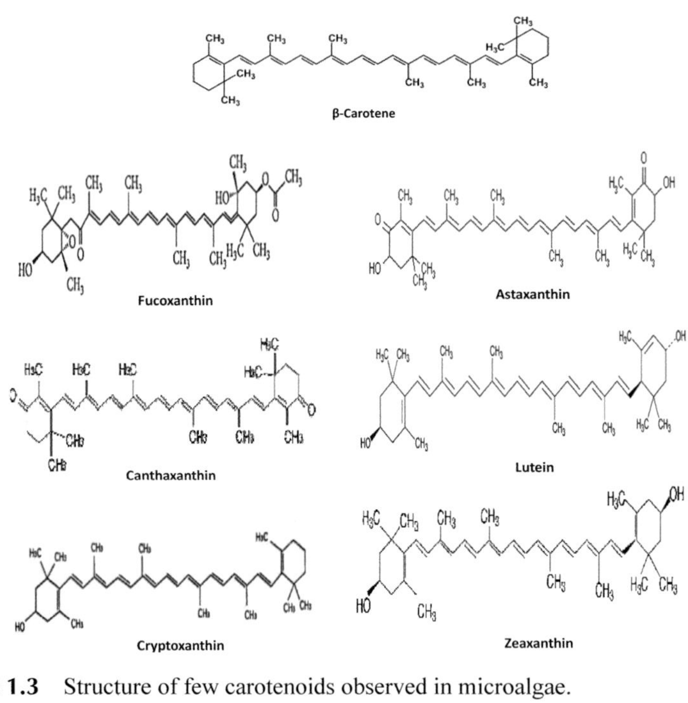

FIGURE 1.3 Structure of few carotenoids observed in microalgae.

1.10 β-CAROTENE FROM *DUNALLIELLA* SP.

β-Carotene ($C_{40}H_{55}$) is a carotenoid of great demand with a wide variety of market applications such as food-coloring agent, as pro-vitamin A (retinol) in food and animal feed, as an additive to cosmetics and multivitamin preparations, and as a health food product with antioxidant benefits (Edge et al., 1997). The health applications and nutraceutical potential of algae-derived β-carotene has been well established and reported (Dufossé et al., 2005; Ravishankar et al., 2008). The major role of β-carotene in microalgae is, as accessory pigment in photosystem, channelizing the excessive photons (light energy) and preventing the photooxidative damage to photosynthetic apparatus (Ben-Amotz & Avron, 1990).

β-Carotene has a major role in stress response of the microalgal cells toward unfavorable growth conditions such as high light irradiation and nutrient-deprived growth conditions. *Dunaliella salina*, a motile, extremophilic green alga, is considered as model organism for carotenoid accumulation, specifically β-carotene (Ben-Amotz, 2004). *Dunaliella* sp. grows in hypersaline environments with high solar radiation such as salt lagoons. *Dunaliella* sp. accumulates β-carotene up to 12% of dry weight which exists as oil globules in interthylakoid spaces of chloroplast. The β-carotene accumulated in *Dunaliella* exists as two major stereoisomers, namely, all-*trans* and 9-*cis*-β-carotene with few other mono-*cis* and di-*cis*-stereoisomers of the β-carotene and no xanthophylls. The ratio of 9-*cis* to all-*trans*-β-carotene depends on light intensity and cell division time, which are in turn determined by the growth conditions. *Dunaliella* as source of β-carotene may be therapeutically better compared to synthetic β-carotene because the synthetic forms contain all-*trans*-β-carotene while *Dunaliella* contains equal amounts of 9-*cis*-β-carotene and minor quantities of other *cis* isomers. The better antioxidant potential of *Dunaliella*-derived β-carotene over synthetic forms was observed when the former had better prevention of methyl linoleneate peroxidation compared to the latter during an *in vitro* study (Levin & Mokady, 1994).

1.11 PRODUCTION OF β-CAROTENE FROM *DUNALIELLA* SP.

Production of β-carotene from *Dunaliella* sp. is already a commercial process. Being a halotolerant organism, *Dunaliella* sp. is grown in very high salt concentrations at 1.5 ± 0.1 M NaCl, giving a selective advantage over competing and predator organisms during growth phase. Bicarbonate is supplied as source of carbon and other nutrients such as nitrate, sulphate, and phosphates are provided during growth phase. The initial growth phase requires 12–14 days in nutrient-replete medium especially rich in nitrate and magnesium-rich medium for chlorophyll accumulation and consequently biomass production. However, for induction of β-carotene accumulation, that is, carotenogenesis, the cells are cultivated in a nitrate-depleted medium under high saline environment (Dufossé et al., 2005).

The induction of carotenogenesis in *Dunaliella* sp. using high light intensity was demonstrated by Vanitha (2007). The vegetative green cells of *Dunaliella* contained only 0.06% β-carotene dry weight, but when cells were exposed to a higher incident light intensity such as 15–20 klx the cells turned yellow and carotenoid content increased to 1.3%, with further increase in light intensity to 30–35 klx the carotenoid content reached to a maximum of 4% dry weight after 4 days of incubation and cells appeared orange in color. In addition to increasing light irradiance, nitrogen deprivation and increase in salinity levels increased β-carotene content in *D. salina*. The β-carotene content increased by fourfold from 1.65 pg/cell to7.05 pg/cell under nitrogen starvation. Similarly, a marked increase in β-carotene content by fivefold was observed under increased salinity levels from 1.5 M NaCl to 4 M NaCl and higher growth temperature of 35°C (Dipak & Lele, 2005). In addition to β-carotene accumulation, under high salinity levels, *Dunaliella* sp. accumulates glycerol as an osmolyte to prevent cellular damage.

1.12 APPLICATIONS OF β-CAROTENE FROM *DUNALIELLA* SP.

The β-carotene obtained from *Dunaliella* sp. finds application as natural food colorant and also as health supplement primarily as an antioxidant (Dufossé et al., 2005). Supplementation of β-carotene from *Dunaliella* sp. improved the egg yolk coloration and has been used as source of vitamin A in animal feed (Gomez & Gonzalez, 2004). The human body converts β-carotene to vitamin A via body tissues, while liver accumulates β-carotene which gets utilized for detoxification. β-Carotene supplementation improves eye health by preventing eye disorders such as night blindness, cataract, and stimulates immune system (Agarwal & Rao, 2000). As an antioxidant, they quench the free radicals such as superoxide anion, hydroxyl radicals, and peroxyl ions, and ameliorate the oxidative stress. Further, the antioxidant property of β-carotene helps in fighting coronary heart disorder, premature aging, arthritis, and cancer (Tornwall et al., 2004). Apart from the antioxidant activity, β-carotene has been shown to possess immune-modulating function such as inhibition of the proliferation of human lymphocytes (Moriguchi et al., 1985), prevention of decrease in antigen expression in human peripheral blood mononuclear cells (Gruner et al., 1986). In addition, Vanitha et al. (2007) reported hepatoprotective activity of β-carotene obtained from *Dunaliella bardawil* (*salina*) against CCl_4-induced toxicity. Raja et al. (2007) have reviewed extensively the health benefits and production of β-carotene from *Dunaliella* sp.

1.13 ASTAXANTHIN FROM *HAEMATOCOCCUS PLUVIALIS*

Astaxanthin (3,3′-dihydroxy-β,β-carotene-4,4′-dione) is a ketocarotenoid with potential nutritional and therapeutic potential. Astaxanthin is a unique carotenoid with the presence of two extra hydroxy and ketone moieties on the ionone rings which confers them a stronger antioxidant property and the ability to form esters with fatty acids. The natural sources of astaxanthin are crustaceans such as krill (Antarctic krill—*Euphausia superba*, Pacific krill—*Euphausia pacifica*, yeast—*Xanthophyllomyces dendrorhous* (formerly *Phaffia rhodozyma*) (Dufossé et al., 2005; Ravishankar et al., 2008). However, among all the natural sources, microalga *Haematococcus pluvialis* accumulates the highest quantity of astaxanthin, up to 4% on dry biomass basis (Kamath, 2007). Astaxanthin has been shown to be a powerful quencher of singlet oxygen from *in vitro* studies (Rangarao, 2011) and has been demonstrated with stronger antioxidant activity compared to β-carotene (10 times higher) and α-tocopherol (500 times higher) and popularly called super vitamin E (Jyonouchi et al., 1994).

H. pluvialis is a motile green alga that is exploited for commercial production of astaxanthin. In general, they accumulate up to 0.2–2.0% (on dry weight basis). Three stable configurational isomers of astaxanthin have been characterized on the basis of the presence of the two hydroxyl groups on the molecule, namely, two enantiomers (3*R*, 3′*R* and 3*S*, 3′*S*) and an achiral diastereomer (3*R*, 3′*S*) or called a mesoform (Turujman et al., 1997). Among them, the 3*S*, 3′*S* stereoisomer is the most abundant in nature (Parajo et al., 1998), and the major form in *H. pluvialis*. As mentioned earlier, astaxanthin exists as esters of fatty acids and among them astaxanthin diesters esterified with oleic acid have highest antioxidant potential followed by astaxanthin monoesters and nonesterified astaxanthin (Rangarao et al., 2014a,b). Astaxanthin accumulation is a unique process where the keto-carotenoid is accumulated in cytoplasm (extra plastidial) unlike β-carotene which is formed in chloroplast. Astaxanthin is accumulated as lipid vesicles or oil globules esterified with fatty acids and triacylglycerol (TAG).

H. pluvialis exist in two phases during its growth cycle. In the vegetative phase and favorable growth conditions, the cells appear green and show motility. During unfavorable growth conditions such as nutrient limiting, oxidative stress, elevated temperature, high light irradiation, and salinity, the cells change their morphology to a nonmotile encysted aplanospore accumulating astaxanthin (Sarada et al., 2002; Jin et al., 2006). During the unfavorable growth conditions, a trilaminar sheath and acetolysis-resistant material is formed around the cell coinciding with astaxanthin accumulation in the extraplastidic lipid vesicles. Astaxanthin prevents cells from photooxidative damage induced due to high light irradiance and consequent oxidative stress.

1.14 PRODUCTION OF ASTAXANTHIN FROM *H. PLUVIALIS*

The commercial production of astaxanthin is performed using either outdoor open raceway pond technology or closed PBRs. Production of astaxanthin-rich *Haematococcus* biomass is a two-step process, where the first step involves production of high-density culture of vegetative cells cultivated under near-optimal growth conditions with careful control of pH, temperature, and nutrient levels. The vegetative phase is carefully cultivated since the cultures are grown in near-neutral pH and are susceptible to predation. Generally, closed PBRs are used for the first phase (Cysewski & Lorenz, 2004; Dufossé et al., 2005). The second stage involves cultivation of cells in stressful environment such as high salinity or nutrient depletion coupled with high light irradiance. For example, nutrient-depleted bold basal medium containing only 1/10th of nitrogen and phosphorus, light irradiance of 60 μmol m^{-2} s^{-1}, NaCl concentration of 17.1 mM, and sodium acetate concentration of 4.4 mM induced astaxanthin accumulation up to 2.45% dry biomass (Vidyavathi, 2008) within 2–3 days of incubation. The carotenoid accumulation was further enhanced by addition of CO_2-mixed air at 2% v/v to the cultures. Among various nitrogen sources such as sodium nitrate, potassium nitrate, ammonium carbonate, ammonium nitrate, and calcium nitrate, supplementation of sodium nitrate at a concentration of 25 mM resulted in maximum biomass (3.3 g L^{-1}) and total carotenoid accumulation of 2.9% with astaxanthin content of 2.5% after 21 days of incubation (Rangarao, 2011). Similarly, among the various inorganic carbon sources like sodium bicarbonate, potassium bicarbonate, ammonium carbonate, potassium carbonate, and sodium carbonate, supplementation of ammonium carbonate at 3 mM resulted in maximum biomass accumulation (2.9 g L^{-1}) with an astaxanthin content of 2.2% dry weight (Rangarao, 2011).

1.15 APPLICATIONS OF ASTAXANTHIN FROM *H. PLUVIALIS*

Astaxanthin is a powerful bioactive antioxidant and has demonstrated efficacy in animal or human models (Snodderly, 1995). The most common application of astaxanthin is its use as fish and crustacean feed for skin coloration (Kamath, 2007). Storebakken (1988) and Torrisen et al. (1989) demonstrated the efficiency of microalgae as source of dietary astaxanthin for flesh pigmentation of Atlantic salmon and rainbow trout. Further, astaxanthin is also considered as a vitamin for salmon, as it is essential for the proper development and survival of juveniles. Astaxanthin-rich *Haematococcus* biomass when fed up to 6% of the diet did not induce any deleterious effect on the growth of rainbow trout (Choubert & Heinrich, 1993). Thus, the algae were concluded to be a safe and effective source of pigment. Natural microalgal astaxanthin has shown superior bioefficacy over the synthetic form in fish pigmentation. The skin color of ornamental koi carp fish increased considerably when fed with diet containing astaxanthin-enriched *H. pluvialis* cells at 25 mg/kg in the feed (Kamath, 2007).

Along with their application as natural colorant, astaxanthin has been promoted as high-value nutraceutical and a health supplement. The nutritional and therapeutic applications of astaxanthin derived from *H. pluvialis* are well documented and reviewed by several authors. Among them, the review by Rangarao et al. (2014a,b) thoroughly describes the complete list of biological activities and disease prevention properties of astaxanthin from *H. pluvialis*. Astaxanthin has been attributed with whole gamut of benefits such as antioxidant, anti-inflammatory, antidiabetic, anticancer, immune modulation, cardioprotective, and hepatoprotective properties. The underlying phenomenon of these varied biological activities is attributed to the superior antioxidant and free radical scavenging activities of astaxanthin. The antioxidant potential of astaxanthin is greater than many other carotenoids such as lutein, β-carotene, lycopene, and α-tocopherol (Naguib, 2000). Astaxanthin has been shown to demonstrate hepatoprotective property by preventing the damage and lipid peroxidation in liver tissue during CCl_4-induced liver toxicity (Kang & Sim, 2008). Further, astaxanthin quenches the singlet oxygen and prevents damage to membranes against free radicals and inhibit free-radical-induced lipid peroxidation (Rangarao, 2011; Rangarao et al., 2013a,b). In addition, astaxanthin-rich *Haematococcus* biomass showed inhibitory activity against the *Helicobacter pylori*-induced ulcer in mice (Wang et al., 2000). This was further proved by Kamath et al. (2008) who demonstrated high antioxidant and ulcer preventive activity of astaxanthin-rich *Haematococcus* biomass. Several reports are available on the anticancer properties of astaxanthin. Tanaka et al. (1995) reported protective effect of astaxanthin against azomethane-induced colon cancer. Anderson (2001) reported suppressive effects of astaxanthin against prostate hyperplasia and prostate cancer. Rangarao (2011) reported antiproliferative property of astaxanthin esters from *H. pluvialis* on human glioma cell lines (LN-229) and liver hepatocellular carcinoma cell line (HepG2). Further, it was demonstrated that astaxanthin inhibits cell cycle progression and induces apoptosis in these cell lines. It was observed that astaxanthin esters showed effective inhibition of skin cancer induced by ultraviolet-7,12-dimethyl benz[α]anthracene (UV-DMBA) in rat models (Rangarao et al., 2013a,b). The anticancer property of astaxanthin esters was attributed to the scavenging of free radicals generated during UV-DMBA treatment and inhibition of tyrosinase enzyme.

1.16 MICROALGAE AS SOURCE OF LUTEIN

Lutein ($C_{40}H_{56}O_2$) is a yellow xanthophyll which is responsible for the yellow color of human retinal macula, egg yolk, and animal fat. The major source of lutein is marigold oleoresin. However, recently, microalgae, *Scenedesmus almeriensis* and *Muriellopsis* sp. are being considered as an alternative source of lutein. Other microalgal sources of lutein include *Chlorella zofingiensis*, *Chlorella fusca*, *Chlorococcum citroforme*, *Neospongiococcum gelatinosum*, and *Dunaliella* sp. (Del Campo et al., 2000). The average lutein content of these microalgae is approximately

500 mg free lutein per 100 g of dry weight, which is 3–6 times higher than marigolds productivity which is approximately 60–100 mg per 100 g (Lin et al., 2015). Among the many microalgae, green-algal species, namely, *S. almeriensis* and *Mureliopsis* sp. have been widely characterized for lutein production.

S. almeriensis is a mesophile with an optimum growth temperature of 35°C and can withstand a wide range of temperature from 10 to 48°C. The strain could tolerate very high irradiance up to 1625 μE m^{-2} s^{-1} without showing any signs of photoinhibition (Sánchez et al., 2008). The optimum pH for the growth of *S. almeriensis* was found to be pH 8; however, maximum yield was obtained at pH 9.5 when cultivated under open outdoor conditions. The strain accumulated lutein up to 0.55% dry weight at these extreme conditions. The maximum volumetric productivity for biomass and lutein was at 0.87 g L^{-1} day^{-1} and 4.77 mg L^{-1} day^{-1}, respectively. Among the various nutrients required for growth, nitrogen (nitrate) plays major role in biomass and lutein accumulation. The optimum nitrate level required for maximum biomass and lutein accumulation was found to be between 3 and 12 mM sodium nitrate. Similarly, maximum productivity was obtained at 85 mM NaCl (i.e., 5 g L^{-1}) (Sánchez et al., 2008; Fernandez-Sevilla et al., 2010). The maximum lutein productivity obtained was 290 mg m^{-2} day^{-1} in open cultivation in tubular PBRs.

Muriellopsis sp., another green microalga, is also a potential source of lutein. The strain accumulates lutein up to 0.55% dry weight. The lutein content of the *Muriellopsis* sp. increased during the exponential phase of growth, with the highest value being recorded in the early stationary phase. The most optimum conditions for maximum lutein accumulation were 20–40 mM $NaNO_3$, 2–100 mM NaCl, 460 μmol photon m^{-2} s^{-1}, pH 6.5, and 28°C. Growth-limiting conditions such as pH values of 6 or 9 and a temperature >30 °C were found to stimulate carotenogenesis in *Muriellopsis* sp. (Del Campo et al., 2000). During the outdoor cultivation of *Muriellopsis* sp. in tubular PBR, a fast increase in lutein content was observed during the early hours of daytime, with maximal lutein content of about 6 mg g^{-1} dry weight recorded at noon, indicating a direct relation between light irradiance and lutein accumulation. A maximum lutein productivity of 180 mg m^{-2} day^{-1} was achieved in outdoor cultivation with tubular PBRs (Del Campo et al., 2001).

Botryococcus braunii, a colonial green alga, has been identified as another potential source of lutein (Rangarao et al., 2014a,b). The strain accumulated approximately 0.29% lutein on dry weight basis. Lutein was accumulated during the linear growth phase of the alga and optimum growth conditions required for biomass production were found to be ideal for lutein accumulation. However, as the cells proceed to stationary phase, *B. braunii* accumulates secondary carotenoids such as xanthophylls—astaxanthin (0.019%), zeaxanthin (0.046%), and β-carotene (0.054%) (Rangarao et al., 2014a,b). Further to these existing carotenoids, *B. braunii* accumulates new class of carotenoids called botryoxanthins (Okada et al., 1996). Among the three races of *B. braunii* species, namely, A, B, and L races, race A accumulates lutein as major pigment while race B accumulates this new class of botryoaxnthin pigments, botryoxanthin A, botryoxanthin B, alpha botryoxanthin, and braunixanthin 1 and 2 (Okada et al., 1997, 1998). The race L has been identified with adonixanthin during stationary phase (Grung et al., 1994).

The nutritional application of lutein is well established in treatment of age-related macular degeneration (AMD) (Ma et al., 2012). Several studies have shown that lutein and its naturally occurring isomer zeaxanthin may provide significant protection against the potential damage caused by light on the retina. Lutein and zeaxanthin have been attributed to filter high-energy wavelengths of visible light and ultraviolet rays and act as antioxidants. They restrict the formation of ROS and subsequent free radicals (Roberts et al., 2009). Rangarao et al. (2013a,b) reported antilipid peroxidation activity of lutein obtained from *B. braunii* biomass in rat model. Conditions that stimulate carotenoid accumulation in microalgae are listed in Table 1.7.

TABLE 1.7 Stimulation of Carotenoid Accumulation in Microalgae.

Carotenoids	Conditions for high yield	Reference
β-Carotene	High light irradiance, high salinity, temperature, nitrate and phosphate deficiency	Dipak and Lele (2005)
Astaxanthin	High light irradiance, high salinity, nitrate and phosphate deficiency, presence of acetate, malonate, and $CaNO_3$	Sarada et al. (2002), Rangarao et al. (2014a,b)
Lutein	Optimum growth conditions	Del Campo et al. (2001), Cordero et al. (2011)
Zeaxanthin	High light irradiance	Jin et al. (2003)
Canthaxanthin	Low light + high salinity	Pelah et al. (2004)

1.17 OTHER MINOR CAROTENOIDS FROM ALGAE

1.17.1 ZEAXANTHIN

Zeaxanthin ($C_{40}H_{56}O_2$) is a yellow pigment found in fruits, vegetables, egg yolk, and corn. It protects the photosynthetic machinery of algae and plants from photodamage. In carotenoid pathway, it is synthesized from β-carotene or β-cryptoxanthin by the addition of hydroxyl group by the enzyme β-carotene hydroxylase. Like other xanthophylls, zeaxanthin is also membrane/protein bound. The main physiological benefit of zeaxanthin is its protective role in vision. Several studies have proven that zeaxanthin is effective against visual abnormalities like AMD and cataract, and a therapeutic candidate for cancer and atherosclerosis treatment (Sajilata et al., 2008; Nishino et al., 2009; Kadian & Garg, 2012). Zeaxanthin-producing microalgae are a mutant strain of *D. salina* (Jin et al., 2003) *Microcystis aeruginosa*, *Spirulina*, and *Synechocystis*. *Flavobacterium* is the one organism with zeaxanthin as its major carotenoid accumulated (Bhosale et al., 2003).

1.17.2 FUCOXANTHIN

Fucoxanthin ($C_{42}H_{58}O_2$) is a major xanthophyll found in brown seaweeds and accounts for almost 10% of the total natural carotenoid production. It is a primary carotenoid found to associate with photosynthetic machinery. It is a potent antioxidant even under anoxic conditions where other carotenoids are inactive. The main source of fucoxanthin is the brown algae, *Laminaria japonica*. The unused part of this alga is used for the commercial extraction of fucoxanthin. Other major sources include *Phaeodactylum tricornutum*, *Eisenia bicyclis*, and *Undaria pinnatifida*. Fucoxanthin exists as isomers, namely, *cis* and *trans* forms, where the *trans* form is more stable but less reactive than *cis* form (Nakazawa et al., 2009). The therapeutic applications of this pigment are well studied, such as anticancer property (Kotake-Nara et al., 2001; Miyashita et al., 2011) and antiobesity (Maeda et al., 2005; Woo et al., 2009).

1.17.3 CANTHAXANTHIN

Canthaxanthin ($C_{40}H_{52}O$) is a xanthophyll formed from β-carotene by addition of keto group by the enzyme β-carotene ketolase. Canthaxanthin finds major application as fish and poultry feed, where they are absorbed on the skin and flesh of the salmonid fish (Brizio et al., 2013) and for the coloration of egg yolk in poultry (Esfahani-Mashhour et al., 2009). In addition to their use in animal feed, canthaxanthin has been demonstrated with high antioxidant activity (Palozza et al., 2000).

1.17.4 CRYPTOXANTHIN

Cryptoxanthin ($C_{40}H_{56}O$) is a xanthophyll formed from β-carotene by addition of one hydroxyl group. Cryptoxanthin is considered as provitamin as they are converted to vitamin A in the animal body. Cryptoxanthin has been reported to possess antioxidant activity and associated with bone homeostasis (Lian et al., 2006; Yamaguchi, 2012). This pigment has a stimulatory effect on bone formation and it prevents bone loss in menopausal of woman.

1.18 EXTRACTION OF CAROTENOIDS

Since carotenoids are lipophilic, they are generally extracted with organic solvents. Cultivation of *Dunaliella* sp. for β-carotene is advantageous for two reasons, namely, for their high saline requirements (extremophilic) during growth which prevents contamination from competing algae and the lack of cell wall which eases the process of extraction. Further, the high hydrophobic nature of β-carotene can be exploited for its synthesis in aqueous/organic biphasic photo-production (León et al., 2001). This process of biphasic extraction uses biocompatible solvents such as decane which forms the organic phase of the biphasic system and carotenoids are easily extracted into the hydrophobic phase continuously during production, this process is called "milking" (Hejazi et al., 2004). β-Carotene could be easily extracted out with a high extraction efficiency and productivity much higher than that of commercial plants (Raja et al., 2007).

However, the extraction of carotenoids are quite challenging in other green microalgae such as *H. pluvialis*, *Chlorella* sp., *Murelliopsis* sp., *S.*

almeriensis, and others, where these algae possess thick cell wall composed of cellulose. Therefore, the preliminary step involves hydrolysis of algal cell wall. Extraction of astaxanthin from encysted *H. pluvialis* biomass was achieved using acid pretreatment for cell wall hydrolysis. Among the different acid treatments, use of hydrochloric acid resulted in a maximum recovery of 80% of total carotenoids (Sarada et al., 2006). However, addition of inorganic acids may not be suitable for bulk extractions since presence of acid residues in the extracts may induce toxic effects when consumed. Therefore, various physical treatments such as autoclaving the biomass for 30 min at 121°C, enzymatic treatment with mix of protease K and driselase, mechanical homogenization, ultrasound sonication, bead milling, and freeze–thawing have been suggested for extraction of astaxanthin from encysted microalgal cells. However, among these pretreatments, autoclaving the biomass followed by solvent extraction or bead milling followed by solvent extraction were most efficient (Mendes-Pinto et al., 2001; Fernandez-Sevilla et al., 2010). Extraction of carotenoids, mainly astaxanthin and lutein, have been achieved using edible vegetable oils such as sunflower, olive, soyabean, corn, and grape seed oil. A highest recovery of 93% for astaxaxnthin was achieved with olive oil (Kang & Sim, 2008). Further, Rangarao et al. (2007, 2014a,b) reported that stabilization of astaxanthin in edible oils resulted in preservation of its antioxidant activity. The principle behind oil-based extraction is that an oil emulsion is formed with biomass slurry and carotenoids are directly absorbed in oil droplets. Generally, large-scale extraction involves Soxhlet extraction with organic solvents repeatedly, followed by vacuum distillation of solvent and drying of the crude extracts (Rangarao et al., 2014a,b). However, accelerated solvent extraction at high pressure and temperature (60–170°C) using solvents such as ethanol and hexane resulted in very high recovery of lutein from *Chlorella* and *Spirulina* biomass. In an another attempt, repeated liquid–liquid counter current extraction with aqueous hydrolysate and hexane for six times resulted in 95% recovery of lutein from *S. almeriensis* biomass (Ceron et al., 2008).

Supercritical CO_2 fluid extraction (SFE) is another widely used method for extraction of carotenoids. SFE involves use of CO_2 at supercritical conditions, where the CO_2 gas has solvent-like properties. Extraction by SFE is ideal for recovery of pharmaceutical or nutraceutical substances since the process is clean and the lack of toxicity of CO_2 as a solvent (Fernandez-Sevilla et al., 2010). Extraction of *Haematococcus* biomass by SFE with ethanol or sunflower oil as cosolvent resulted in up to 80–90% recovery of astaxanthin (Machmudah et al., 2006; Nobre et al., 2006; Wang et al., 2012). However, in the case of lutein extraction, SFE did not result in very high recoveries compared to conventional solvent extraction. Only 40–50% recovery of lutein was obtained with SFE (300–400 bar, 60°C extraction temperature) for *Chlorella* sp., *D. salina*, and *Nannochloropsis gaditana* compared to conventional solvent extraction with hexane, ethanol, or DMF (dimethyl formamide) (Macías-Sánchez et al., 2005, 2009; Kitada et al., 2009). The main disadvantage with SFE was co-extraction of chlorophyll which was again needed to be separated using organic solvents (Macías -Sanchez et al., 2009).

1.19 POLYUNSATURATED FATTY ACIDS FROM MICROALGAE

Lipids are inherent components of cells and play major role in cell membrane development, energy storage, stress response, and act as secondary messengers in many metabolic pathways. Lipids are esters of glycerol and fatty acids, where the fatty acid structure and composition determine the function of a particular lipid compound. Among the many types of fatty acids present or identified in biological systems, PUFAs have gained great importance owing to their multitude health benefits.

PUFAs are methylene-interrupted polyenes with two or more double bonds. They are classified based on the occurrence of double bond position from the methyl (omega) end of the carbon chain into omega-3 (ω-3), omega-6 (ω-6), and so on. Examples of ω-3 include as alpha-linolenic acid (ALA, C-18:3), eicosapentaenoic acid (EPA, 20:5), and docosahexaenoic acid (DHA, 22:6), while examples of ω-6 include linoleic acid (LA, C-18:2), gamma-linolenic acid (GLA, C-18:3), and arachidonic acid (AA, 20:4). Though PUFAs are not classified under secondary metabolites they have huge role in the physiology of the producing organism such as cold tolerance, carbon metabolism, secondary messengers, and environmental adaptation (Guschina & Harwood, 2009). Further, PUFAs have been attributed with various biological functions in human physiology and metabolism like cardiovascular protection, prevention of age-associated cognition disorders, fetal neurodevelopment, and anti-inflammatory properties (Doughman et al., 2007). Considering the importance of PUFAs in human metabolism, WHO/FAO has recommended a minimum dietary intake of 250 mg day^{-1} ω-3 PUFAs while American Heart Association has recommended up to 500 mg day^{-1} for a healthy adult (Lichtenstein et al., 2006). PUFAs are considered as essential fatty acids since they cannot be synthesized in humans and are derived from food sources. The common food sources of PUFAs are fish and shellfish, flaxseed (linseed), hemp oil, soya oil, canola (rapeseed) oil, chia seeds, pumpkin seeds, sunflower seeds, leafy vegetables, and walnuts. However, fish oil obtained from marine fish such as salmon, mackerel, and herring are the major commercial sources of therapeutically important EPA and DHA (Guedes et al., 2011).

The increasing health consciousness of people and surge in the demand for nutracueticals have increased the PUFA production to 24.87 kt in 2013 and the demand is further expected to grow at 13.7% from 2014 to 2020 (www.iffo.net). This increased demand of PUFAs will exert a huge pressure on world's marine fishery resources which are depleting due to excessive fish capture and intense fishery practices (FAO, 2011). In addition, many toxic chemicals such as methyl mercury, dioxins, and polychlorinated biphenyl are found in fish oil due to the increasing pollution levels of oceanic ecosystem caused by anthropogenic activities. These toxic contaminants are hydrophobic in nature and bind to the lipid deposits in fish causing bioaccumulation down the food chain (Storelli et al., 2004). Further, fish oil has unpleasant odor, high cholesterol levels (Melanson et al., 2005), and the proportion of constituent fatty acids are difficult to control.

Microalgae, as a source of PUFAs, have advantage over other plant-based sources as they have higher photosynthetic and surface area productivity (10-fold) than terrestrial crop plants (Rittman, 2008). They can be cultivated in nonarable land in open outdoor ponds throughout the year (Borowitzka, 1999) with minimal nutritional input requirement. Further, microalgae accumulate lipids as an adaptive response to stressful environmental conditions such as nutrient deprivation, cold shock, and others. Hence, they can be targeted for industrial production of PUFAs (Vidyashankar et al., 2015).

1.20 PUFA-PRODUCING MICROALGAE

Microalgae have a class-specific fatty acid composition which can be used as a chemotaxonomic marker for identifying the algal species (Sahu et al.,

2013). PUFAs are majorly distributed in the polar lipid fractions of microalgae and contribute to the chloroplast development. Although PUFAs have specific physiological role in cellular metabolism, their composition differ in different class of algae. Cyanobacteria (blue-green algae), the prokaryotic group, are generally low lipid producers with palmitic and oleic acid as major fatty acids. However, *Spirulina* sp., belonging to the Oscillatoriales order, is an exception synthesizing GLA (C-18:3, n-6) (Hu, 2004). The ubiquitously occurring Chlorophyceae (green algae) members, such as *Chlorella* sp., *Scenedesmus* sp., *Dunaliella* sp., synthesize ALA as the prominent PUFA, however, with certain exceptions such as *Chlorella minutissima, Tetraselmis* sp. that accumulate EPA and *Parietochloris incisa* synthesizing ARA (Bigogno et al., 2002; Vidyashankar et al., 2015). Diatoms (Bacillariophyceae), a commonly occurring group of marine algae, generally synthesize high amounts of palmitoleic acid (C16:1) and EPA, for example, *Phaeodactylum tricornutum*, *Thalassiosira pseudonana*, and *Nitzschia laevis* (Tonon et al., 2002; Patil and Raghavarao, 2007; Wen & Chen, 2003). Chrysophyceae (golden algae) produce a wide variety of fatty acids such as palmitoleic acid (C-16:1), EPA, and DHA. Among them, strains of *Isochrysis* sp. have been reported to contain high amounts of PUFAs (Patil and Raghavarao, 2007; Fernandez-Sevilla et al., 1998). Dinoflagellates such as *Gyrodinium* and *Crypthecodinium cohnii* produce unusual fatty acids such as C-18:4, C-18:5 along with EPA and DHA (de Swaaf et al., 2003; Couto et al., 2010). *Porphyridium* sp., a popular red alga (Rhodophyta), is known to synthesize linoleic acid (C18:2), ARA, and EPA (Arad & Richmond, 2004). Thraustochytrids are another class of PUFA-accumulating marine microheterotrophs, taxonomically aligned with heterokont algae (Lewis et al., 1999; Raghukumar, 2008). Certain species belonging to this heterotrophic class of algae like, *Schizochytrium* sp. and *Thrausochytrium* sp., accumulate DHA-rich lipids up to 40% of the dry weight and have been exploited for commercial DHA production (Adarme-Vega et al., 2012). Vidyashankar et al. (2015) reported that marine microalgae species such as *Nannochloropsis* sp. and *Chorella* sp. could be a potential source of EPA under autotrophic growth conditions with an average productivity of 2.0–2.5 mg L^{-1} day^{-1} in a batch cultivation. The PUFA content of some oleaginous microalgae are listed in Table 1.8.

TABLE 1.8 List of PUFA-Accumulating Microalgae.

Microalgae	Class	Cultivation condition	Content of major PUFA(s) (% composition of total fatty acids)	Reference
Nannochloropsis sp.	Euastigophyceae	Autotrophic/ Heterotrophic	EPA—23–33	Patil and Raghavarao (2007), Zhukova and Aizdaicher (1995), Huang et al. (2013), Bellou and Aggelis (2012), Huerlimann et al. (2010), Andrich et al. (2005)
Pavlova sp.	Haptophyceae	Autotrophic/ Heterotrophic	EPA—15–29 DHA—7.5–10	Lang et al. (2011), Zhukova and Aizdaicher (1995), Griffiths et al. (2012), Huang et al. (2013)
Isochrysis sp.	Haptophyceae	Autotrophic/ Heterotrophic	DHA—7–15	Patil and Raghavarao (2007), Renaud et al. (2002), Huerlimann et al. (2010), Vidyashankar et al. (2015)
Porphyridium cruentum	Rhodophyceae	Autotrophic	ARA—17–27 EPA—20–38	Cohen et al. (1988), Zhukova and Aizdaicher (1995), Tsuzuki et al. (1990)
Parietochloris incisa	Chlorophyceae	Autotrophic	ARA—14 EPA—4.3	Lang et al. (2011)
Phaeodactylum tricornutum	Bacillariophyceae	Autotrophic/ Heterotrophic	EPA—14–33	Tonon et al. (2002), Patil and Raghavarao, (2007), Zhukova and Aizdaicher (1995), Griffiths et al. (2012)
Nitzschia laevis	Bacillariophyceae	Heterotrophic	EPA—10–15	Wen and Chen (2003)
Tetraselmis sp.	Chlorophyceae	Autotrophic	EPA—6–8	Vidyashankar et al. (2015)
Schizochytrium sp.	Thraustochytrids	Heterotrophic	DHA—57	Chang et al. (2013)
Crypthecodinium cohnii	Dinophyceae	Heterotrophic	DHA—72	Couto et al. (2010)
Prorocentrum sp.	Dinophyceae	Autotrophic	DHA—20–22	Makri et al. (2011)

1.21 PRODUCTION OF PUFA FROM ALGAL BIOMASS

PUFAs are generally present in polar lipid fractions which constitute the membrane lipids, while saturated fatty acids and monounsaturated fatty acids constitute the neutral lipids, mainly TAG, in autotrophic microalgae (Hu et al., 2008; Guschina & Harwood, 2009). However, certain microalgae like *P. incisa*, *Pavlova lutheri*, *Nannochloropsis oculata*, *T. pseudonana*, and *P. tricornutum* accumulate TAG containing PUFA (Bigogno et al., 2002; Khozin-Goldberg et al., 2002). Among these algae, *P. incisa* accumulates highest amounts of PUFA (AA)-containing TAGs (Bigogno et al., 2002). Targeting microalgae that accumulate PUFA-rich TAG would be beneficial since the TAG metabolism is greatly dependent on growth environment which can be controlled (Hu et al., 2008). The three major factors that determine the growth and lipid metabolism of microalgae are nutrient availability, medium pH, and carbon supplies. Lipid synthesis and fatty acid profiles are particularly affected by nutrient availability specifically nitrogen, phosphate, sulphate, or silica limitation, occurring with culture age (Roessler, 1988; Khozin-Goldberg & Cohen, 2006; Guiheneuf and Stengel, 2013; Vidyashankar et al., 2013). Cultivation temperature plays an important role in the fatty acid composition of microalgae. When the incubation temperature was decreased from 25°C to 10°C, the cultures of haptophyte *Isochrysis galbana* showed an increased accumulation of DHA and ALA with corresponding decrease in linoleic and oleic acid (Zhu et al., 1997). Several studies have demonstrated the effect of carbon supplementation on the fatty-acid profiles of microalgae. Tang et al. (2011) reported increased lipids and PUFA (ALA, EPA) levels under higher CO_2 supplementation (up to 50% v/v CO_2) in *Scenedesmus obliquus* and *Chlorella pyrenoidosa*. Similarly, Vidyashankar et al. (2013) reported a twofold increase in the ALA content of an indigenous *Scenedesmus dimorphus* strain when supplemented with CO_2 at 10% v/v. However, the response to CO_2 supplementation varies among different species where increased CO_2 supplementation decreased PUFA levels in the lipids. For example, the strains of *Chlamydomonas reinhardtii*, *D. salina*, and *Chlorococcum littorale* showed decreased PUFA accumulation during increasing CO_2 supplementation (Ota et al., 2009).

Nannochloropsis sp., belonging to the class Eumastigophyceae, has been widely cultivated for production of EPA and has been exploited as live feed in aquaculture (Sukenik, 1999). Several reports on the effect of environmental factors on the total lipid and EPA content of *Nannochloropsis* sp. are available. The biomass production, total fatty acids, and EPA content of *Nannochloropsis* decreased with increasing light irradiance, high salinity levels, and nitrogen depletion (Pal et al., 2011; Hu & Gao, 2006); however, at reduced salinity levels to that of brackish water and lower light intensities, the EPA content increased. This decrease in PUFA content at higher light irradiance and salinity levels could be attributed to the formation of ROS. The stressed culture utilize PUFAs to counter negative effects of ROS and quench the free radicals generated during unfavorable growth conditions (Sukenik et al., 1989, 2009).

1.22 HETEROTROPHIC PRODUCTION

The heterotrophic growth approach of algal biomass and PUFA production has several distinct advantages over phototrophic culture such as higher cell densities, readily available large-scale bioreactors, control over process, and lesser cost. The production of heterotrophic algal biomass using the standardized technology cost less than US$5 kg^{-1} (Gladue and Maxey, 1994), whereas autotrophic algae production can be two times higher (Benemann, 1992). The higher cell density obtained with heterotrophic production further reduces the difficulty in harvesting process since autotrophic production in open raceway or PBRs maintain a low cell density such as 0.5–1.0 g L^{-1} requiring energy-intensive harvesting process (Vandamme et al., 2013).

Heterotrophic production of PUFAs has been commercially established with several strains of algae such as *Schizochytrium* sp., *C. cohnii*, *Thraustochytrium aureum*, *Skeletonema costatum*, and others (Sijtsma & de Swaaf, 2004; Adarme-Vega et al.,2012). Some of the main factors affecting the growth and PUFA accumulation during heterotrophic cultivation are carbon source, carbon/nitrogen ratio, and phosphorus concentration, while silicate concentration in case of diatoms. The list of essential growth criteria required for production of DHA from Thraustochytrids is summarized in Table 1.9. Currently, Thraustochytrids are exploited for production of DHA at industrial level (Raghukumar, 2008). Among the essential criteria, carbon source is the primary parameter that provides the energy and carbon skeletons for cell growth since heterotrophic production involves dark cultivation of microalgae. The most commonly used sources are acetate or glucose. In some cases mono-, di-, and polysaccharides such as fructose, sucrose, lactose, and starch have been reported to promote growth and PUFA accumulation (Wen & Chen, 2003). Pyle et al. (2008) reported production of DHA by *Schizochytrium limnacium* under crude glycerol supplementation derived from algal biodiesel production. In addition to carbon source, carbon/nitrogen (C/N) ratio influences lipid accumulation by controlling the protein and lipid synthesis. A high C/N ratio favors lipid accumulation where the cells are supplemented with low nitrogen levels (Gordillo et al., 1998). However, for accumulation of PUFA a low C/N ratio is required (Chen and Johns, 1991). The most commonly used nitrogen sources for microalgal growth are nitrate and urea. Yongmanitchai and Ward (1991) reported increased growth and EPA accumulation in *P. tricornutum* with nitrate, while Wen and Chen (2001) reported increased productivity in *N. laevis* with urea as nitrogen source. Use of ammonia as sole nitrogen source inhibited growth and PUFA production in diatoms (Wen & Chen, 2001). This could be attributed to pH fluctuations during assimilation of ammonium ions. Complex nitrogen sources such as yeast extract, tryptone, and corn steep liquor could promote algal growth since they supply reduced form of nitrogen sources such as amino acids and other growth factors such as vitamins (Wen & Chen, 2003). Phosphorus is essential for the growth and metabolic processes such as energy transfer in cells, biosynthesis of nucleic acids, and more importantly for biosynthesis of phospholipids and membrane development. Since the PUFAs are generally distributed in polar lipids (phospholipids), phosphorus concentration in growth medium determines PUFA production turnover (Guschina & Harwood, 2009). Higher phosphorus concentration in growth medium resulted in higher EPA yield in *P. tricornutum* (Yongmanitchai & Ward, 1991). Similarly, silica supplementation is essential for the growth and cell wall formation in diatoms. Several authors have reported that depletion of silica and nitrate in the growth medium resulted in increased lipid synthesis and consequently PUFA accumulation (Roessler, 1988; Ramachandra et al., 2009). Wen and Chen (2000) reported an increase in the EPA content of *N. laevis* when silicate became the limiting factor. The increased lipid/PUFA accumulation in silicate-limited cultures could be attributed to that the cell tends to alter their metabolism and divert energy toward lipid storage which otherwise would have been allocated for silicate uptake (Wen & Chen, 2003). The PUFA productivity of few oleaginous microalgae is listed in Table 1.10.

TABLE 9.1 Essential Growth Requirements for Production of DHA from Thrausochytrids.

Parameter	Requirement
Carbon source	Glucose, fructose, glycerol, lactose, starch
Nitrogen source	Ammonium salt (for *Schizochytrium*) and organic nitrogen (peptone or glutamate) for others; nitrate and urea for mixotrophic cultivation
Macroelements	Seawater or a Na^+ salt and other major cations and anions in seawater; KH_2PO_4
Microelements	Various trace elements of seawater
Vitamins	Thiamine, biotin, riboflavin, and cobalamin depending on the organism
C/N ratio	Based on cell ratio for biomass production; excess C/N for lipid accumulation; low C/N ratio for PUFA accumulation
Temperature	25–30°C for biomass production; ~15°C for lipid accumulation
Salinity	50–100% of that of seawater
pH	5–8
Oxygen levels	>4% for growth and <3% for DHA production (for *Schizochytrium*)
Preferred culture mode	Fed batch
Age of culture	Stationary phase

Source: Chen and Johns (1991), Wen and Chen (2003), Raghukumar 2008; Pyle et al. (2008).

TABLE 1.10 PUFA Productivity of Oleaginous Microalgae.

Microalgae	Production system	PUFA productivity (mg L−1 day−1)	References
Phaeodactylum tricornutum	Autotrophic—glass tanks, continuous mixotrophic—glass vessels, batch culture	EPA—19–25 EPA—33.5	Yongmanitchai and Ward (1991, 1992) Garcia et al. (2000)
Nitszchia laevis	Heterotrophic—perfusion cell bleeding	EPA—175	Wen and Chen (2001)
Isochrysis galbana	Autotrophic—cylindrical PBRs, continuous culture	EPA—24	Fernandez-Sevilla et al. (1998)
Nannochloropsis sp.	Autotrophic—Tubular PBRs, continuous culture	EPA—32	Zittelli et al. (1999)
Monodus subterraneus	Autotrophic—Flat plate PBRs, semi-continuous	EPA—58.9	Qiang et al. (1997)
Schizochytrium sp. SR21	Heterotrophic—bioreactors	DHA—3300*	Yaguchi et al. (1997)
Crypthecodinium cohnii	Heterotrophic—bioreactors	DHA—1200*	de Swaaf et al. (2003)
Thraustochytrium sp. strain G13	Heterotrophic—bioreactors	DHA—900*	Bowles et al. (1999)

*Calculated based on the values expressed as mg L^{-1} h^{-1}

1.23 EXTRACTION OF LIPIDS

Extraction and purification of lipids and PUFAs from microalgae have been considerably challenging due to their complex cell wall composition and to certain extent their morphology. Oleaginous strains belonging to Chlorophyceae have thick cellulosic cell wall while diatoms have silicified cell walls which decrease the extractability of lipids from algal cell. Cell disruption is an important preliminary step in recovering intracellular products. The most commonly practiced method for oil/lipid extraction is use of organic solvents such as hexane, ethanol, and others (Adarme-Vega et al., 2012). However, before organic solvent extraction, a preliminary pretreatment step for cell wall degradation is essential (Mata et al., 2010). The conventional pretreatment processes involve soaking of dry or wet biomass in diluted inorganic acids as sulfuric, phosphoric, or hydrochloric acids, and heating at high temperature for a certain period of time followed by neutralization with alkali (NaOH). This treatment leads to pores formation that reduces the cellulose crystallinity which weakens cell wall. Although this process is efficient in degrading tough lignocellulose layers, it may not be recommended for oil extraction for commercial applications (specially nutritional applications) as there may be residues of these toxic and corrosive solvents in the extracts and leftover biomass (Taher et al., 2014a).

Several physical/mechanical pretreatment methods, such as for oil extraction, have been proposed, for example, expelling at high pressures like French press, high-pressure homogenization, wet milling, ultrasonication, bead beating, microwave extraction, autoclaving at a high temperature and pressure (Mercer & Armenta, 2011; Halim et al., 2012; Kim et al., 2013; Taher et al., 2014a; Li et al., 2014). These mechanical pretreatments disrupt the cell wall integrity and result in increased oil/lipid recovery. However, these methods have high capital and high operational costs and require external cooling since they generate excess heat during extraction. Further, the heat generated during extraction may affect the properties of the extracted lipids and may generate degraded toxic intermediates. Therefore, mild extraction methods are essential that do not compromise on oil recovery and the quality.

Mild extraction methods in terms of use of chemicals and lesser energy requirements are economical and effective in oil production for nutritional or therapeutic applications. Enzyme pretreatment is a commonly preferred method for milder operation conditions. Enzymes such as lysozyme that act on peptidoglycan layer (*n*-acetyl glucosamine polymer) can be used for breaking the cell walls of cyanobacteria and cellulase can be used for hydrolyzing cellulose layers of chlorphycean algae. The main advantage of enzyme extraction is that they operate at low temperatures with high selectivity and fewer side products (Taher et al., 2014a). Further, enzyme extraction can be applied to wet biomass, thus avoiding the energy-intensive step of drying microlgal biomass. The enzymatic treatment individually or in combination with other physical disruption methods such as sonication can improve lipid recovery and fasten the extraction process. However, the major disadvantage of this process is its scalability and high process cost (Halim et al., 2012).

Several nontoxic methods have been developed/proposed for extraction of lipids from microalgae such as supercritical CO_2 extraction, osmotic shock-mediated extraction, and use of biocompatible solvents (Sahena et al., 2009). Among these, supercritical fluid extraction (SFE) has been widely practiced as the end product is devoid of any trace of solvent. SFE involves the use of substances that have properties of both liquids and gases (i.e., CO_2) when exposed to increased temperatures and pressures. This property allows them to act as an extracting solvent, leaving no residues behind when the system is brought back to atmospheric pressure and RT. Taher et al. (2014b) reported highest recovery of crude lipids and TAG from *Scenedesmus* biomass compared to conventional solvent extraction (chloroform: methanol, hexane, Soxhlet extraction, etc.). Further, SFE has been used to extract both polar and nonpolar compounds, such as phenolics (polar), fucoxanthin, and PUFAs (nonpolar), from algal biomass (*Sargassum muticum*) by varying the extraction conditions, namely, temperature and pressure (Conde et al., 2015). Li et al. (2014) reported improved recovery of total lipids and PUFAs (EPA and DHA) from *Tetraselmis* sp. compared to conventional solvent extraction method.

The final step in PUFA production involves enrichment of PUFAs from crude lipid extracts. The enrichment process mainly utilizes the specific physicochemical properties (molecular weight, boiling points/freezing points, polarity) of PUFAs. Some of the commonly practiced methods of enrichment are molecular distillation, molecular sieving (membrane process), winterization, urea complexation, column chromatography, etc. Molecular distillation involves purification of fatty acid esters in vacuum based on the different boiling points of the fatty acid esters and molecular sieving process involves separation of fatty acid ester based on membrane permeability and selectivity. Urea complexation involves addition of urea and ethanol to saturation point to lipid extract, heating the mixture, and cooling it to crystal formation. The PUFA crystals are purified by filtration. Winterization involves reduction of temperature rendering the saturated fats insoluble, while the PUFAs which do solidified are separated and dried to purity (Adarme-Vega et al., 2012). Belarbi et al. (2000) reported use of argentated silica gel chromatography for purification of EPA from *P. tricornutum* PUFA concentrates by eluting the crude extracts with mixture of acetone and hexane, further the use of silica helped in removal of chlorophyll contamination from the crude extracts. Further, PUFAs such as EPA, DHA, and AA can be purified by reverse-phase chromatography from PUFA concentrates of the microalgae. The EPA and DHA esters from PUFA concentrates obtained from *I. galbana*, *P. tricornutum*, and *Porphyridium cruentum* were separated and purified by this technique with purity >90% (Molina Grima et al., 2004).

1.24 MAJOR APPLICATIONS OF MICROALGAL PUFAS

Microalgae as source of PUFAs are primarily exploited in aquaculture where the various oleaginous species of microalgae are used as food source (live feed) for commercial rearing of marine organisms such as molluscs, crustaceans, and some species of fish (Brett & Muller–Navarra, 1997; Patil and Raghavarao, 2007). Among the oleaginous microalgae, diatom species such as *S. costatum*, *T. pseudonana*, and *Chaetoceros mulleri*; haptophytes such as *Isochrysis* sp.; chrysophyte such as *Pavlova* sp.; and parsinophytes like *Platymonas* sp., have been used as live feed for penaeid shrimp larvae, bivalve mollusc larvae, freshwater prawn larvae, bivalve mollusc post larvae, abalone larvae, marine rotifers (*Brachionus*), brine shrimp (*Artemia*), saltwater copepods, and freshwater zooplanktons (De Pauw & Persoone, 1988; Brown et al., 1989; Muller-Feuga, 2004). In addition to their application as aquaculture feed, PUFA-rich microalgae are targeted as single cell oil and vegetarian source of PUFAs (Cohen & Khozin-Goldberg, 2005). Doughman et al. (2007) and Bellou and Aggelis (2012) extensively reviewed the health benefits of PUFAs (EPA and DHA) obtained from oleaginous microalgae, such as thraustochytrids especially *Schizochytrium* sp., which are utilized for commercial production of DHA. Dried *Schizochytrium* has GRAS status for use as feed to broiler chickens and laying hen feed in order to enhance DHA in the meat and eggs (Barclay, 1994; Ward & Singh, 2005; Raghukumar, 2008). The EPA/DHA obtained from these microalgae finds application in infant food formulations and as adult health supplements (Ravishankar et al., 2012).

1.25 CONCLUSIONS

This chapter compiles the developments in microalgal biotechnology in the last two decades especially focussing on value-added metabolites such as PBPs, carotenoids, and PUFAs. Attempts have been made to bring out the latest developments in the cultivation of algae, stimulation of the metabolite accumulation, downstream processing, and their application. As indicated in this chapter, the higher productivity of microalgal forms can be utilized for a sustainable production of nutraceuticals which can prevent overexploitation of the existing natural resources. Further, development of cost-effective and innovative production systems with minimal energy inputs and ecofriendly downstream processes would pave the way for wider acceptance of microalgae as industrial feedstock.

KEYWORDS

- **microalgae**
- **phycobiliproteins**
- **carotenoids**
- **polyunsaturated fatty acids**
- **autotrophy**
- **heterotrophy**

REFERENCES

Abalde, J.; Betancourt, L.; Torres, E.; Cid, A.; Barwell, C. Purification and Characterization of Phycocyanin from the Marine Cyanobacterium *Synechococcus* sp. IO9201. *Plant Sci.* **1998,** *136*, 109–120.

Adarme-Vega, T. C.; Lim, D. K. Y.; Timmins, M.; Vernen, F.; Li, Y.; Schenk, P. M. Microalgal Biofactories: A Promising Approach Towards Sustainable Omega-3 Fatty Acid Production. *Microb. Cell Factories* **2012,** *11*(1), 1.

Agarwal, S.; Rao, A. V. Tomato Lycopene and its Role in Human Health and Chronic Diseases. *Canadian Med. Assoc. J.* **2000,** *163*, 739–744.

Anderson, M. Method of Inhibiting 5-Alpha Reductase with Astaxanthin to Prevent and Treat Benign Prostate Hyperplasia (BPH) and Prostate Cancer in Human Males. US Patent 6277417, 2001.

Andrade, M. R.; Costa, J. A. V. Mixotrophic Cultivation of Microalga *Spirulina platensis* Using Molasses as Organic Substrate. *Aquaculture* **2007,** *264*(1–4), 130–134.

Andrich, G.; Nesti, U.; Venturi, F.; Zinnai, A.; Fiorentini, R. Supercritical fluid extraction of bioactive lipids from the microalga *Nannochloropsis* sp. *Eur. J. Lipid Sci. Technol.* **2005,** *107*(6), 381–386.

Antelo, F. S.; Costa, J. A. V.; Kalil, S. J. Purification of C-Phycocyanin from *Spirulina platensis* in Aqueous Two-phase Systems Using an Experimental Design. *Braz. Arch. Biol. Technol.* **2015,** *58*, 1–11.

Arad, S.; Cohen, E.; Ben-Amotz, A. Accumulation of Canthaxanthin in *Chlorella emersonii*. *Physiol. Plant.* **1993,** *87*, 232–236.

Arad, S.; Richmond, A. Industrial Production of Microalgal Cell-mass and Secondary Products—Major Industrial Species. *Porphyridium* sp. In *Hand Book of Microalgal Culture: Biotechnology and Applied Phycology*; Richmond, A., Ed.; Blackwell Science: Oxford, UK, 2004; pp 289–298.

Bai, S.; Dai, J.; Xia, M.; Ruan, J.; Wei, H.; Yu, D.; Li, R.; Jing, H.; Tian, C.; Song, L.; et al. Effects of Intermediate Metabolite Carboxylic Acids of TCA Cycle on Microcystis with Overproduction of Phycocyanin. *Environ. Sci. Pollut. Res.* **2014,** 1–7.

Barclay, W. R. Food Product having High Concentrations of omega-3 Highly Unsaturated Fatty Acids. US Patent 5,340,594, 1994.

Barsanti, L.; Gualtieri, P. *Algae Anatomy, Biochemistry and Biotechnology*; CRC Press: Florida, USA, 2006; pp 15–27.

Becker, W. Microalgae in Human and Animal Nutrition. In *Hand Book of Microalgal Culture: Biotechnology and Applied Phycology*; Richmond, A. Ed.; Blackwell Science, 2004; pp 312–352.

Belarbi, H.; Molina, E.; Chisti, Y. A Process for High and Scaleable Recovery of High Purity Eicosapentaenoic Acid Esters from Microalgae and Fish Oil. *Enzyme Microb. Technol.* **2000,** *26*, 516–29.

Bellou, S.; Aggelis, G. Biochemical Activities in *Chlorella* sp. and *Nannochloropsis salina* During Lipid and Sugar Synthesis in a Lab-scale Open Pond Simulating Reactor. *J. Biotechnol.* **2012,** *164*(2), 318–329.

Ben-Amotz, A. Effect of Low Temperature on the Steroisomer Composition of β-Carotene in the Halotolerant Alga *Dunaliella bardawil* (Chloropyta). *J. Phycol.* **1996,** *32*, 272–275.

Ben-Amotz, A. Industrial Production of Microalgal Cell-mass and Secondary Products—Major Industrial Species *Dunaliella*. In *Hand Book of Microalgal Culture: Biotechnology and Applied Phycology*; Richmond, A. Ed.; Blackwell Science: Oxford, UK, 2004; pp 273–280.

Ben-Amotz, A.; Avron, M. *Dunaliella bardawil* Can Survive Especially High Irradiance Levels by the Accumulation of β-Carotene. *Trends Biotechnol.* **1990,** *8*, 121–126.

Benedetti, S.; Rinalducci, S.; Benvenuti, F.; Francogli, S.; Pagliarani, S.; Giorgi, L.; Micheloni, M.; D'Amici, G. M.; Zolla, L.; Canestrari, F. Purification and Characterization of Phycocyanin from the Blue-Green Alga *Aphanizomenon flos-aquae*. *J. Chromatogr. B* **2006,** *833*, 12–18.

Benemann, J. R. Microalgae Aquaculture Feeds. *J. Appl. Phycol.* **1992,** *4*, 233–245.

Bertolin, T. E.; Farias, D.; Guarienti, C.; Petry, F. T. S.; Colla, L. M.; Costa, J. A. V. Antioxidant Effect of Phycocyanin on Oxidative Stress Induced with Monosodium Glutamate in Rats. *Braz. Arch. Biol. Technol.* **2011,** *54*, 733–738.

Bhat, V. B.; Madyastha, K. M. C-Phycocyanin: A Potent Peroxyl Radical Scavenger In Vivo and In Vitro. *Biochem. Biophys. Res. Commun.* **2000,** *275*, 20–25.

Bhosale, P.; Ermakov, I. V.; Ermakova, M. R.; Gellermann, W.; Bernstein, P. S. Resonant Raman Quantification of Zeaxanthin Production from *Flavobacterium multivorum*. *Biotechnol. Lett.* **2003,** *25*, 1007–1011.

Bigogno, C.; Khozin-Goldberg, I.; Boussiba, S.; Vonshak, A.; Cohen, Z. Lipid and Fatty Acid Composition of the Green Oleaginous Alga *Parietochloris incisa*, The Richest Plant Source of Arachidonic Acid. *Phytochemistry* **2002,** *5*; 497–503.

Binder, A.; Wilson, K.; Zuber, H. C-Phycocyanin from the Thermophilic Blue-green Alga *Mastigocladus laminosus*, Isolation, Characterization and Subunit Composition. *FEBS Lett.* **1972,** *20*, 111–116.

Borowitzka, M. A. Commercial Production of Microalgae: Ponds, Tanks, Tubes and Fermenters. *J. Biotechnol.* **1999,** *70*, 313–321.

Boussiba, S.; Richmond, A. E. C-Phycocyanin as a Storage Protein in the Blue-green Alga *Spirulina platensis*. *Arch. Microbiol.* **1980,** *125*, 143–147.

Boussiba, S.; Richmond, A. E.; Isolation and Characterization of Phycocyanins from the Blue-green Alga *Spirulina platensis*. *Arch. Microbiol.* **1979,** *120,* 155–159.

Bowles, R. D.; Hunt, A. E.; Bremer G. B.; Duchars, M. G.; Eaton, R. A. Long-chain-omega-3 Polyunsaturated Fatty Acid Production by Members of the Marine Protistan Group the Thraustochytrids: Screening the Isolates and Optimisation of DHA Production. *J. Biotechnol.* **1999,** *70*, 193–202.

Brett, M.; Muller-Navarra, D. The Role of Highly Unsaturated Fatty Acids in Aquatic Food Web Processes. *Freshwater Biol.* **1997,** *38*, 483–499.

Brizio, P.; et al. Astaxanthin and Canthaxanthin (Xanthophyll) as Supplements in Rainbow Trout Diet: In Vivo Assessment of Residual Levels and Contributions to Human Health. *J. Agric. Food Chem.* **2013,** *61*, 10954–10959.

Brown, M. R.; Jeffrey, S. W.; Garland, C. D. Nutritional Aspects of Microalgae Used in Mariculture: A Literature Review. *CSIRO Mar. Lab. Rep.* **1989,** *205*, 44 pp.

Bryant, D. A.; Guglielmi, G.; de Marsac, N. T.; Castets, A.-M.; Cohen-Bazire, G. The Structure of Cyanobacterial Phycobilisomes: A Model. *Arch. Microbiol.* **1979,** 123, 113–127.

Cai, C.; Wang, Y.; Li, C.; Guo, Z.; Jia, R.; Wu, W.; Hu, Y.; He, P. Purification and Photodynamic Bioactivity of Phycoerythrin and Phycocyanin from *Porphyra yezoensis Ueda*. *J. Ocean Univ. China* **2014,** *13*, 479–484.

Carlozzi, P. Dilution of Solar Radiation Through "Culture" Lamination in Photobioreactor Rows Facing South–North: A Way to Improve the Efficiency of Light Utilisation of Cyanobacteria (*Arthrospira platensis*). *Biotechnol. Bioeng.* **2003,** *81*, 305–315.

Ceron, M. C.; Campos, I.; Sánchez, J. F.; Acien, F. G.; Molina, E.; Fernandez-Sevilla, J. M. Recovery of Lutein from Microalgae Biomass: Development of a Process for *Scenedesmus almeriensis*. *J. Agric. Food Chem.* **2008,** *56*, 11761–11766.

Chang, G.; Luo, Z.; Gu, S.; Wu, Q.; Chang, M.; Wang, X. Fatty Acid Shifts and Metabolic Activity Changes of *Schizochytrium* sp. S31 Cultured on Glycerol. *Bioresour. Technol.* **2013,** *142*, 255–260.

Chen, F.; Johns, M. R. Effect of C/N Ratio and Aeration on the Fatty Acid Composition of Heterotrophic *Chlorella sorokiniana*. *J. Appl. Phycol.* **1991,** *3*, 203–209.

Chen, F.; Zhang, Y. High Cell Density Mixotrophic Culture of *Spirulina platensis* on Glucose for Phycocyanin Production Using a Fed-Batch System. *Enzyme Microb. Technol.* **1997,** *20*, 221–224.

Chen, F.; Zhang, Y.; Guo, S. Growth and Phycocyanin Formation of *Spirulina platensis* in Photoheterotrophic Culture. *Biotechnol. Lett.* **1996,** *18*, 603–608.

Chen, H.-W.; Yang, T.-S.; Chen, M.-J.; Chang, Y.-C.; Eugene, I.; Wang, C.; Ho, C.-L.; Lai, Y.-J.; Yu, C.-C.; Chou, J.-C.; et al.. Purification and Immunomodulating Activity of C-Phycocyanin from *Spirulina platensis* Cultured Using Power Plant Flue Gas. *Process Biochem.* **2014,** 49, 1337–1344.

Chen, T.; Wong, Y.-S.; Zheng, W. Purification and Characterization of Selenium-Containing Phycocyanin from Selenium-enriched *Spirulina platensis*. *Phytochemistry* **2006,** *67*, 2424–2430.

Chen, T.; Zheng, W.; Yang, F.; Bai, Y.; Wong, Y-S. Mixotrophic Culture of High Selenium-enriched *Spirulina platensis* on Acetate and the Enhanced Production of Photosynthetic Pigments. *Enzyme Microb. Technol.* **2006,** *39*, 103–107.

Chisti, Y. Biodiesel from Micro-algae. *Biotechnol. Adv.* **2007,** *25*, 294–306.

Chiu, H.-F.; Yang, S.-P.; Kuo, Y.-L.; Lai, Y.-S.; Chou, T.-C. Mechanisms Involved in the Antiplatelet Effect of C-Phycocyanin. *Br. J. Nutr.* **2006,** *95*, 435–440.

Chojnacka, K.; Noworyta, A. Evaluation of *Spirulina* sp. Growth in Photoautotrophic, Heterotrophic and Mixotrophic Cultures. *Enzyme Microb. Technol.* **2004,** *34*, 461–465.

Choubert, G.; Heinrich, O. Carotenoid Pigments of Green-algae *Haematococcus pluvialis*: Assay on Rain Bow Trout, *Oncorhynchus mykiss* Pigmentation in Comparison with Synthetic Astaxanthin and Canthaxanthin. *Aquaculture* **1993,** *112*(2–3), 217–226.

Cohen, Z.; Khozin-Goldberg, I. Searching for PUFA Rich Microalgae. In *Single Cell Oils*; Cohen Z., Ratledge C., Eds.; AOCS Press: Urbana, IL, 2005; pp 53–52.

Cohen, Z.; Vonshak, A.; Richmond, A. Effect of Environmental Conditions on Fatty Acid Composition of the Red Alga *Porphyridium cruentum*: Correlation to Growth Rate. *J. Phycol.* **1988,** *24*(3), 328–332.

Conde, E.; Moure, A.; Dominguez, H. Supercritical CO_2 Extraction of Fatty Acids, Phenolics and Fucoxanthin from Freeze-dried *Sargassum muticum*. *J. Appl. Phycol.* **2015,** *27*(2), 957–964.

Cordero, B. F.; et al. Enhancement of Lutein Production in *Chlorella sorokiniana* (Chorophyta) by Improvement of Culture Conditions and Random Mutagenesis. *Mar. Drugs* **2011,** *9*, 1607–1624.

Couto, R. M.; Simões, P. C.; Reis, A.; Da Silva, T. L.; Martins, V. H.; Sánchez-Vicente, Y. Supercritical Fluid Extraction of Lipids from the Heterotrophic Microalga *Crypthecodinium cohnii*. *Eng. Life Sci.* **2010,** *10*(2), 158–164.

Cysewski, G. R.; Lorenz, R. T. Industrial Production of Microalgal Cell-mass and Secondary products—Major Industrial Species *Haematococcus*. In *Hand Book of Microalgal Culture: Biotechnology and Applied Phycology*; Richmond, A., Ed.; Blackwell Science: Oxford, UK, 2004; pp 281–288.

Dangeard, P. Sur une algue bleue alimentaire pour l'homme: *Arthrospira platensis* (Nordst.) Gomont. Actes Soc. Linn. Boreaux Extr. *Proces-verbaux* **1940,** *91*, 39–41, 122.

De Marsac, N. T.; Cohen-Bazire, G. Molecular Composition of Cyanobacterial Phycobilisomes. *Proc. Natl. Acad. Sci.* **1977,** *74*, 1635–1639.

De Pauw, N.; Persoone, G. Micro-algae for Aquaculture. In *Micro-algal Biotechnology*; Borowitzka, M. A., Borowitzka, L. J., Eds.; Cambridge University Press, Cambridge, U.K., 1988; pp 197–221.

de Swaaf, M. E.; Pronk, J. T.; Sijtsma, L. Fed-batch Cultivation of the Docosahexaenoic Acid Producing Marine Alga *Crypthecodinium cohnii* on Ethanol. *Appl. Microbiol. Biotechnol.* **2003,** *61*, 40–43.

Del Campo, J. A.; Moreno, J.; Rodriguez, H.; Vargas, M. A.; Rivas, J.; Guerrero, M. G. Carotenoid Content of Chlorophycean Microalgae: Factors Determining Lutein Accumulation in *Muriellopsis* sp. (Chlorophyta). *J. Biotechnol.* **2000,** *76*, 51–59.

Del Campo, J. A.; Rodriguez, H.; Moreno, J.; Vargas, M. A.; Rivas, J.; Guerrero M. G. Lutein Production by *Muriellopsis* sp. in an Outdoor Tubular Photobioreactor. *J. Biotechnol.* **2001,** 85, 289–295.

Dipak, S. P.; Lele, S. S. Carotenoid production from microalga, *Dunaliella salina*. *Indian J. Biotechnol.* **2005,** *4*, 476–483.

Domonkos, I.; Kis, M.; Gombos, Z.; Ughy, B. Carotenoids, Versatile Components of Oxygenic Photosynthesis. *Prog. Lipid Res.* **2013,** *52*, 539–561.

Doughman, S. D.; Krupanidhi, S.; Sanjeevi, C. B. Omega-3 Fatty Acids for Nutrition and Medicine: Considering Microalgae Oil as a Vegetarian Source of EPA and DHA. *Curr. Diabetes Rev.* **2007,** *3*, 198–203.

Dufossé, L.; Galaup, P.; Yaron, A.; Arad, S. M.; Blanc, P.; Chidambara Murthy, K. N.; Ravishankar, G. A. Microorganisms and Microalgae as Sources of Pigments for Food Use: A Scientific Oddity or Industrial Reality? *Trends Food Sci. Technol.* **2005,** *16*, 389–406.

Edge, R.; McGarvey, D. J.; Truscott, T. G. The Carotenoids as Anti-oxidants—A Review. *J. Photochem. Photobiol. B* **1997,** *41*, 189–200.

Eriksen, N. T. Production of Phycocyanin—A Pigment with Applications in Biology, Biotechnology, Foods and Medicine. *Appl. Microbiol. Biotechnol.* **2008,** *80*, 1–14.

Esfahani-Mashhour, M.; Moravej, H.; Mehrabani-Yeganeh, H.; Razavi, S. H.; et al. Evaluation of Coloring Potential of *Dietzia natronolimnaea* Biomass as Source of Canthaxanthin for Egg Yolk Pigmentation. *Asian-Austral. J. Anim. Sci.* **2009,** *22*, 254.

FAO. Review of the State of World Marine Fishery Resources. *Fisheries and Aquaculture Technical Paper No. 569*. FAO: Rome, Italy, 2011.

Farooq, S. M.; Asokan, D.; Kalaiselvi, P.; Sakthivel, R.; Varalakshmi, P. Prophylactic Role of Phycocyanin: A Study of Oxalate Mediated Renal Cell Injury. *Chem.-Biol. Interact.* **2004,** *149*, 1–7.

Farooq, S. M.; Boppana, N. B.; Asokan, D.; Sekaran, S. D.; Shankar, E. M.; Li, C.; Gopal, K.; Bakar, S. A.; Karthik, H. S.; Ebrahim, A. S. C-Phycocyanin Confers Protection against Oxalate-Mediated Oxidative Stress and Mitochondrial Dysfunctions in MDCK Cells. *PLoS ONE* **2014,** *9*, e93056.

Fernández-Rojas, B.; Hernández-Juárez, J.; Pedraza-Chaverri, J. Nutraceutical Properties of Phycocyanin. *J. Funct. Foods* **2014a,** *11*, 375–392.

Fernández-Rojas, B.; Medina-Campos, O. N.; Hernández-Pando, R.; Negrette-Guzmán, M.; Huerta-Yepez, S.; Pedraza-Chaverri, J. C-Phycocyanin Prevents Cisplatin-induced Nephrotoxicity through Inhibition of Oxidative Stress. *Food Funct.* **2014b,** *5*, 480–490.

Fernandez-Sevilla, J. M.; Acien Fernandez, F. G.; Molina Grima, E. Biotechnological Production of Lutein and Its Applications. *Appl. Microbiol. Biotechnol.* **2010,** *86*, 27–40.

Fernandez-Sevilla, J. M.; Molina Grima, E.; Carcia Camacho, F.; Acien Fernandez, F. G.; Sanchez Perez, J. A. Photolimitation and Photoinhibition as Factors Determining Optimal Dilution Rate to Produce Eicosapentaenoic Acid from Cultures of the Microalga *Isochrysis galbana*. *Appl. Microbiol. Biotechnol.* **1998,** *50*, 199–205.

Gantar, M.; Dhandayuthapani, S.; Rathinavelu, A. Phycocyanin Induces Apoptosis and Enhances the Effect of Topotecan on Prostate Cell Line LNCaP. *J. Med. Food* **2012,** *15*, 1091–1095.

Gantt, E.; Conti, S. F. Granules Associated with the Chloroplast Lamellae of *Porphyridium cruentum*. *J. Cell Biol.* **1966,** *29*, 423–434.

Garcia, M. C. C.; Sevilla, J. M. F.; Fernandez, F. G. A.; Grima, E. M.; Camacho, F. G. Mixotrophic Growth of *Phaeodactylum tricornutum* on Glycerol: Growth Rate and Fatty Acid Profile. *J. Appl. Phycol.* **2000,** 12, 239–248.

Garcia-Gonzalez, M.; Moreno, J.; Manzano, J. C.; Florencio, F. J.; Guerrero, M. G. Production of *Dunaliella salina* Biomass Rich in 9-Cis-β-carotene and Lutein in a Closed Tubular Photobioreactor. *J. Biotechnol.* **2005,** *115*, 81–90.

Gardeva, E.; Toshkova, R.; Yossifova, L.; Minkova, K.; Ivanova, N.; Gigova, L. Antitumor Activity of C-Phycocyanin from *Arthronema africanum* (Cyanophyceae). *Braz. Arch. Biol. Technol.* **2014,** *57*(5), 675–684.

Ge, B.; Tang, Z.; Zhao, F.; Ren, Y.; Yang, Y.; Qin, S. Scale-up of Fermentation and Purification of Recombinant Allophycocyanin Over-expressed in *Escherichia coli*. *Process Biochem.* **2005,** *40*, 3190–3195.

Gladue, R. M.; Maxey, J. E. Microalgal Feeds for Aquaculture. *J. Appl. Phycol.* **1994,** *6*, 131–141.

Glazer, A. N. Phycobiliproteins—A Family of Valuable, Widely Used Fluorophores. *J. Appl. Phycol.* **1994,** *6*, 105–112.

Gomez, P. I.; Gonzalez, M. A. Genetic Variation among Seven Strains of *Dunaliella salina* (Chlorophyta) with Industrial Potential, Based on RAPD Banding Patterns and on Nuclear ITS rDNA Sequences. *Aquaculture* **2004,** *233*, 149–162.

Gordillo, F. J. L.; Goutx, M.; Figueroa, F. L.; Niell, F. X. Effects of Light Intensity, CO_2 and Nitrogen Supply on Lipid Class Composition of *Dunaliella viridis*. *J. Appl. Phycol.* **1998,** *10*, 135–144.

Griffiths, M. J.; van Hille, R. P.; Harrison, S. T. Lipid Productivity, Settling Potential and Fatty Acid Profile of 11 Microalgal Species Grown under Nitrogen Replete and Limited Conditions. *J. Appl. Phycol.* **2012,** *24*(5), 989–1001.

Grobbelaar, J. U. Photosynthetic Characteristics of *Spirulina platensis* Grown in Commercial-Scale Open Outdoor Raceway Ponds: What the Organisms Tell Us? *J. Appl. Phycol.* **2007,** *19*, 591–598.

Gross, W.; Schnarrenberger, C. Heterotrophic Growth of Two Strains of the Acido-Thermophilic Red Alga *Galdieria sulphuraria*. *Plant Cell Physiol.* **1995,** *36*, 633–638.
Gruner, S.; Volk, H.; Falck, P.; von Baehr, R. The Influence of Phagocytic Stimuli on the Expression of HLA-DR Antigen: Role of Reactive Oxygen Intermediates. *Eur. J. Immunol.* **1986,** *16*, 212–215.
Grung, M.; Metzger, P.; Liaaen Jensen, S. Algal Carotenoids 53; Secondary Carotenoids of Algae 4; Secondary Carotenoids in the Green Alga *Botryococcus braunii*, Race L, New Strain. *Biochem. Syst. Ecol.* **1994,** *22*, 25–29.
Guedes, A. C.; Amaro, H. M.; Barbosa, C. R.; Pereira, R. D.; Malcata, F. X. Fatty Acid Composition of Several Wild Microalgae and Cyanobacteria, with a Focus on Eicosapentaenoic, Docosahexaenoic and α-Linolenic Acids for Eventual Dietary Uses. *Food Res. Int.* **2011,** *44*, 2721–2729.
Guedes, A. C.; Amaro, H. M.; Malcata, F. X. Microalgae as Source of Carotenoids. *Mar. Drugs* **2011,** 9, 625–644.
Guiheneuf, F.; Stengel, D. B. LC-PUFA-Enriched Oil Production by Microalgae: Accumulation of Lipid and Triacylglycerols Containing n-3 LC-PUFA Is Triggered by Nitrogen Limitation and Inorganic Carbon Availability in the Marine Haptophyte *Pavlova lutheri*. *Mar. Drugs* **2013,** *11*, 4246–4266.
Guschina, I. A.; Harwood, J. L. Algal Lipids and Effect of the Environment on their Biochemistry. In *Lipids in Aquatic Ecosystems*; Arts, M. T., Brett, M. T., Kainz, M., Eds.; Springer: New York, 2009; pp 1–24.
Halim, R.; Danquah, M. K.; Webley, P. A. Extraction of Oil from Microalgae for Biodiesel Production: A Review. *Biotechnol. Adv.* **2012,** *30*, 709–732.
Hayashi, O.; Ono, S.; Ishii, K.; Shi, Y. H.; Hirahashi, T.; Katoh, T. Enhancement of Proliferation and Differentiation in Bone Marrow Hematopoietic Cells by *Spirulina (Arthrospira) platensis* in Mice. *J. Appl. Phycol.* **2006,** *18*, 47–56.
Hejazi, M. A.; Holwerda, E.; Wiffels, R. H. Milking Microalga *Dunaliella salina* for β-Carotene Production in Two-phase Bioreactors. *Biotechnol. Bioeng.* **2004,** *85*, 475–481.
Hsiao, G.; Chou, P.-H.; Shen, M.-Y.; Chou, D.-S.; Lin, C.-H.; Sheu, J.-R. C-Phycocyanin, a Very Potent and Novel Platelet Aggregation Inhibitor from *Spirulina platensis*. *J. Agric. Food Chem.* **2005,** *53*, 7734–7740.
Hu, H.; Gao, K. Response of Growth and Fatty Acid Compositions of *Nannochloropsis* sp. to Environmental Factors under Elevated CO_2 Concentration. *Biotechnol. Lett.* **2006,** *28*, 987–992.
Hu, Q. Industrial Production of Microalgal Cell-Mass and Secondary Products—Major Industrial Species. *Arthospira (Spirulina) platensis*. In *Hand Book of Microalgal Culture: Biotechnology and Applied Phycology*; Richmond, A., Ed.; Blackwell Science: Oxford, UK, 2004; pp 264–273.
Hu, Q.; Sommerfeld, M.; Jarvis, E.; Ghirardi, M.; Posewitz, M.; Seibert, M.; Al Darzins. Microalgal Triacylglycerols as Feedstocks for Biofuel Production: Perspectives and Advances. *Plant J.* **2008,** *54*, 621–639.
Huang, X.; Huang, Z.; Wen, W.; Yan, J. Effects of Nitrogen Supplementation of the Culture Medium on the Growth, Total Lipid Content and Fatty Acid Profiles of Three Microalgae (*Tetraselmis subcordiformis*, *Nannochloropsis oculata* and *Pavlova viridis*). *J. Appl. Phycol.* **2013,** *25*(1), 129–137.

Huerlimann, R.; de Nys, R.; Heimann, K. Growth, Lipid Content, Productivity, and Fatty Acid composition of Tropical Microalgae for Scale-up Production. *Biotechnol. Bioeng.* **2010,** *107*, 245–257.

Hussein, M. M.; Ali, H. A.; Ahmed, M. M. Ameliorative Effects of Phycocyanin against Gibberellic Acid Induced Hepatotoxicity. *Pest. Biochem. Physiol.* **2015,** *119*, 28–32.

Hwang, J.-H.; Chen, J.-C.; Chan, Y.-C. Effects of C-Phycocyanin and *Spirulina* on Salicylate-induced Tinnitus, Expression of NMDA Receptor and Inflammatory Genes. *PLoS ONE* **2013,** *8*, e58215.

Ibañez, E.; Herrero, M.; Mendiola, J. A.; Castro-Puyana, M. Extraction and Characterization of Bioactive Compounds with Health Benefits from Marine Resources: Macro and Micro Algae, Cyanobacteria, and Invertebrates, In *Marine Bioactive Compounds*; Springer, 2012; pp 55–98.

Ip, P. F.; Chen, F. Production of Astaxanthin by the Green Microalga *Chlorella zofingiensis* in the Dark. *Process Biochem.* **2005,** *40*, 733–738.

Jensen, G. S.; Attridge, V. L.; Beaman, J. L.; Guthrie, J.; Ehmann, A.; Benson, K. F. Antioxidant and Anti-Inflammatory Properties of an Aqueous Cyanophyta Extract Derived from *Arthrospira platensis*: Contribution to Bioactivities by the Non-phycocyanin Aqueous Fraction. *J. Med. Food* **2015**.

Jian-Feng, N. I. U.; Guang-Ce, W.; Lin, X.; Zhou, B.-C. Large-Scale Recovery of C-Phycocyanin from *Spirulina platensis* using Expanded Bed Adsorption Chromatography. *J. Chromatogr. B* **2007,** *850*, 267–276.

Jiménez, C.; Cossío, B. R.; Labella, D.; Niell, F. X. The Feasibility of Industrial Production of *Spirulina* (*Arthrospira*) in Southern Spain. *Aquaculture* **2003,** *217*, 179–190.

Jin, E.; Feth, B.; Melis. A. A Mutant of the Green Alga *Dunaliella salina* Constitutively Accumulates Zeaxanthin under All Growth Conditions. *Biotechnol. Bioeng.* **2003,** *81*(1), 115–124.

Jin, E.; Lee, C. G.; Polle, J. E. W. Secondary Carotenoid Accumulation in *Haematococcus* (chlorophyceae): Biosynthesis, Regulation and Biotechnology. *J. Microbiol. Biotechnol.* **2006,** *16*, 821–831.

Juin, C.; Chérouvrier, J.-R.; Thiéry, V.; Gagez, A.-L.; Bérard, J.-B.; Joguet, N.; Kaas, R.; Cadoret, J.-P.; Picot, L. Microwave-Assisted Extraction of Phycobiliproteins from *Porphyridium purpureum*. Appl. Biochem. Biotechnol. **2014,** *175*, 1–15.

Jyonouchi, H.; Sun, S.; Gross, M. Effect of Carotenoids on In Vitro Immunoglobulin Production by Human Peripheral Blood Mononuclear Cells: Astaxanthin, a Carotenoid without Vitamin A Activity, Enhances In Vitro Immunoglobulin Production in Response to a T-dependent Stimulant and Antigen. *Nutr. Cancer* **1994,** *23*, 171–183.

Kadian, S. S.; Garg, M. Pharmacological Effects of Carotenoids: A Review. *Int. J. Pharm. Sci. Res.* **2012,** *3*, 42–48.

Kamath, B. S.; Srikanta, B. M.; Dharmesh, S. M.; Sarada, R.; Ravishankar, G. A. Ulcer Preventive and Antioxidative Properties of Astaxanthin from *Haematococcus pluvialis*. *Eur. J. Pharmacol.* **2008,** *590*, 387–395.

Kamath, S. B. Biotechnological Production of Micro-algal Carotenoids with Reference to Astaxanthin and Evaluation of Its Biological Activity. Ph. D. Thesis. University of Mysore: Mysore, India, 2007.

Kang, C. D.; Sim. S. J. Direct Extraction of Astaxanthin from *Haematococcus* Culture Using Vegetable Oils. *Biotechnol. Lett.* **2008,** *30*, 441–444.

Khozing-Goldberg, I.; Cohen, Z. The Effect of Phosphate Starvation on the Lipid and Fatty Acid Composition of the Fresh Water Eustigmatophyte *Monodus subterraneus*. *Phytochemistry* **2006,** *67*, 696–701.

Khozin-Goldberg, I.; Bigogno, C.; Shrestha, P.; Cohen, Z. Nitrogen Starvation Induces the Accumulation of Arachidonic Acid in the Freshwater Green Alga *Parietochloris incisa* (Trebouxiophyceae). *J. Phycol.* **2002,** *38*, 991–994.

Kim, J.; Yoo, G.; Lee, H.; Lim, J.; Kim, K.; Kim, C. W.; Park, M. S.; Yang, J-W. Methods of Downstream Processing for Production of Biodiesel from Microalgae. *Biotech. Adv.* **2013,** *31*, 862–876.

Kitada, K.; Machmudah, S, Sasaki, M.; Goto, M.; Nakashima, Y.; Kumamoto, S.; Hasegawa, T. Supercritical CO_2 Extraction of Pigment Components with Pharmaceutical Importance from *Chlorella vulgaris*. *J. Chem. Technol. Biotechnol.* **2009,** *84*(5), 657–661.

Kotake-Nara, E.; et al. Carotenoids Affect Proliferation of Human Prostate Cancer Cells. *J. Nutr.* **2001,** *131*, 3303–3306.

Kuddus, M.; Singh, P.; Thomas, G.; Al-Hazimi, A. Recent Developments in Production and Biotechnological Applications of C-Phycocyanin. *BioMed Res. Int.* **2013**1-9.

Kulkarni, S. U.; Badakere, S. S.; Oswald, J.; Kamat, M. Y. Fluorescent Phycocyanin from *Spirulina platensis. Application for Diagnosis. Biotech. Lab Int. 1996*; *September–October*, 1996; pp 14–16.

Kumar, D.; Dhar, D. W.; Pabbi, S.; Kumar, N.; Walia, S. Extraction and Purification of C-Phycocyanin from *Spirulina platensis* (CCC540). *Indian J. Plant Physiol.* **2014,** *19*, 184–188.

Kumari, R. P.; Anbarasu, K. Protective Role of C-Phycocyanin against Secondary Changes during Sodium Selenite Mediated Cataractogenesis. *Nat. Prod. Bioprospect.* **2014,** *4*, 81–89.

Kumari, R. P.; Ramkumar, S.; Thankappan, B.; Natarajaseenivasan, K.; Balaji, S.; Anbarasu, K. Transcriptional Regulation of Crystallin, Redox, and Apoptotic Genes by C-Phycocyanin in the Selenite-induced Cataractogenic Rat Model. *Mol. Vis.* **2015,** *21*, 26.

Kupka, M.; Scheer, H. Unfolding of C-Phycocyanin Followed by Loss of Non-covalent Chromophore–protein Interactions: 1. Equilibrium Experiments. *Biochim. Biophys. Acta (BBA)—Bioenerget.* **2008,** *1777*, 94–103.

Lang, I.; Hodac, L.; Friedl, T.; Feussner, I. Fatty Acid Profiles and their Distribution Patterns in Microalgae: A Comprehensive Analysis of More Than 2000 Strains from the SAG Culture Collection. *BMC Plant Biol.* **2011,** *11*(1), 124.

Larkum, A. W. D. Limitation and Prospects of Natural Photosynthesis for Bio-energy Production. *Curr. Opin. Biotechnol.* **2010,** *21*, 271–276.

Lee, T. L. Marino, G. E. Microalgae for "Healthy" Foods—Possibilities and Challenges. *Compr. Rev. Food Sci. Food Saf.* **2010,** *9*, 655–675.

Lee, Y-K. Commercial Production of Microalgae in the Asia Pacific Rim. *J. Appl. Phycol.* **1997,** *9*, 403–411.

León, R.; Garbayo, I.; Hernández, R.; Vigara Vílchez, C. Organic Solvent Toxicity in Photoautotrophic Unicellular Microorganisms. *Enzyme Microb. Technol.* **2001,** *29*, 173–180.

Leonard, J.; Compere, P. *Spirulina platensis* (Gom.) Geitl., algue bleue de grande valeur alimentaire par sa richesse en proteines. *Bull Jard. Bot. Nat. Belg.* **1967,** *37*(Suppl.), 1–23.

Levin, G.; Mokady, S. Antioxidant Activity of 9-Cis Compared to All Trans-beta-Carotene In Vitro. *Free Rad. Biol. Med.* **1994,** *17*, 77–82.

Lewis, T. E.; Nichols, P. D.; McMeekin, T. A. The Biotechnological Potential of Thrausochytrids. *Mar. Biotechnnol.* **1999,** *1*(6), 580–587.

Li, B.; Gao, M.-H.; Chu, X.-M.; Teng, L.; Lv, C.-Y.; Yang, P.; Yin, Q.-F. The Synergistic Antitumor Effects of All-trans Retinoic Acid and C-Phycocyanin on the Lung Cancer A549 Cells In Vitro and In Vivo. *Eur. J. Pharmacol.* **2015,** *749*, 107–114.

Li, X.; Ma, L.; Zheng, W.; Chen, T. Inhibition of Islet Amyloid Polypeptide Fibril Formation by Selenium-containing Phycocyanin and Prevention of Beta Cell Apoptosis. *Biomaterials* **2014,** *35*, 8596–8604.

Li, Y.; Naghdi, F. G.; Garg, S.; Adarme-Vega, T. C.; Thurecht, K. J.; Ghafor, W. A.; Tannock, S.; Schenk, P. M. A Comparative Study: The Impact of Different Lipid Extraction Methods on Current Microalgal Lipid Research. *Microb. Cell Factories* **2014,** 13, 14.

Lian, F.; Hu, K.-Q.; Russell, R. M.; Wang, X.-D. β-Cryptoxanthin Suppresses the Growth of Immortalized Human Bronchial Epithelial Cells and Non-small-cell Lung Cancer Cells and Up-regulates Retinoic Acid Receptor β-Expression. *Int. J. Cancer* **2006,** *119*, 2084–2089.

Liao, H. H.; Medwid, R. D.; Heefner, D. L.; Sniff, K. S.; Hassler, R. A.; Yarus, M. J. inventors. Carotenoid Producing Culture Using *Nespongiococcum excentricum*. U.S. Patent 5,437,997, 1995.

Lichtenstein AH, Appel LJ, Brands M, Carnethon M, Daniels S, Franch HA Diet and Lifestyle Recommendations Revision 2006—A Scientific Statement From the American Heart Association Nutrition Committee. *Circulation* **2006,** *114*, 82–96.

Lin J.-H.; Lee D.-J.; Chang J.-S. Lutein Production from Biomass: Marigold Flowers versus Microalgae. Bioresour. Technol. **2015,** 184, 421–428.

Liu, L.-N.; Chen, X.-L.; Zhang, X.-Y.; Zhang, Y.-Z.; Zhou, B.-C. One-step Chromatography Method for Efficient Separation and Purification of R-Phycoerythrin from *Polysiphonia urceolata*. *J. Biotechnol.* **2005,** *116*, 91–100.

Liu, Q.; Wang, Y.; Cao, M.; Pan, T.; Yang, Y.; Mao, H.; Sun, L.; Liu, G. Anti-allergic Activity of R-Phycocyanin from *Porphyra haitanensis* in Antigen-Sensitized Mice and Mast Cells. *Int. Immunopharmacol.* **2015** *25*(2), 465–473.

Lorenz, R. T.; Cysewski, G. R. Commercial Potential for *Haematococcus* Microalgae as a Natural Source of Astaxanthin. *Trends Biotechnol.* **2000,** *18*, 160–167.

Ma, L.; Yan, S.-F.; Huang, Y.-M.; Lu, X.-R.; Qjan, F.; Pang, H.-L.; Xu, X.-R.; Zou, Z.-Y.; Dong, P.-C.; Xiao, X.; Wang, X.; Sun, T.-T.; Dou, H.-L.; Lin, X.-M. Effect of Lutein and Zeaxanthin on Macular Pigment and Visual Function in Patients with Early Age Related Macular Degeneration. *Opthamology* **2012,** *119*(11), 2290–2297.

Machmudah, S.; Shotipruk, A.; Goto, M.; Sasaki, M.; Hirose, T. Extraction of Astaxanthin from *Haematococcus pluvialis* Using Supercritical CO_2 and Ethanol as Entrainer. *Ind. Eng. Chem. Res.* **2006,** *45*, 3652–3657.

Macías-Sánchez, M. D.; Mantell Serrano, C.; Rodríguez Rodríguez, M.; Martínez de la Ossa, E. Kinetics of the Supercritical Fluid Extraction of Carotenoids from Microalgae with CO_2 and Ethanol as Cosolvent. *Chem. Eng. J.* **2009,** *150*, 104–113.

Macías-Sánchez, M. D.; Mantell, C.; Rodríguez, M.; Martínez de la Ossa, E.; Lubián, L. M.; Montero, O. Supercritical Fluid Extraction of Carotenoids and Chlorophyll *a* from *Nannochloropsis gaditana*. *J. Food Eng.* **2005,** *66*, 245–251.

Madhyastha, H. K.; Radha, K. S.; Sugiki, M.; Omura, S.; Maruyama, M. Purification of C-Phycocyanin from *Spirulina fusiformis* and Its Effect on the Induction of Urokinase-type Plasminogen Activator from Calf Pulmonary Endothelial Cells. *Phytomedicine* **2006,** *13*, 564–569.

Madhyastha, H. K.; Sivashankari, S.; Vatsala, T. M. C-Phycocyanin from *Spirulina fussiformis* Exposed to Blue Light Demonstrates Higher Efficacy of *In Vitro* Antioxidant Activity. *Biochem. Eng. J.* **2009,** *43*, 221–224.

Maeda, H.; Hosokawa, M.; Sashima, T.; Funayama, K.; Miyashita, K. Fucoxanthin from Edible Seaweed, *Undaria pinnatifida*, Shows Antiobesity Effect Through UCP1 Expression in White Adipose Tissues. *Biochem. Biophys. Res. Commun.* **2005,** *332*, 392–397.

Makri, A.; Bellou, S.; Birkou, M.; Papatrehas, K.; Dolapsakis, N. P.; Bokas, D.; et al. Lipid Synthesized by Micro-algae Grown in Laboratory and Industrial-Scale Bioreactors. *Eng. Life Sci.* **2011,** *11*(1), 52–58.

Mallick, N.; Mohn, F. H. Reactive Oxygen Species: Response of Algal Cells. *J. Plant Physiol.* **2000,** *157*, 183–193.

Marquez, F. J.; Nishio, N.; Nagai, S. Enhancement of Biomass and Pigment Production During Growth of *Spirulina platensis* in Mixotrophic Culture. *J. Chem. Tech. Biotechnol.* **1995,** *62*, 159–164.

Marquez, F. J.; Sasaki, K.; Kakizono, T.; Nishio, N.; Nagai, S. Growth Characterization of *Spirulina platensis* in Mixotrophic and Heterotrophic Conditions. *J. Ferment. Bioeng.* **1993,** *76*, 408–410.

Martelli, G.; Folli, C.; Visai, L.; Daglia, M.; Ferrari, D. Thermal Stability Improvement of Blue Colorant C-Phycocyanin from *Spirulina platensis* for Food Industry Applications. *Process Biochem.* **2014,** *49*, 154–159.

Mata, T. M.; Martins, A. A.; Caetano, N. S. Micro-algae for Biodiesel Production and Other Applications: A Review. *Renew. Sustain. Energy Rev.* **2010,** *14*, 217–232.

Melanson SF, Lewandrowski EL, Lewandrowski KB Measurement of Organochlorines in Commercial Over-the-counter Fish Oil Preparations. *Arch. Pathol. Lab. Med.* **2005,** 129:74–77.

Mendes-Pinto, M. M.; Raposo, M. F. J.; Bowen, J.; Young, A. J.; Morais, R. Evaluation of Different Cell Disruption Processes on Encysted Cells of *Haematococcus pluvialis*: Effects on Astaxanthin Recovery and Implications for Bioavailability. *J. Appl. Phycol.* **2001,** *13*, 19–24.

Mercer, P.; Armenta, R. E. Developments in Oil Extraction from Microalgae. *Eur. J. Lipid Sci. Technol.* **2011,** *113*, 539–547.

Minkova, K. M.; Tchernov, A. A.; Tchorbadjieva, M. I.; Fournadjieva, S. T.; Antova, R. E.; Busheva, M. C. Purification of C-Phycocyanin from *Spirulina (Arthrospira)* fusiformis. *J. Biotechnol.* **2003,** *102*, 55–59.

Miyashita, K.; *et al.* The Allenic Carotenoid Fucoxanthin, A Novel Marine Nutraceutical from Brown Seaweeds. *J. Sci. Food Agric.* **2011,** *91*, 1166–1174.

Molina Grima, E.; Acien Fernandez, F. G.; Medina, A. R. Downstream Processing of Cell-mass and Products. In *Hand Book of Microalgal Culture: Biotechnology and Applied Phycology*; Richmond, A., Ed.; Blackwell Science: Oxford, UK, 2004; pp 215–251.

Moon, M.; Mishra, S. K.; Kim, C. W.; Suh, W. I.; Park, M. S.; Yang, J.-W. Isolation and Characterization of Thermostable Phycocyanin from *Galdieria sulphuraria*. *Kor. J. Chem. Eng.* **2014,** *31*, 490–495.

Moraes, C. C.; Kalil, S. J. Strategy for a Protein Purification Design Using C-Phycocyanin Extract. *Bioresour. Technol.* **2009,** *100*, 5312–5317.

Moraes, C. C.; Sala, L.; Ores, J. da C.; Braga, A. R. C.; Costa, J. A. V.; Kalil, S. J. Expanded and Fixed Bed Ion Exchange Chromatography for the Recovery of C-phycocyanin in a Single Step by Using Lysed Cells. *Can. J. Chem. Eng.* **2015,** *93*, 111–115.

Moreno, J.; Vargas, M. A.; Rodríguez, H.; Rivas, J.; Guerrero, M. G. Outdoor Cultivation of a Nitrogen-fixing Marine Cyanobacterium, *Anabaena* sp. ATCC 33047. *Biomol. Eng.* **2003,** *20*, 191–197.

Moriguchi, S.; Jackson, J. C.; Watson, R. R. Effect of Retinoids on Human Lymphocyte Functions In Vitro. *Hum. Toxicol.* **1985,** *4*, 365–378.

Muller-Feuga, A. Microalgae for Aquaculture—the Current Global Situation and Future Trends. In *Hand Book of Microalgal Culture: Biotechnology and Applied Phycology*; Richmond, A., Ed.; Blackwell Science: Oxford, UK, 2004; pp 352–365.

Naguib, Y. M. A. Antioxidative Activities of Astaxanthin and Related Carotenoids. *J. Agric. Food Chem.* **2000,** *48*, 1150–1154.

Naidu, K. A.; Sarada, R.; Manoj, G.; Khan, M. Y.; Swamy, M. M.; Viswanatha, S.; Murthy, K. N.; Ravishankar, G. A.; Srinivas, L. Toxicity Assessment of Phycocyanin—A Blue Colorant from Blue Green Alga *Spirulina platensis*. *Food Biotechnol.* **1999,** *13*, 51–66.

Nakazawa, Y.; Sashima, T.; Hosokawa, M.; Miyashita, K. Comparative Evaluation of Growth Inhibitory Effect of Stereoisomers of Fucoxanthin in Human Cancer Cell Lines. *J. Funct. Foods* **2009,** *1*, 88–97.

Nield, J.; Rizkallah, P. J.; Barber, J.; Chayen, N. E. The 1.45\AA Three-dimensional Structure of C-Phycocyanin from the Thermophilic Cyanobacterium *Synechococcus elongatus*. *J. Struct. Biol.* **2003,** *141*, 149–155.

Nishino, H.; Murakoshi, M.; Tokuda, H.; Satomi, Y. Cancer Prevention by Carotenoids. *Arch. Biochem. Biophys.* **2009,** *483*, 165–168.

Nobre, B.; Marcelo, F.; Passos, R.; Beiro, L.; Palavra, A.; Gouveia, L.; Mendes, R. Supercritical Carbon Dioxide Extraction of Astaxanthin and Other Carotenoids from the Microalga *Haematococcus pluvialis*. *Eur. Food Res. Technol.* **2006,** *223*, 787–790.

Okada, S.; Mastuda, M.; Murakami, M.; Yamaguchi, K. Botryoxanthin-A a New Member of the New Class of Carotenoids From Green Microalga *Botryococcus braunii*, Berkley. *Tetrahedron Lett.* **1996,** *37*, 1065–1068.

Okada, S.; Tonegawa, I.; Mastuda, M.; Murakami, M.; Yamaguchi, K. Botryoxanthin-B and Alpha Botryoxanthin-A from Green Microalga *Botryococcus braunii*. *Phytochemistry* **1998,** *47*, 1111–1115.

Okada, S.; Tonegawa, I.; Mastuda, M.; Murakami, M.; Yamaguchi, K. Braunixaxnthins 1 and 2, New Carotenoids from the Green Microalga *Botryococcus braunii*. *Tetrahedron* **1997,** *53*, 11307–11316.

Ota, M.; Kato, Y.; Watanabe, M.; Sato, Y.; Smith, R. L.; Inomata, H. Fatty Acid Production from a Highly CO_2 Tolerant Alga, *Chlorococcum llittorale*, in the Presence of Inorganic Carbon and Nitrate. *Bioresour. Technol.* **2009,** *100*, 5237–5242.

Ou, Y.; Zheng, S.; Lin, L.; Jiang, Q.; Yang, X. Protective Effect of C-Phycocyanin Against Carbon Tetrachloride-induced Hepatocyte Damage In Vitro and In Vivo. *Chemico-biol. Interact.* **2010,** *185*, 94–100.

Pal D.; Khozin-Goldberg, I.; Cohen, Z.; Boussiba, S. The Effect of Light Salinity and Nitrogen Availability on Lipid Production by *Nannochloropsis* sp. *Appl. Microbiol. Biotechnol.* **2011,** *90*, 1429–1441.

Palozza, P.; Calviello, G.; Emilia De Leo, M.; Serini, S.; Bartoli, G. M. Canthaxanthin Supplementation Alters Antioxidant Enzymes and Iron Concentration in Liver of Balb/c Mice. *J. Nutr.* **2000,** *130*, 1303–1308.

Pangestuti, R.; Kim, S.-K. Biological Activities and Health Benefit Effects of Natural Pigments Derived from Marine Algae. *J. Funct. Foods* **2011,** *3*, 255–266.

Parajo, J. C.; Santos, V.; Vazquez, M. Production of Carotenoids by *P. rhodozyma* Growing on Media Made from Hemi-cellulosic Hydrolysates of *Eucalyptus globulus* wood. *Biotechnol. Bioeng.* **1998,** *59*, 501–506.

Park, E. K.; Lee, C. G. Astaxanthin Production by *Haematococcus pluvialis* under Various Light Intensities and Wavelengths. *J. Microbiol. Biotechnol.* **2001,** *11*, 1024–1030.

Patel, A.; Mishra, S.; Pawar, R.; Ghosh, P. K. Purification and Characterization of C-Phycocyanin from Cyanobacterial Species of Marine and Freshwater Habitat. *Protein Express. Purif.* **2005**, *40*, 248–255.

Patil, G.; Chethana, S.; Madhusudhan, M. C.; Raghavarao, K. Fractionation and Purification of the Phycobiliproteins from *Spirulina platensis*. *Bioresour. Technol.* **2008,** *99*, 7393–7396.

Patil, G.; Raghavarao, K. Aqueous Two Phase Extraction for Purification of C-Phycocyanin. *Biochem. Eng. J.* **2007**, 34, 156–164.

Patil, V.; Källqvist, T.; Olsen, E.; Vogt, G.; Gislerød, H. R. Fatty Acid Composition of 12 Microalgae for Possible Use in Aquaculture Feed. *Aquacult. Int.* **2007,** *15*(1), 1–9.

Pelah, D.; Sintov, A.; Cohen, E. The Effect of Salt Stress on the Production of Canthaxanthin and Astaxanthin by *Chlorella zofingiensis* Grown under Limited Light Intensity. *World J. Microbiol. Biotechnol.* **2004,** *20*, 483–486.

Pleonsil, P.; Soogarun, S.; Suwanwong, Y. Anti-oxidant Activity of Holo- and Apo-C-Phycocyanin and their Protective Effects on Human Erythrocytes. *Int. J. Biol. Macromol.* **2013,** *60*, 393–398.

Posten, C. Design Principles of Photo-bioreactors for Cultivation of Micro-algae. *Eng. Life Sci.* **2009,** *9*(3), 165–177.

Pulz, O. Photobioeractors: Production Systems for Phototrophic Microorganisms. *Appl. Microbiol. Biotechnol.* **2001,** *57*, 287–293.

Pushparaj, B.; Pelosi, E.; Tredici, M. R.; Pinzani, E.; Materassi, R. An Integrated Culture System for Outdoor Production of Microalgae and Cyanobacteria. *J. Appl. Phycol.* **1997,** *9*, 113–119.

Pyle, D. J.; Garcia, R. A.; Wen, Z. Producing Docosahexanoic Acid (DHA)-rich Algae from Biodiesel Derived Crude Glycerol: Effects of Impurities on DHA Production and Algal Biomass Composition. *J. Agric. Food Chem.* **2008,** *56*(11), 3933–3939.

Qiang, H.; Zhengyu, H.; Cohen, Z.; Richmond, A. Enhancement of Eicosapentaenoic Acid (EPA) and Gamma-Linolenic Acid (GLA) Production by Manipulating Algal Density of Outdoor Cultures of *Monodus subterraneus* (Eustigmatophyta) and *Spirulina platensis* (Cyanobacteria). *Eur. J. Phycol.* **1997,** *32*, 81–86.

Raghukumar, S. Thraustochytrid Marine Protists: Production of PUFAs and Other Emerging Technologies. *Mar. Biotechnol.* **2008,** *10*, 631–640.

Raja. R.; Hemaiswarya. S.; Rengasamy, R. Exploitation of *Dunaliella* for β-Carotene Production. *Appl. Microbiol. Biotechnol.* **2007,** 74, 517–523.

Ramachandra, T. V.; Mahapatra, D. M.; Karthick, B. Milking Diatoms for Sustainable Energy: Biochemical Engineering versus Gasoline Secreting Diatom Solar Panels. *Ind. Eng. Chem. Res.* **2009,** *48*, 8769–8788.

Rangarao, A. Production of Astaxanthin from Cultured Green Alga *Haematococcus pluvialis* and its Biological Activities. Ph. D. Thesis, University of Mysore: Mysore, India, 2011.

Rangarao, A.; Baskaran, V.; Sarada, R.; Ravishankar, G. A. In Vivo Bioavailability and Antioxidant Activity of Carotenoids from Microalgal Biomass—A Repeated Dose Study. *Food Res. Int.* **2013a,** *54*, 711–717.

Rangarao, A.; Moi, P. S.; Sarada, R.; Ravishankar, G. A. Astaxanthin: Sources, Extraction, Stability, Biological Activities and Its Commercial Applications—A Review. *Mar. Drugs*, **2014a,** *12*, 128–152.

Rangarao, A.; Sarada, R.; Ravishankar, G. A. Stabilization of Astaxanthin in Edible Oils and Its Use as an Antioxidant. *J. Sci. Food Agric.* **2007,** *87*, 957–965.

Rangarao, A.; Sarada, R.; Ravishankar, G. A.; Phang S. M. Industrial Production of Microalgal Cell-mass and Bioactive Constituents from Green Microalga *Botryococcus braunii*. In *Recent Advances in Microalgal Biotechnology*; Liu, J., Sun, Z., Gerken, H.; Omics group Ebooks: California, USA, **2014b**.

Rangarao, A.; Sindhuja, H. N.; Dharmesh, S. M.; Sankar, K. U.; Sarada, R.; Ravishankar, G. A. Effective Inhibition of Skin Cancer, Tyrosinase and Antioxidative Properties by Astaxanthin and Astaxanthin Esters from the Green Alga *Haematococcus pluvialis*. *J. Agric. Food Chem.* **2013b,** *61*, 3842–3851.

Rathnasamy, S.; Debora, J. J.; Hari, B. V. Extraction and Purification of C-Phycocyanin from *Spirulina platensis* Using Aqueous Two Phase Extraction and Its Applications. *Asian J. Chem.* **2014,** *26*, 3729–3732.

Ravishankar, G. A.; Sarada, R.; Kamath, B. S.; Namitha, K. K. Food Applications of Algae. In *Food Biotechnology*; Shetty, K., Paliyath, G., Pometto, A., Levin, R. E., Eds.; CRC Press New York, 2008; pp 491–524.

Ravishankar, G. A.; Sarada, R.; Vidyashankar, S.; VenuGopal, K. S.; Kumudha, A. Cultivation of Micro-algae for Lipids and Hydrocarbons, and Utilization of Spent Biomass for Livestock Feed and for Bio-active Constituents. In *Biofuel Co-products as Livestock Feed—Opportunities and Challenges*; Makkar, H. P. S., Ed.; FAO: Rome, 2012; pp 423–446.

Reddy, C. M.; Bhat, V. B.; Kiranmai, G.; Reddy, M. N.; Reddanna, P.; Madyastha, K. M. Selective Inhibition of Cyclooxygenase-2 by C-Phycocyanin, a Biliprotein from *Spirulina platensis*. *Biochem. Biophys. Res. Commun.* **2000,** *277*, 599–603.

Reddy, M. C.; Subhashini, J.; Mahipal, S. V. K.; Bhat, V. B.; Reddy, P. S.; Kiranmai, G.; Madyastha, K. M.; Reddanna, P. C-Phycocyanin, a Selective Cyclooxygenase-2 Inhibitor, Induces Apoptosis in Lipopolysaccharide-stimulated RAW 264.7 Macrophages. *Biochem. Biophys. Res. Commun.* **2003,** 304, 385–392.

Reis, A.; Mendes, A.; Lobo-Fernandes, H.; Empis, J. A.; Novais, J. M. Production, Extraction and Purification of Phycobiliproteins from *Nostoc* sp. *Bioresour. Technol.* **1998,** *66*, 181–187.

Renaud, S. M.; Thinh, L. V.; Lambrinidis, G.; Parry, D. L. Effect of Temperature on Growth, Chemical Composition and Fatty Acid Composition of Tropical Australian Microalgae Grown in Batch Cultures. *Aquaculture* **2002,** *211*, 195–214.

Richmond, A.; Grobbelaar, J. U. Factors Affecting the Output Rate of *Spirulina platensis* with Reference to Mass Cultivation. *Biomass* **1986,** *10*, 253–264.

Richmond, A.; Lichtenberger, E.; Stahl, B.; Vonshak, A. Quantitative Assessment of the Major Limitations on Productivity of *Spirulina platensis* in Open Raceways. *J. Appl. Phycol.* **1990,** *2*, 195–206.

Rittmann, E. B. Opportunities for Renewable Bioenergy Using Microorganisms. *Biotechnol. Bioeng.* **2008,** *100*, 203–212.

Roberts, R. L.; Green, J.; Lewis, B. Lutein and Zeaxanthin in Eye and Skin Health. *Clin. Dermatol.* **2009,** *27*(2), 196–201.

Roessler, P. G. Effects of Silicon Deficiency on Lipid Composition and Metabolism in the Diatom *Cyclotella cryptica. J. Phycol.* **1988,** *24*(3), 394–400.

Roman, B. R.; Alvarez-Pez, J. M.; Fernandez, A.; Molina Grima, E. Recovery of Pure B-Phycoerythrin from the Microalga *Porphyridium cruentum*. *J. Biotechnol.* **2002,** *93*, 73–85.

Romay, C.; Ledón, N.; González, R. Further Studies on Anti-inflammatory Activity of Phycocyanin in Some Animal Models of Inflammation. *Inflamm. Res.* **1998,** *47*, 334–338.

Roy, K. R.; Nishanth, R. P.; Sreekanth, D.; Reddy, G. V.; Reddanna, P. C-Phycocyanin Ameliorates 2-Acetylaminofluorene Induced Oxidative Stress and MDR1 Expression in the Liver of Albino Mice. *Hepatol. Res.* **2008,** *38*, 511–520.

Sabarinathan, K. G.; Ganesan, G. Antibacterial and Toxicity Evaluation of C-phycocyanin and Cell Extract of Filamentous Freshwater Cyanobacterium. *Eur. Rev. Med. Pharmacol. Sci.* **2008,** *12*, 79–82.

Sahena, F.; Zaidul, A. S. M.; Jinap, S.; Karim, A. A.; Abbas, K. A.; Norulaini, N. A. N.; Omar, A. K. M. Application of Supercritical CO_2 in Lipid Extraction—A Review. *J. Food Eng.* **2009,** *95*, 240–253.

Sahu A.; Pancha I.; Jain D.; Paliwal C.; Ghosh T.; Shailesh P.; Bhattacharya S.; Mishra S. Fatty acids as Biomarkers of Algae. *Phytochemistry* **2013,** *89*, 53–58.

Saini, M. K.; Sanyal, S. N. Piroxicam and C-Phycocyanin Prevent Colon Carcinogenesis by Inhibition of Membrane Fluidity and Canonical Wnt/β-Catenin Signaling while Up-regulating Ligand Dependent Transcription Factor PPARγ. *Biomed. Pharmacother.* **2014b,** *68*(5), 537–550.

Saini, M. K.; Sanyal, S. N. Targeting Angiogenic Pathway for Chemoprevention of Experimental Colon Cancer using C-Phycocyanin as Cyclooxygenase-2 Inhibitor. *Biochem. Cell Biol.* **2014a,** *92*(3), 206-218.

Sajilata, M. G.; Singhal, R. S.; Kamat, M. Y. The Carotenoid Pigment Zeaxanthin—A Review. *Compr. Rev. Food Sci. Food Saf.* **2008,** 7, 29–49.

Sánchez, F.; Fernández, J. M.; Acien, F. G.; Rueda, A.; Perez-Parra, J.; Molina, E. Influence of Culture Conditions on the Productivity and Lutein Content of the New Strain *Scenedesmus almeriensis*. Process Biochem. **2008,** *43*(4), 398–405.

Sánchez, F.; Fernández, J. M.; Acien, F. G.; Rueda, A.; Perez-Parra, J.; Molina, E. Influence of Culture Conditions on the Productivity and Lutein Content of the New Strain *Scenedesmus almeriensis*. *Process Biochem.* **2008,** *43*(4), 398–405.

Sandeep, K. P.; Shukla, S. P.; Vennila, A.; Purushothaman, C. S.; Manjulekshmi, N. Cultivation of *Spirulina (Arthrospira) platensis* in Low Cost Seawater Based Medium for Extraction of Value Added Pigments. *Indian J. Mar. Sci.* **2015,** *44*, 3.

Santiago-Santos, M. C.; Ponce-Noyola, T.; Olvera-Ramírez, R.; Ortega-López, J.; Cañizares-Villanueva, R. O. Extraction and Purification of Phycocyanin from *Calothrix* sp. *Process Biochem.* **2004,** 39, 2047–2052.

Sarada, R.; Pillai, M. G.; Ravishankar, G. A. Phycocyanin from *Spirulina* sp.: Influence of Processing of Biomass on Phycocyanin Yield, Analysis of Efficacy of Extraction Methods and Stability Studies on Phycocyanin. *Process Biochem.* **1999,** *34*, 795–801.

Sarada, R.; Usha, T.; Ravishankar, G. A. Influence of Stress on Astaxanthin Production in *Haematococcus pluvialis* Grown under Different Culture Conditions. *Process Biochem.* **2002,** *37*, 623–637.

Sarada, R.; Vidhyavathi, R.; Usha, D.; Ravishankar, G. A. An Efficient Method for Extraction of Astaxanthin from Green Alga *Haematococcus pluvialis*. *J. Agric. Food Chem.* **2006,** *54*, 7585–7588.

Satyantini, W. H.; Harris, E.; Utomo, N. B. P.; et al.. Administration of *Spirulina* Phycocyanin Enhances Blood Cells, Phagocytic Activity and Growth in Humpback Grouper Juvenile. *J. Vet.* **2014,** *15*, 46–56.

Scoglio, S.; Benedetti, Y.; Benvenuti, F.; Battistelli, S.; Canestrari, F.; Benedetti, S. Selective Monoamine Oxidase B Inhibition by an *Aphanizomenon flos-aquae* Extract and by Its Constitutive Active Principles Phycocyanin and Mycosporine-like Amino Acids. *Phytomedicine* **2014,** *21*, 992–997.

Sekar, S.; Chandramohan, M. Phycobiliproteins as a Commodity: Trends in Applied Research, Patents and Commercialization. *J. Appl. Phycol.* **2008,** *20*, 113–136.

Setyaningsih, I.; Bintang, M.; Madina, N. Potentially Antihyperglycemic from Biomass and Phycocyanin of *Spirulina Fusiformis Voronikhin* by In Vivo Test. *Procedia Chem.* **2015,** *14*, 211–215.

Shahidi, F.; Metusalach, Brown, J. A. Carotenoid Pigments in Sea Foods and Aquaculture. *Crit. Rev. Food Sci.* **1998,** *38*, 1–67.

Shi, X.; Zhengyun, W.; Chen, F. Kinetic Model of Lutein Production by Heterotrophic *Chlorella* at Various pH and Temperature. *Mol. Nutr. Food Res.* **2006,** *50*, 763–768.

Siegelman, H. W.; Kycia, J. H. Algal biliproteins. In *Handbook of Phycological Methods, Physiological and Biochemical Methods*; Hellebust, J. A., Craigie, J. S., Eds.; Cambridge University Press: Cambridge, 1978; pp 71–79.

Sijtsma, L.; de Swaaf, M. E. Biotechnological Production and Applications of ω-3 Poly Unsaturated Fatty Acid DHA. *Appl. Microbiol. Biotechnol.* **2004,** *64*, 146–153.

Silveira, S. T.; Burkert, J. F. M.; Costa, J. A. V.; Burkert, C. A. V.; Kalil, S. J. Optimization of Phycocyanin Extraction from *Spirulina platensis* Using Factorial Design. *Bioresour. Technol.* **2007,** *98*, 1629–1634.

Singh, N. K.; Parmar, A.; Madamwar, D. Optimization of Medium Components for Increased Production of C-Phycocyanin from *Phormidium ceylanicum* and Its Purification by Single Step Process. *Bioresour. Technol.* **2009,** *100*, 1663–1669.

Sloth, J. K.; Wiebe, M. G.; Eriksen, N. T. Accumulation of Phycocyanin in Heterotrophic and Mixotrophic Cultures of the Acidophilic Red Alga *Galdieria sulphuraria*. *Enzyme Microb. Technol.* **2006,** *38*, 168–175.

Snodderly, D. M. Evidence for Protection Against Age-related Macular Degeneration by Carotenoids and Antioxidant Vitamins. *Am. J. Clin. Nutr.* **1995,** *62*, 1448S–1461S.

Sonani, R. R.; Singh, N. K.; Kumar, J.; Thakar, D.; Madamwar, D. Concurrent Purification and Antioxidant Activity of Phycobiliproteins from *Lyngbya sp.* A09DM: An Antioxidant and Anti-aging Potential of Phycoerythrin in *Caenorhabditis elegans*. *Process Biochem.* **2014,** *49*, 1757–1766.

Soni, B.; Kalavadia, B.; Trivedi, U.; Madamwar, D. Extraction, Purification and Characterization of Phycocyanin from *Oscillatoria quadripunctulata* Isolated from the Rocky Shores of Bet-Dwarka, Gujarat, India. *Process Biochemi.* **2006,** *41*, 2017–2023.

Spolaore, P.; Joannis-Cassan, C.; Duran, E.; Isambert, A. Commercial Applications of Microalgae. *J. Biosci. Bioeng.* **2006,** *101*, 87–96.

Storebakken, T. Krill as a Potential Feed Source for Salmonids. *Aquaculture* **1988,** *70*, 193–205.

Storelli, M. M.; Storelli, A.; Marcotrigiano, G. O. Polychlorinated Biphenyls, Hexa Chloro Benzene, Hexachloro Cyclohexane Isomers, and Pesticide Organo Chlorine Residues in Cod-liver Oil Dietary Supplements. *J. Food Protect.* **2004,** 67, 1787–1791.

Su, C.-H.; Liu, C.-S.; Yang, P.-C.; Syu, K.-S.; Chiuh, C.-C. Solid–Liquid Extraction of Phycocyanin from *Spirulina platensis*: Kinetic Modeling of Influential Factors. *Sep. Purif. Technol.* **2014,** *123*, 64–68.

Sukenik, A. Production of EPA by *Nannochloropsis*. In *Chemicals from Microalgae*; Cohen, Z., Ed.; Taylor and Francis, London, 1999; pp 41–56.

Sukenik, A.; Beardall, J.; Kromkamp, J. C.; Kopeck, J.; Masojídek, J.; van Bergeijk, S.; Gabai, S.; Shaham, E.; Yamshon, A. Photosynthetic Performance of Outdoor *Nannochloropsis* Mass Cultures under a Wide Range of Environmental Conditions. *Aquat. Microb. Ecol.* **2009,** *56*, 297–308.

Sukenik, A.; Carmeli, Y.; Berner, T. Regulation of Fatty Acid Composition by Irradiance Level in the Eustigmatophyte *Nannochloropsis* sp. *J. Phycol.* **1989**, *25*, 686–692.

Taher, H.; Al-Zuhair, S.; Al-Marzouqi, A.; Haik, Y.; Farid, M. Effective Extraction of Microalgae Lipids from Wet Biomass for Biodiesel Production. *Biomass Bioenergy* **2014a,** *66*, 159–167.

Taher, H.; Al-Zuhair, S.; Al-Marzouqi, A.; Haik, Y.; Farid, M.; Tariq, S. Supercritical Carbon Dioxide Extraction of Microalgae Lipid: Process Optimization and Laboratory Scale-up. *J. Supercrit. Fluids* **2014b,** *86*, 57–66.

Tanaka, T.; Kawamori, T.; Ohnishi, M.; Makita, H.; Mori, H.; Satoh, K.; Hara, A. Suppression of Azoxymethane-induced Rat Colon Carcinogenesis by Dietary Administration of Naturally Occurring Xanthophylls Astaxanthin and Canthaxanthin During the Postinitiation Phase. *Carcinogenesis* **1995,** *16*, 2957–2963.

Tang, D.; Han, W.; Li, P.; Miao, X.; Zhong, J. CO_2 Biofixation and Fatty Acid Composition of *Scenedesmus obliquus* and *Chlorella pyrenoidosa* in Response to Different CO_2 Levels. *Bioresour. Technol.* **2011,** *102*, 3071–3076.

Tantirapan, P.; Suwanwong, Y. Anti-proliferative Effects of C-Phycocyanin on a Human Leukemic Cell Line and Induction of Apoptosis via the PI3K/AKT Pathway. *J. Chem. Pharm. Res.* **2014,** *6*, 1295–1301.

Thangam, R.; Suresh, V.; Asenath Princy, W.; Rajkumar, M.; SenthilKumar, N.; Gunasekaran, P.; Rengasamy, R.; Anbazhagan, C.; Kaveri, K.; Kannan, S. C-Phycocyanin from *Oscillatoria tenuis* Exhibited an Antioxidant and *In Vitro* Antiproliferative Activity Through Induction of Apoptosis and *G sub 0 sub G sub 1 sub* Cell Cycle Arrest. *Food Chem.* **2013,** *140*, 262–272.

Tonon, T.; Harvey, D.; Larson, T. R.; Graham, I. A. Long Chain Polyunsaturated Fatty Acid Production and Partitioning to Triacylglycerols in Four Microalgae. *Phytochemistry* **2002,** *61*, 15–24.

Tornwall, M. E.; Virtamo, J.; Korhonen, P. A.; Virtanen, M. J.; Taylor, P. R.; Albanes, D.; et al. Effect of a-Tocopherol and b-Carotene Supplementation on Coronary Heart Disease During the 6-Year Post-Trial Follow-Up in the ATBC Study. *Eur. Heart J.* **2004,** *25*, 1171–1178.

Torrisen, O. J.; Hardy, W. H. & Shearer, K. D. Pigmentation of Salmonids—Carotenoid Deposition and Metabolism. *Rev. Aquat. Sci.* **1989,** *1*, 209–227.

Tredici, M. R.; Carlozzi, P.; Zittelli, G. C.; Materassi, R. A Vertical Alveolar Panel (VAP) for Outdoor Mass Cultivation of Microalgae and Cyanobacteria. *Bioresour. Technol.* **1991,** *38*, 153–159.

Tsuzuki, M.; Ohnuma, E.; Sato, N.; Takaku, T.; Kawaguchi, A. Effects of CO_2 concentration During Growth on Fatty Acid Composition in Microalgae. *Plant Physiol.* **1990,** *93*, 851–856.

Turujman, S. A.; Wamer, W. G.; Wei, R. R.; Albert, R. H. Rapid Liquid Chromatographic Method to Distinguish Wild Salmon from Aquacultured Salmon Fed Synthetic Astaxanthin. *J. Am. Oil Chem. Soc.* **1997,** *80*, 622–632.

Vandamme, D.; Foubert, I.; Muylaert, K. Flocculation as a Low-Cost Method for Harvesting Microalgae for Bulk Biomass Production. *Trends Biotechnol.* **2013,** *31*(4), 233–239.

Vanitha, A.; Chidambara Murthy, K. N.; Vinod Kumar, Sakthivelu, G.; Jyothi, M. V.; Saibaba, P.; Ravishankar, G. A. Effect of Carotenoid-Producing Alga, *Dunaliella bardawil*, on CCl4 Induced Toxicity in Rats. *Int. J. Toxicol.* **2007,** *26*, 159–167.

Vanitha. A. Cultivation of *Dunaliella bardawil* rich in Carotenoids and Studies on Nutritional and Biological Activities. Ph. D. Thesis, University of Mysore: Mysore, India, 2007.

Venkataraman, L. V. and Becker, E. W. *Biotechnology and Utilization of Algae: The Indian Experience*. Department of Science and Technology: New Delhi, 1985, 257.

Vidyashankar, S.; Deviprasad, K.; Chauhan, V. S.; Ravishankar, G. A.; Sarada, R. Selection and Evaluation of CO_2 Tolerant Indigenous Microalga *Scenedesmus dimorphus* for Unsaturated Fatty Acid Rich Lipid Production under Different Culture Conditions. *Bioresour. Technol.* **2013,** *144*, 28–37.

Vidyashankar, S.; Sireesha, E.; Chauhan, V. S.; Sarada, R. Evaluation of Microalgae as Vegetarian Source of Dietary Polyunsaturated Fatty Acids under Autotrophic Growth Conditions. *J. Food Sci. Technol.* **2015,** *52*(11), 7070–7080.

Vidyavathi, R. Molecular and Biochemical Studies of Astaxanthin Biosynthesis in *Haematococcus pluvialis*. Ph. D. Thesis, University of Mysore: Mysore, India, 2008.

Vonshak, A.; Tomaselli, L. *Arthrospira* (*Spirulina*): Systematics and Ecophysiology. In *The Ecology of Cyanobacteria*; Whitton, B. A., Potts, M. Eds. Kluwer Academic Publishers, The Netherlands, 2000; pp 505–22.

Wang, L.; Qu, Y.; Fu, X.; Zhao, M.; Wang, S.; Sun, L. Isolation, Purification and Properties of an R-Phycocyanin from the Phycobilisomes of a Marine Red Macroalga *Polysiphonia urceolata*. *PLoS ONE* **2014,** *9*, e87833.

Wang, L.; Yang, B.; Yan B.; Yao, X. Supercritical Fluid Extraction of Astaxanthin from *Haematococcus pluvialis* and its Antioxidant Potential in Sunflower Oil. *Innov. Food Sci. Emerg. Technol.* **2012,** *13*, 120–127.

Wang, X.; Willen, R.; Wadstorm, T. Astaxanthin Rich Algal Meal and Vitamin C Inhibit *Helicobacter pylori* Infection in BALB/CA Mice. *Antimicrob. Agents Chemother.* **2000,** 44, 2452–2457.

Ward, O. P.; Singh, A. Omega-3/6 Fatty Acids: Alternative Sources of Production. *Process Biochem.* **2005,** *40*, 3627–3652.

Wen, Z. Y.; Chen, F. Heterotrophic Production of Eicosapentaenoic Acid by the Diatom *Nitzschia laevis*: Effects of Silicate and Glucose. *J. Ind. Microbiol. Biotechnol.* **2000,** *25*, 218–224.

Wen, Z. Y.; Chen, F. Heterotrophic Production of Eicosapentaenoic Acid by Microalgae. *Biotechnol. Adv.* **2003,** *21*, 273–294.

Wen, Z. Y.; Chen, F. Optimization of Nitrogen Sources for Heterotrophic Production of Eicosapentaenoic Acid by the Diatom *Nitzschia laevis*. *Enzyme Microb. Technol.* **2001,** *29*, 341–347.

Woo, M.-N. et al. Anti-obese Property of Fucoxanthin is Partly Mediated by Altering Lipid-Regulating Enzymes and Uncoupling Proteins of Visceral Adipose Tissue in Mice. *Mol. Nutr. Food Res.* **2009,** *53*, 1603–1611.

Yaguchi, T.; Tanaka, S.; Yokochi, T.; Nakahara, T.; Yaguchi, T. Production of High Yields of docodahexaenoic Acids by *Schizochytrium* sp. strain SR21. *J. Am. Oil. Chem. Soc.* **1997,** 74, 1431–1434.

Yamaguchi, M. Role of Carotenoid β-cryptoxanthin in Bone Homeostasis. *J. Biomed. Sci.* **2012,** *19*, 36.

Yan, M.; Liu, B.; Jiao, X.; Qin, S. Preparation of Phycocyanin Microcapsules and Its Properties. *Food Bioproducts Process.* **2014,** *92*, 89–97.

Yang, F.; Li, B.; Chu, X.-M.; Lv, C.-Y.; Xu, Y.-J.; Yang, P. Molecular Mechanism of Inhibitory Effects of C-Phycocyanin Combined with All-trans-retinoic Acid on the Growth of HeLa Cells In Vitro. *Tumor Biol.* **2014,** *35*(6), 5619–5628.

Yongmanitchai, W.; Ward, O. P. Growth and Eicosapentaenoic Acid Production by *Phaeodactylum tricornutum* in Batch and Continuous Culture Systems. *J. Am. Oil Chem. Soc.* **1992,** *69*, 584–590.

Yongmanitchai, W.; Ward, O. P. Growth of and Omega-3 Fatty Acid Production by *Phaeodactylum tricornutum* under Different Culture Conditions. *Appl. Environ. Microbiol.* **1991,** 57, 419–425.

Zhang, L.-X.; Cai, C.-E.; Guo, T.-T.; Gu, J.-W.; Xu, H.-L.; Zhou, Y.; Wang, Y.; Liu, C.-C.; He, P.-M. Anti-cancer Effects of Polysaccharide and Phycocyanin from *Porphyra yezoensis*. *J. Mar. Sci. Technol.* **2011,** *19*, 377–382.

Zhang, X.; Zhang, F.; Luo, G.; Yang, S.; Wang, D. Extraction and Separation of Phycocyanin from *Spirulina* using Aqueous Two-Phase Systems of Ionic Liquid and Salt. *J. Food Nutr. Res.* **2015,** *3*, 15–19.

Zhao, L.; Peng, Y.; Gao, J.; Cai, W. Bioprocess Intensification: An Aqueous Two-phase Process for the Purification of C-Phycocyanin from Dry *Spirulina platensis*. *Eur. Food Res. Technol.* **2014,** *238*, 451–457.

Zhu, C. J.; Lee, Y. K.; Chao, T. M. Effects of Temperature and Growth Phase on Lipid and Biochemical Composition of *Isochrysis galbana* TK1. *J. Appl. Phycol.* **1997,** *9*, 451–457.

Zhu, L.; Yan, S.; Lv, A. Efficient Purification and Active Configuration Investigation of R-Phycocyanin from *Polysiphonia urceolata*. In *Advances in Applied Biotechnology*; Springer: Berlin Heidelberg, 2015; pp 489–496.

Zhukova, N. V.; Aizdaicher, N. A. Fatty Acid Composition of 15 Species of Marine Microalgae. *Phytochemistry* **1995,** *39*(2), 351–356.

Zittelli, G. C.; Lavista, F.; Bastianini, A.; Rodolfi, L.; Vincenzini, M.; Tredici, M. R. Production of Eicosapentaenoic Acid by *Nannochloropsis* sp. Cultures in Outdoor Tubular Photobioreactors. *J. Biotechnol.* **1999,** *70*, 299–312.

CHAPTER 2

MEDICINAL PLANTS IN PREVENTIVE AND CURATIVE ROLE FOR VARIOUS AILMENTS

R. SINGH[1], K. K. PRASAD[1], MOHAMMED WASIM SIDDIQUI[2], and KAMLESH PRASAD[3*]

[1]*Department of Gastroenterology, Post Graduate Institute of Medical Education and Research, Chandigarh 160012, India*

[2]*Department of Food Science and Postharvest Technology, Bihar Agricultural University, Sabour, Bhagalpur 813210, Bihar, India*

[3]*Department of Food Engineering and Technology, SLIET, Longowal 148106, Punjab, India*

[*]*Corresponding author, E-mail: profkprasad@gmail.com*

CONTENTS

ABSTRACT

Plants which can be employed for therapeutic purpose or used to extract active components and bearing the curative medicinal properties are referred to as medicinal plants according to World Health Organization (WHO). Plant parts or the extracts have been found to be used since time immemorial for therapeutic benefits throughout the globe as preventive or curative purposes against various ailments and are depicted broadly in well-recognized literatures related to Ayurveda, Chinese, and Unani medicines. Archeologists found the corroboration in support of medicinal plants utilization by Neanderthal man more than 50,000 years ago. Since prehistoric periods, people of Europe, Asia, and American subcontinents knew regarding the usefulness of medicinal herbs. Also, historians depicted the existence of herbal medicine in garden present in Pisa and Italy during the 16th century. It is evident that as the first nation, the Canadians used numerous of plant species for their pharmacological roles in various diseases and still advocate the utilization of pharmaceutical plants in modern times. Majority of the people still believe the importance of herbs for their pharmacological benefits.

2.1 INTRODUCTION

Plants which can be employed for therapeutic purpose or used to extract active components and bearing the curative medicinal properties are referred to as medicinal plants according to World Health Organization (WHO). Plant parts or the extracts have been found to be used since time immemorial for therapeutic benefits throughout the globe as preventive or curative purposes against various ailments and are depicted broadly in well-recognized literatures related to Ayurveda, Chinese, and Unani medicines. Archeologists found the corroboration in support of medicinal plants utilization by Neanderthal man more than 50,000 years ago. Since prehistoric periods, people of Europe, Asia, and American subcontinents knew regarding the usefulness of medicinal herbs. Also, historians depicted the existence of herbal medicine in garden present in Pisa and Italy during the 16th century. It is evident that as the first nation, the Canadians used numerous of plant species for their pharmacological roles in various diseases and still advocate the utilization of pharmaceutical plants in modern times. Majority of the people still believe the importance of herbs for their pharmacological benefits.

It is reported that over 20% of Canadians have regularly used pharmaceutical plants and around 25% of current pharmaceutical medicines contain plant products. In the United States, about 1800 pharmacological active plant species are available. According to historians, more than 13,000 herbal species are employed as traditional medicines by different ethnicities. At present, over 50,000 plant species are found to have pharmacologically active biomaterial properties. Alternative medicine is the term which deals with the use of herbs for their pharmacological benefits in disease state. Therapy through alternative medicine is thus getting popularity worldwide. The active element of plants is extracted in order to obtain various drugs for lifestyle disorders especially diabetes, cardiovascular disease, neoplasms, cardiovascular and neurological disorders (Prasad et al., 2009, 2010; Haq & Prasad, 2015; Gull et al., 2015; Ankita & Prasad, 2015a,b).

Thus, the use of plants in medicine has a promising future as large number of plant species which are still the subject of investigations for their pharmacological claims and their medicinal activities could be having the decisive role in treating the present or future ailments. The emergence of new concept of complementary and alternative medicine (CAM) flood of employing phytotherapy will bestow the field of medicine worldwide.

2.2 MEDICINAL PLANTS USED IN DIABETES

Diabetes has emerged as the biggest challenge to healthcare consultants. Diabetes mellitus (DM), or diabetes, is a metabolic disorder that leads to imbalance of insulin hormone, which is responsible for the breakdown of carbohydrate of food materials into energy. Affected population will reach to 300 million by 2025 as per WHO. Current management for diabetes includes the supply of insulin, sulfonylureas, metformin, glinides, which are also associated with adverse drug reactions and moreover highly expensive. Natural products (NPs) on the other hand provide an alternative, apart from the fact that traditional medicines have an edge as per availability and associated with lesser adverse effects. Moreover, the plant products have other phytochemicals such as flavonoids, alkaloids, and glycosides, which are having multifaceted role apart from antidiabetic effects. Many NPs are depicted in various literatures as having antihyperglycemic activity by induction of insulin output or inhibit the intestinal absorption of glucose (Table 2.1).

TABLE 2.1 Medicinal Plants in Diabetes.

Medicinal plant	Family	Action	Model
Acacia arabica	Leguminosae	Hypoglycemic effect	Alloxan-induced diabetic rabbits
Aegle marmelos	Rutaceae	Antidiabetic effect	STZ-induced diabetic rats
Agrimonia eupatoria	Rosaceae	Antidiabetic activity	Pancreatic beta-cell lines
Alangium salviifolium	Alangiaceae	Antihyperglycemic	Dexamethasone-induced insulin-resistant rats
Allium sativum	Alliaceae	Hypoglycemic effect	Alloxan-induced diabetic rats or rabbits
Aloe vera	Liliaceae	Hypoglycemic effect	Diabetic rats
Azadirachta indica	Meliaceae	Antihyperglycemic, antidyslipidemic	STZ-induced diabetes in rats, rabbits, *in vitro* rat pancreas
Berberis aquifolium	Berberidaceae	Insulinotropic	Diabetic rats
Biophytum sensitivum	Oxalidaceae	Hypoglycemic effect	Diabetic rabbits
Brassica juncea	Brassicaceae	Hypoglycemic, antihyerlipidemic	STZ-induced diabetic rat
Citrullus colocynthis	Cucurbitaceae	Antidiabetic activity	STZ-induced diabetic rats
Coccinia indica	Cucurbitaceae	Antidiabetic activity	STZ-induced diabetic rats
Eucalyptus	Myrtaceae	Antihyperglycemic	STZ-induced mice and rats
Eugenia jambolana	Myrtaceae	Antihyperglycemic, antihyperlidemic	STZ-induced diabetic rats
Gymnema sylvestre	Asclepiadaceae	Hypolipidemic, hypoglycemic	MIN6 β-cell lines, alloxan-induced diabetes
Hibiscus rosa-sinensis	Malvaceae	Hypoglycemic, antidyslipidemic	STZ-induced diabetic mice
Ocimum sanctum	Lamiaceae	Antidiabetic	STZ-induced diabetic rats
Trigonella foenum -graceum	Fabaceae	Antihyperglycemic	STZ-induced diabetic rats, alloxan induced diabetic rats

2.2.1 ACACIA ARABICA (LEGUMINOSAE)

Acacia arabica, usually called as babul or kikar which is depicted in the Asian scriptures, was used to treat DM. It is cultivated in the arid regions around the world. It is credited with a number of medicinal properties. Traditionally, its gum proved to be beneficial in many skin-related problems like rashes, inflammation, and burns, and its bark is used in dysentery.

Study showed that the seed of *A. arabica* was able to produce hypoglycemic effect by inducing insulin production in rats and also powdered seeds showed hypoglycemic effect in rabbits (Singh, 2011).

2.2.2 AEGLE MARMELOS (RUTACEAE)

Aegle marmelos (bael) belongs to Rutaceae family. Bael trees are profoundly found in Asian regions specifically in Southeast Asian regions. Antidiabetic properties of this plant had been discovered in rats, which might be the result of its ability to enhance glucose utilization (Ayodhya et al., 2010).

2.2.4 AGRIMONIA EUPATORIA (ROSACEAE)

Agrimonia eupatoria is spread throughout China and has been a part of Chinese medicine in many ailments like bacterial infection, tumors, trichomonas vaginitis, and others. Presence of flavonoids, lactone, tannin, and glycosides might be associated with its therapeutic activity. Under *in vitro* condition, *Agrimonia* effectively supports in the production of insulin, which is helpful in its antidiabetic activity (Bnouham et al., 2006).

2.2.5 ALANGIUM SALVIIFOLIUM (ALANGIACEAE)

Alangium salviifolium is also referred to as Ankola. It is native to Indian regions and its pharmacological properties has also mentioned in Ayurveda. It showed many beneficial medicinal activities like anti-inflammatory, antimicrobial, antifertility, and cardiovascular activities. *Alangium* has shown beneficial effect when tested in insulin-resistant rats induced by dexamethasone. Methanolic extract was observed to produce antihyperglycemic effects in these strains which could be useful in clinical settings (Kshirsagar et al., 2010).

2.2.6 *ALLIUM SATIVUM* (ALLIACEAE)

Commonly referred to as garlic, it is the part of our life since 5000 years as documented. It has rich values, being used as food as well as in medicines. *Allium* is found to influence the insulin production from parietal cell of pancreas by lowering the blood sugar level in rats under observations (Chauhan et al., 2010). Allicin, a biologically active element of garlic, has also shown hypoglycemic effect. The effect is associated with increased insulin secretion and hepatic activity. *S*-allyl cystein sulfoxide is the precursor of allicin, also able to enhance the increased production of insulin (Bnouham et al., 2006; Modak et al., 2007). It is known that having garlic in routine use can increase insulin level in the body.

2.2.7 *ALOE VERA* (XANTHORRHOEACEAE)

Since time immemorial, *Aloe vera* (Ghritkumari) has been recommended as herbal medicine and also used in cosmetics on a large scale. *Aloe vera* extract has shown to stimulate beta-cells of Langerhans to enhance insulin secretion. When *Aloe vera* was administered in murines, insulin release from β-cells of pancreas depicted hypoglycemic effect (Singh, 2011).

2.2.8 *AZADIRACHTA INDICA* (MELIACEAE)

It is frequently referred as "Neem" and is mainly found in tropical areas. *Azadirachta indica* is determined as an antihyperglycemic and anti-dyslipidemic agent, which maintains the glucose level in diabetes-induced rats. *A. indica* inhibited the utilization of glucose by suppressing adrenaline in diabetes-induced rabbits (Chattopadhyay, 1996). The components extracted from leaves were also observed to obliterate the serotonin inhibitory effect on insulin production.

2.2.9 *BERBERIS AQUIFOLIUM* (BERBERIDACEAE)

Berberine is mainly an isoquinoline quaternary alkaloid. The compound is isolated from different types of pharmacologically beneficial herbs such as Oregon grape, barberry, or tree turmeric, having multiple positive effects in numerous of disease pathologies like in various gastrointestinal, hepatological disorder, heart disease, and others. Berberine induced the insulin release in rat pancreatic cells, thereby proving its potential effect in diabetes. When observed in adipocytes cell lines, it increased insulin-signaling pathway, thereby increasing uptake of glucose (Ko et al., 2005).

2.2.10 *BIOPHYTUM SENSITIVUM* (OXALIDACEAE)

Commonly found in south-eastern regions of Asia like in India, Nepal, and others, *Biophytum sensitivum* is referred to as life plant and *Lakshmana* in India. Various pharmacological characteristics like antioxidants, anti-inflammatory, as well as the antineoplasm activities of the plant, make it a good choice in different disease conditions. In diabetes too, when observed in fasting and diabetic rabbits, it showed hypoglycemic effect by insulin release (Puri, 2001).

2.2.11 *BRASSICA JUNCEA* (BRASSICACEAE)

Mainly known as Indian mustard green, it is cultivated in variety of forms throughout the globe. Its seeds, stems, roots, and oil are edible and also show medicinal values. *Brassica juncea* showed significant medicinal values when studied in diabetes-induced rats. Hypoglycemic effects were reported on different doses which attributed toward its effective nature to activate glycogen synthetase and suppression of gluconeogenic enzymes (Thirumalai et al., 2011).

2.2.12 *CITRULLUS COLOCYNTHIS* (CUCURBITACEAE)

Citrullus colocynthis, also referred as bitter cucumber or bitter apple, is known to have important pharmacological effects. Preclinical study on rats, which were induced with diabetes using alloxan when administered with *C. colocynthis*, showed a significant antihyperglycemic and insulino-tropic potential of the medicinal herb. Various animal studies using different extracts like aqueous and alcoholic administered in different doses 50, 100, 300 mg/kg showed positive results (Dallak et al., 2009).

2.2.13 *COCCINIA INDICA* (CUCURBITACEAE)

Coccinia indica, also referred as little gourd, is used for its beneficial value in numerous of ailments like asthma, bronchitis, and also found to be having effect in diabetes. Aqueous extracts showed antidiabetic activity while alcoholic extract normalized glucose as well as lipid levels (Ajay, 2009). In clinical settings, dried extract enhanced glucose level in the blood (Balaraman et al., 2010).

2.2.14 *EUCALYPTUS GLOBULUS* (MYRTACEAE)

Eucalyptus globulus is commonly called blue gum tree and mainly found in Australian subcontinents. Since the ancient times, it has been used as antidiabetic agent in the land of South America and Africa. When used in diabetic mice, it showed antihyperglycemic effect. It also stimulates the insulin production (Gray & Flatt, 1998). *Eucalyptus* partially restores activity of β-cells in pancreas and reduces streptozotocin-induced damage (Mahmoudzadeh et al., 2010).

2.2.15 *EUGENIA JAMBOLANA* (MYRTACEAE)

Eugenia jambolana, also referred as Indian blackberry, observed to contain antihyperglycemic and antihyperlidemic potential when studied in different preclinical settings (Ravi et al., 2004). It inhibits insulinase function, thereby increasing insulin secretion (Patelet al., 2012).

2.2.16 *GYMNEMA SYLVESTRE* (ASCLEPIADACEAE)

Gymnema sylvestre is a plant which is mainly cultivated in Asian regions like in Sri Lanka and India. Antidiabetic potential of the plant is depicted in Ayurveda. It leads to the increase in insulin production in mouse and also in human β-cells with type-2 diabetes. Increased level of insulin might be associated with rejuvenation of the cells of the pancreas. *G. sylvestre* extract also causes the release of insulin in MIN6 β-cell lines induced diabetic rats (Liu et al., 2009). Aqueous extract of the leaves was observed to be hypolipidemic as well as hypoglycemic effect in diabetes-induced rats by alloxan (Mall et al., 2009).

2.2.17 *HIBISCUS ROSA-SINENSIS* (MALVACEAE)

Hibiscus, commonly known as China rose or shoe flower, plays a vital role against neoplasm but recently found useful in diabetes. *Hibiscus* extract has shown hypoglycemic effect in diabetes-induced rats (Venkatesh et al., 2008). The root extract has also reported to have hypoglycemic, anti-dyslipidemic as well as antioxidant potential in various preclinical studies (Kumar et al., 2013).

2.2.18 *OCIMUM SANCTUM* (LAMIACEAE)

The plant, better referred to as *tulsi* in India, is an important part of medicinal herbs used since time immemorial. Its useful roles are also depicted in Atharvaveda and in Ayurveda and in other old literatures also. The whole plant can be utilized because of its pharmacological activity. An antidiabetic property of the plant has already been observed in previous studies. Extract of *Ocimum sanctum* significantly reduced blood glucose and increased glycogen level in diabetic rats and also showed significant decrease in diabetogenic and dyslipidemia parameters (Husain et al., 2015).

2.2.19 *TRIGONELLA FOENUM-GRAECUM* (FABACEAE)

Trigonella foenum-graecum is popularly referred as fenugreek and in India known as *methi* (Ankita & Prasad, 2015b). Presence of glucosides and alkaloids adds to its antidiabetic properties. *T. foenum-graecum* decreases the somatostatin level and increases glucagon level, thereby increasingly leading to antihyperglycemic effect (Ribes et al., 1986). The herb reduces the maltase activity in diabetes-induced rats to produce antidiabetic effect and also has the potential to reduce renal toxicity in diabetes-induced rats with alloxan (Kumar et al., 2005).

2.3 MEDICINAL PLANTS USED IN NEOPLASM

Neoplasm is among the prime cause of death worldwide after heart diseases. Pharmacological plants showed as potential agents in anticancer therapy. Studies conducted in the regions of America, East Asia, and Europe depicted that about 45% of drugs employed against neoplasm in the period from 1940 to 2006 were extracted from plants. The important lead compounds utilized in neoplasm therapy like vinblastine, vincristine, podophyllotoxin, paclitaxel, and camptothecin are derived from NPs. The possible model and mode of actions for respective medicinal plants are presented in Table 2.2.

TABLE 2.2 Medicinal Plants in Neoplasm.

Medicinal plant	Family	Action	Model
Andrographis paniculata	Acanthaceae	Increases apoptosis by increasing caspase substrate, increases the sensitivity of cancerous cells toward chemotherapeutic agents	NSCLC, breast cancer cells, ovarian cancer cell lines
Centella asiatica	Apiaceae	Biochemical modulator, decreases tumor size, increases tumor apoptosis	MCF-7 cell lines, DMBA-induced rat mammary cancer model, colon cancer cell line
Curcuma longa	Zingiberaceae	Downregulates the expression of a COX2, NF-kβ, EGR-1, NOS, TNF, cyclins, and growth factor receptors	MCF-7, rat and human bladder cells, lung neoplasms cells
Phyllanthus amarus	Euphorbiaceae	Induced the expression of p53 and p45NFE2 and decreased the expression of Bcl-2 in the spleen of infected mice, inhibits DNA topoisomerase II of *Saccharomyces cerevisiae* mutant cell cultures	Erythroleukemia in mice, 20-methylcholanthrene (20-MC)-induced sarcoma

TABLE 2.2 *(Continued)*

Medicinal plant	Family	Action	Model
Podophyllum hexandrum	Berberidaceae	Increases the tumor doubling time	Breast cancer cell line (MCF-7), Ehrlich ascites tumorous mice
Tinospora cordifolia	Menispermaceae	Increases apoptosis of cancer cell, increases survival of tumor free cell	HeLa cells, Ehrlich ascites carcinoma transplanted mice
Withania somnifera	Solanaceae	Inhibited vimentin in breast cancer cell, inhibited lung collagen hydroxyproline	Breast cancer cells, B16F1 melanoma in C57BL mice, B16F-10 melanoma-induced metastasis in mice
Ziziphus jujube	Rhamnaceae	Inhibiting cell cycle and increase in apoptosis via the mitochondria transduction pathway, antioxidative property and the flavonoid content	Human breast cancer cell line MCF-7, HeLa cervical cancer cells and A549 lung cancer cells

2.3.1 *ANDROGRAPHIS PANICULATA* (ACANTHACEAE)

Andrographis paniculata mainly grows in south Asian countries. Roots and leaves are generally employed for pharmacological purpose. Andrographolide is the pharmacological active component of the plant, used in different disease conditions. It can also activate antigen-specific and nonspecific immune response, and it is considered as important infectious and oncogenic agent (Puri et al., 1993).

Andrographolide is potent in non-small-cell lung cancer (NSCLC) by increasing HLJ1, a caspase-3 substrate level, thereby inhibiting tumorigenesis (Lai et al., 2013). Andrographolide suppressed STAT3 phosphorylation by IL-6 and nuclear translocation in tumor cells, which in turn sensitize these cells toward doxorubicin-induced apoptosis (Zhou et al., 2010). Andrographolide also inhibits matrix metalloproteinase-9 expression that ultimately affects tumorigenesis and invasion in breast cancer cells (Chao et al., 2013). Andrographolide when administered along with cisplatin increased the programmed cell death in ovarian cancer cells (Yunos et al., 2013).

2.3.2 *CENTELLA ASIATICA* (APIACEAE)

Centella asiatica is a herb native to the wetlands of Asia and mentioned in Ayurveda for its beneficial role in ulcer, diarrhea, asthma, bacterial infection, and wound healing. Asiaticoside, an extract of *C. asiatica,* is a biochemical modulator which increases programmed cell death in MCF-7 cell line and *in vivo* in 7,12-dimethylbenz(*a*)anthracene-induced rat mammary cancer model. It has a capacity to decrease tumor sizes (Al-Saeedi, 2014). In colon cancer tissue, asiaticoside induced cancer apoptosis via mitochondrial death apoptosis cascade. So, it is determined that the herb has potential chemopreventive, antitumor activity, and anti-inflammatory effects.

2.3.3 *CURCUMA LONGA* (ZINGIBERACEAE)

Popularly known as turmeric, *Curcuma longa* is cultivated in southeast India. The herbaceous plant is mainly used in cooking. In medicine, it is employed for both cancer prevention and treatment. The antineoplasm activity of *C. longa* is attributed due to its ability to block proliferation in numerous of tumor cells. *C. longa* decreases the expression of a cyclooxygenase 2 (COX2), nuclear factor-κB (NF-kβ), epidermal growth receptor-1, tumor necrosis factor (TNF), cyclins, and growth factor receptors, and others, thereby producing anticancerous activity (Aggarwal et al., 2003; Shao et al., 2002; Surh et al., 2001). Numerous studies of *Curcuma* on neoplasms have been established for its efficacy in the disease state. *Curcuma* inhibited MCF-7 cells (human mammary epithelial neoplasms cells), thereby proving its beneficial role in breast cancer (Ströferet al., 2011). Similarly in prostate cancer, *Curcuma* down-regulated the genes expression that could support tumor-cell growth. When administered along with anti-Ki-67 siRNA to rat and human bladder cells, *Curcuma* decreased growth rate of neoplasms cells and increased rate of apoptosis (Pichu et al., 2012). *Curcuma* also efficiently restricted the movement and invasion of pulmonary neoplasm cells by inhibiting Rac1, a protein which contributes in tumor cell growth and migration (Chen et al., 2014).

2.3.4 PHYLLANTHUS AMARUS (EUPHORBIACEAE)

Phyllanthus amarus is an annual herb that mainly grows in tropical and subtropical regions. *P. amarus* by suppressing expression of Bcl-2, regulator protein of apoptosis and enhancing level of p53, a tumor-suppressor protein, proved to be effective in erythroleukemia developed in BALB/c mice (Harikumar et al., 2009). It is a potent inhibitor of the cell cycle and carcinogens, thereby producing antineoplasm effect (Rajeshkumar et al., 2002).

2.3.5 PODOPHYLLUM HEXANDRUM (BERBERIDACEAE)

Podophyllum hexandrum, mainly cultivated in India, Pakistan, China, and Afghanistan, is observed to be a potent antineoplastic agent. When administered in tumorous mice, it significantly increases the tumor doubling time from 1.94 ± 0.26 days to 19.1 ± 2.5 days (Goel et al., 1998). In MCF-7 cell lines (breast cancer cell line), the herb inhibited the neoplasm growth by 50%. Moreover, its early use provides better result in neoplasm (Chattopadhyay et al., 2004).

2.3.6 TINOSPORA CORDIFOLIA (MENISPERMACEAE)

Tinospora cordifolia commonly referred as giloya or heartleaf grows mainly in Asian lands like India, Sri Lanka, and China. Root as well as stem extracts are associated with antioxidant, anti-inflammatory, antiallergic, antiarthritic properties. Different doses of *T. cordifola* are able to inhibit HeLa cell growth. On the other side, micronuclei induction took place in treated samples which proved the tumor-free cell-survival activity, suggesting its potential as an anticancer agent. Another study demonstrated the potential role of *T. cordifolia* in *Ehrlich ascites* neoplasm transplanted mice. The herb increased the survival of tumor-free cells by reducing glutathione and increasing lipid peroxidation, thereby killing tumor cells (Jagetia & Rao, 2006). *In vitro* study on human epithelial neoplasm tissue revealed that *Tinospora* leads to chemotherapeutic drug-mediated toxicity in these cells by blocking ABC-G2 and ABC-B1 transporters (Maliyakkal et al., 2015).

2.3.7 *WITHANIA SOMNIFERA (SOLANACEAE)*

Withania somnifera is popularly called ashwagandha, Indian ginseng. Withaferin A is the potent constituent of *W. somnifera*. In mammary cancer cells, Withaferin A inhibited vimentin, a type of protein which promotes cancer cell growth. Withaferin A decreased the level of thermotolerance in B16F1 melanoma in black mice (Kalthur et al., 2009). Withanolide inhibited the melanoma growth in lungs by reducing the levels of hexosamines, uronic acids, and sialic acid (Leyon & Kuttan, 2004).

2.3.8 *ZIZIPHUS JUJUBE (RHAMNACEAE)*

Ziziphus jujube, also called as *ber* or Chinese date, is mainly distributed in southern-Asian regions. Betulinic acid, extracted from *jujube* fruits, significantly inhibits MCF-7 cell line. The effect is initiated by inhibiting cell cycle and increase in programmed cell death through mitochondria transduction pathway (Sun et al., 2013). The antioxidative property and the flavonoid content have antineoplastic nature against HeLa cells and A549 lung cancer cells.

2.4 MEDICIANL PLANTS USED IN CARDIOVASCULAR DISEASE

Cardiovascular disease is one of the prevalent non-communicable diseases worldwide. As per the estimate, stroke, heart attack, and failure are the main cause of death in developed countries. Since the last few decades, different strategies are being designed depending on the pathogenesis of heart-related ailments but morbidity and mortality still persist in clinical settings. But the use of plants in cardiovascular disease showed some hopes (Table 2.3). Different herbs have been used in different pathology of heart-related diseases.

TABLE 2.3 Medicinal Plants in Cardiovascular Disease.

Medicinal plant	Family	Action	Studied on
Allium sativum	Liliaceae	Effective in arrhythmias, hypertrophy, ischemia–reperfusion injury, myocardial infarction (MI)	–
Azadirachta indica	Meliaceae	Decreased the apoptosis rate; restored biochemical and hemodynamic properties	–
Crataegus oxyacantha	Rosaceae	Decreased inflammatory and apoptotic markers, decreasing NF-κβ expression and increasing G-protein-coupled receptor kinase 2 expression	Isoproterenol-induced MI in a rat model; acute myocardial ischemia/reperfusion in anesthetized dogs
Ginkgo biloba	Ginkgoaceae	Improves vasomotor function; alters signal transduction	Human aortic endothelial cells (HAECs), coronary ligated rats
Inula racemosa	Asteraceae	Adrenergic beta-blocking activity	Adrenaline-induced hyperglycemia in rats, isoproterenol-induced myocardial injury in rats, experimental atherosclerosis in guinea-pigs
Magnolia officinalis	Magnoliaceae	Effective in atherosclerosis, inhibits ventricular arrhythmia	HAECs, coronary ligated rats
Panax ginseng	Araliaceae	PPAR activity restored	STZ-induced rat
Terminalia arjuna	Combretaceae	Antianginal, decongestive and hypolipidemic effect, antithrombotic	STZ-induced diabetic rats; isoproterenol-induced myocardial fibrosis

2.4.1 *ALLIUM SATIVUM* (LILIACEAE)

It is popularly called garlic. It is, used since time immemorial and mainly cultivated in central Asia but used all around the world. Many observations depictedvital role of *Allium* in cardiovascular ailments. It has beneficiary effects on lipids, platelets, blood pressure along with acting as antioxidant and fibrinolytic help in disease state (Mikaili et al., 2013). *Allium* is useful in arrhythmias, hypertrophy, ischemia–reperfusion injury as well as myocardial infarction (MI). Many of its important properties like cytochrome P450 inhibition, ion channels control, histone deacetylase inhibition are important in disease-related conditions (Khatua et al., 2013).

2.4.2 AZADIRACHTA INDICA (MELIACEAE)

A. indica, also known as Neem, is useful as cardio-protective agent. Its potent activity in MI, cardiotoxicity, and hypertension has been depicted. It decreased the apoptosis rate, restored biochemical and hemodynamic properties which evident in histopathological studies (Peer et al., 2008).

2.4.3 CRATAEGUS OXYACANTHA (ROSACEAE)

Commonly referred to as Hawthorn, it has been employed in many cardio-vascular-related studies. Beneficial effect of *Crataegus* was determined in isoproterenol-induced MI, in which lactate dehydrogenase along with creatine kinase and all other inflammatory and apoptotic markers finally controlled (Vijayan et al., 2012). In case of acute myocardial ischemia, *Crataegus* showed to decrease inflammation by decreasing NF-κβ expression and increasing G-protein-coupled receptor kinase 2 expressions (Fu et al., 2013). *C. oxyacantha* is potent in chronic heart failure but proved satisfactory in many other cardiac-related ailments also.

2.4.4 GINKGO BILOBA (GINKGOACEAE)

Ginkgo biloba is mainly cultivated in China and has beneficial medicinal value. In cardiovascular disease, it shows myocardial protective effect by lowering the level of oxygen-free radicals and enhances the antioxidant potential, improves vasomotor function, and alters signal transduction. Furthermore, *G. biloba* showed positive effect on the myocardial cells when bone-marrow-derived stem cells were transplanted into cardiomyocytes (Liu et al., 2014).

2.4.5 *INULA RACEMOSA (ASTERACEAE)*

Inula racemosa, known commonly as *Pushkarmoola*, is mainly cultivated in the hilly regions in the north-western Himalayas, and is useful in dyspnea and angina. *Inula* was found to decrease in the level of insulin and glucose which showed its β-blocking activity in rats (Tripathi et al., 1988). It can also act as antioxidant which helps in ventricular functions of the heart in MI (Ojha et al., 2011). It provides an edge in artherosclerosis by reducing cholesterol, low-density lipoprotein cholesterol, and increased superoxide dismutase and glutathione peroxidase (Mangathayaru et al., 2009).

2.4.6 *MAGNOLIA OFFICINALIS (MAGNOLIACEAE)*

Magnolol, one of the active constituent of *Magnolia officinalis*, has both antioxidative and anti-inflammatory properties and is helpful in athero-sclerosis and inflammatory responses (Liang et al., 2014). Magnolol also prevents ischemic injury in myocardial cell and also plays preventive role in ischemia and reperfusion (Ho & Hong, 2012).

2.4.7 *PANAX GINSENG (ARALIACEAE)*

Panax, commonly referred to as ginseng, is helpful in preventing cardiac hypertrophy by blocking calcineurin activation, thereby helping in remodeling (Deng et al., 2009). Reduction in cardiac PPAR delta which is a receptor activated by proliferation of peroxisomes, leads to fall in the cardiac output leading to increased incidence of heart failure. But ginseng is found to restore PPAR delta activity and ultimately restores the cardiac contractility of STZ-rats (Tsai et al., 2014).

2.4.8 *TERMINALIA ARJUNA (COMBRETACEAE)*

Teminalia arjuna, mainly referred as *Arjun* tree, is mainly found on the river banks in India. In Streptozotocin, (STZ)-induced diabetic Wistar Albino rats, it ameliorated autonomic neuropathy in cardiovascular system by improving autonomic control, left ventricular function, and myocardial remodeling in chronic heart failure (Khaliq et al., 2013).

In myocardial changes, it helps in myocardial fibrosis and in oxidative stress. The antithrombotic activity of *T. arjuna* helped in many coronary artery diseases (Dwivedi, 2007). It is also useful in angina (Bharani et al., 2002). In chronic smokers, *Arjuna* reduces oxidative stress in order to improve endothelial abnormalities (Bharani et al., 2004).

2.5 MEDICINAL PLANTS USED IN GASTROINTESTINAL DISORDERS

The gastrointestinal system always faces some of the common problems like dysentery, diarrhea, stomach-ache, worms (anti-helminthic), constipation, vomiting, bowel movements, ulcer, and others, due to food habits. Apart from these common ailments, inflammatory bowel syndrome (IBS), inflammatory bowel disease (IBD), colon cancer, stomach cancer, and others, are the real threats to gastrointestinal system. Man always relies on plants not only for food but also for medicines. There are many medicinal plants which showed promising effect in gastrointestinal disease (Table 2.4).

2.5.1 AEGLE MARMELOS (RUTACEAE)

Aegle marmelos (AM), commonly referred as *Bael*, is a spiny tree, found mainly in Asia. In colitis-induced rodents, *Aegle* helps in reducing inflammation and mucosal damage (65). It also showed antibacterial and antioxidant activities. It is also beneficial in ulcer as depicted by biochemical parameters (Ilavarasan et al., 2002).

2.5.2 ALLIUM SATIVUM (LILIACEAE)

Garlic is found to be effective in colorectal cancer because of its property to induce apoptosis, carcinogen metabolizing enzymes (Chung et al., 2004; Chen et al., 1998), and restrain COX2 (Sengupta et al., 2004) and lipid peroxidation (Sengupta et al., 2003). It is beneficial against *H. pylori* infection as it acts as antibiotic agent (Sivam et al., 1997). Garlic is effective in diarrhea as it increases smooth muscle tension of the gastric tract when given in low doses.

TABLE 2.4 Medicinal Plants in Gastrointestinal Disease.

Name	Family	Action	Disease state
Aegle marmelos	Rutaceae	Anti-inflammatory, antioxidant	IBD, peptic ulcer
Allium sativum	Liliaceae	Antibiotic, antiproliferative, antioxidant	Colorectal cancer, *H. pylori*, antidiarrheic
Amaranthus tricolor	Amaranthaceae	Antioxidant, antiproliferative	Peptic ulcer, stomach cancer, colon cancer
Andrographis paniculata	Acanthaceae	Antioxidant, antiproliferative	Peptic ulcer, colon cancer
Aloe vera	Aloeaceae	Antisecretory, anti-inflammatory	Gastric secretion, ulcerative colitis, peptic ulcer
Artemisia annua	Asteraceae	Immunomodulatory, anti-inflammatory	HCT-116, TNBS colitis model
Azadirachta indica	Meliaceae	Antisecretory, proton pump inhibitor; anticlastogenic	Peptic ulcer, infection
Boswellia serrata	Burseraceae	Anti-inflammatory, antioxidant effects, cancer cell cycle inhibitor	Ulcerative colitis, colon cancer
Centella asiatica	Apiaceae	Antioxidant, anti-inflammatory	Acetic acid-induced gastric ulcer, intestinal tumorigenesis, DNBS-induced colitis
Chelidonium majus	Papaveraceae	Anti-inflammatory, apoptosis in neoplasm cell	Colorectal cancer, pancreatic cancer, ulcerative colitis
Coptis chinensis	Ranunculaceae	Anti-inflammatory, reduce apoptosis in normal cell	Antiulcer, antineoplastic, antidiarrheic
Croton cajucara	Euphorbiaceae	Anti-inflammatory, antiseptic	Gastric ulcer, antineoplastic
Curcuma longa	Zingiberaceae	Sensitize cancerous cell toward chemotherapeutic drugs	IBD, pancreatic cancer, colon cancer, peptic ulcer
Elettaria cardamomum	Zingiberaceae	Antioxidant, anti-inflammatory, anti-proliferative, apoptosis	Colon cancer, gastric ulcer

TABLE 2.4 *(Continued)*

Name	Family	Action	Disease state
Lippia sidoides	Verbenaceae	Antiseptic, anti-inflammatory	Ethanol-induced damage in experimental model
Panax ginseng	Araliaceae	Anti-inflammatory	Ethanol-induced gastric damage, neoplasm
Phyllanthus amarus	Euphorbiaceae	Clastogenic, stop cell cycle in neoplasm cells	Stomach cancer
Rosmarinus officinalis	Lamiaceae	Anti-inflammatory, antioxidant, vaso-dilator, antiproliferative	Colon cancer, pancreatic cancer, gastric ulcer
Solanum tuberosum	Solanaceae	Anti-inflammatory, antioxidant, reduce toxicity	Colitis
Withania somnifera	Solanaceae	Anti-inflammatory, mucorestorative, Immunomodulatory	Colon cancer, IBD, gastric ulcer

2.5.3 AMARANTHUS TRICOLOR (AMARANTHACEAE)

Amaranthus tricolor is referred as red amaranth, cultivated mainly in south-east Asian regions. *A. tricolor* is traditionally used for diarrhea, dysentery, cough, liver ailments, and tumor. Presence of flavonoids, antioxidants, and polysaccharides provided antiulcer properties in preclinical studies (Gopal et al., 2013). These properties also add to its antiproliferative nature in stomach as well as in colon neoplasms (Alam et al., 2013).

2.5.4 ANDROGRAPHIS PANICULATA (ACANTHACEAE)

Andrographis paniculata is used widely in Chinese medicine for the cure of viral infections and inflammatory disease. Andrographolide (Andro) isolated from *A. paniculata* showed beneficial medicinal properties in many disease states. Anti-inflammatory as well as antioxidant properties of *A. paniculata* minimized the lesions in ulcer (Wasman et al., 2011). It promotes programmed cell death under *in vitro* condition and generates synergistic effect when administered with chemotherapeutic agents (Yunos et al., 2013).

Recently a placebo-controlled, double-blind clinical trial of HMPL-004 (*A. paniculata* extract) has shown its efficacy in mild-to-moderate category of ulcerative colitis (UC) by inducing remission and mucosal healing (Sandborn et al., 2013).

2.5.5 ALOE VERA (XANTHORRHOEACEAE)

Aloe vera undoubtedly is effective in diabetes and cardiovascular diseases. But it has a significant effect in gastric lesions, peptic ulcer, and colitis. *Aloe* decreases the expression of nitric oxide (NO) synthase enzyme and matrix metalloproteinase-9 in gastric mucosa to produce healing effect (Park et al., 2011). It has antisecretory activity due to inhibition of gastric acid production in a dose-dependent manner. It suppresses TNF-α and increases IL-10 level and induces peptic ulcer healing which proved in histological examinations (Eamlamnam et al., 2006).

In UC, *Aloe* was found to lower the concentration of TNF-α and leukotriene B4 and significantly reduced inflammation (Park et al., 2011).

2.5.6 *ARTEMISIA ANNUA* (ASTERACEAE)

Artemisia annua is better used for fever treatment in conventional Chinese medicine. Artemisinin, extracted from *A. annua*, is used widely as antimalarial. Dihydroartemisinin (DHA), the semisynthetic derivative of artemisinin, has shown to promote programmed cell loss in HCT-116 cells (colorectal cancer cells) (Lu et al., 2015). In TNBS colitis model, the immunomodulatory and anti-inflammatory potential of artesunate inhibited TNF-α expression and T-helper responses, thereby showing gastroprotective effect (Yang et al., 2012a).

2.5.7 *AZADIRACHTA INDICA* (MELIACEAE)

The worthy role of *Azadirachta* or Neem has been also depicted in Ayurveda as it is useful in ulcer and gastric discomfort. The beneficial role of Neem in gastrointestinal disease is because of its property to eliminate harmful bacteria and toxic agents thereby maintaining healthy digestive system. The antisecretory, as well as proton-pump blocking by Neem, helps in controlling increased acidity level (Bandyopadhyay et al., 2004). It also exerts anticlastogenic effects by modulating antioxidant defense and oxidative stress (Gangar et al., 2010).

2.5.8 *BOSWELLIA SERRATA* (BURSERACEAE)

Boswellia is a potent agent in diarrhea, UC, Crohn's disease, and tumor in gastrointestinal tract. In UC, *B. serrata* is effective in decreasing inflammation level as it is antioxidant and anti-inflammatory in nature. All the biochemical parameters and histological examinations proved a beneficial effect of *Boswellia* in UC (Yadav et al., 2012).

Boswellia is efficacious in colorectal cancer in mice. Its extract lowered down NF-κB, COX2, bcl-2, and cyclin D1 level which in turn suppresses tumor growth (Guo et al., 2004).

2.5.9 *CENTELLA ASIATICA* (APIACEAE)

Centella asiatica is referred to as Asiatic pennywort or Indian pennywort. It has a mention in Ayurveda as well as in Chinese medicine and is native to Asian wetlands. Anti-inflammatory property was observed to be beneficial in gastric ulcer in rats. *C. asiatica* inhibited NO output and finally reduced ulcers (Bunpo et al., 2004). When administered in rats having intestinal neoplasm, it significantly decreased the abnormal crypt foci and modified cell proliferation resulting in chemo-preventive effect. In dinitrobenzene sulphonic acid-induced colitis rat, it reduced various pro-inflammatory cytokines and suggested beneficial role of *C. asiatica* in IBD (di Paola et al., 2010).

2.5.10 *CHELIDONIUM MAJUS* (PAPAVERACEAE)

Chelidonium majus is mainly native to Eastern Asia. Its therapeutic effect is well known as it has analgesics, antibiotic, and oncostatic effects. Numerous clinical and preclinical studies have been conducted by using Ukrain, an extract of *C. majus,* in colorectal and pancreatic cancer. It is also pharmacologically active in neoplasms. It suppressed the neoplasms at molecular stage by interfering in synthesis of protein and inhibited oxygen consumption and induced apoptosis (Nadova et al., 2008). In UC, it is responsible to attenuate the level of COX2 which is an inducible enzyme, upregulated during tissue damage during inflammation and produce prostaglandins which ultimately mediate inflammation. It also decreases the level of hypoxia-induced factor-1α (HIF-1α) that is involved in inflammation (Kim et al., 2012).

2.5.11 *COPTIS CHINENSIS* (RANUNCULACEAE)

Coptis chinensis is a herb, native to China, used extensively in Chinese medicines. *Coptis* is employed for abdominal pain and diarrhea but also useful in IBS (Yang et al., 2012b) and gastric inflammation. The rhizomes of the herb can reduce visceral pain raised during IBS by decreasing the levels of serotonin and cholecystokinin in rats (Tjong et al., 2011). Alkaloids of *Coptis* are also useful in *H. pylori* lipopolysaccharide-induced gastric inflammation in rodents by inhibiting apoptosis in epithelial tissues and inflammatory response (Lu et al., 2007).

2.5.12 CROTON CAJUCARA (EUPHORBIACEAE)

Different essential oils of *Croton cajucara* have medicinal values as reported in previous studies. In ulcers, *C. cajucara* decreased inflammation as depicted in histopathological evaluations. The anti-inflammatory effect of *C. cajucara* is due to the heat shock protein-70 and sulfhydryl compounds activities present in it (Rozza et al., 2011).

Croton lechleri another form of Croton best known as Sangre de Drago is used widely in wounds, fractures, hemorrhoids, gastrointestinal ulcers, and also for the empirical cure of cancers. Under *in vitro* condition, it is also observed to block colorectal cancer cell growth, supporting its effectiveness as antineoplastic agent (Montopoli et al., 2012).

2.5.13 CURCUMA LONGA (ZINGIBERACEAE)

Curcumin, active component extracted from *C. longa,* is found to be effective in IBD, pancreatic cancer, and colorectal cancer. It also inhibits NF-kβ and therefore the pathways which lead to the progress of esophageal adenocarcinoma (Hartojo et al., 2010; Tian et al., 2008). Curcumin helped to attain improvement in premalignant patients which was shown in histological studies. When curcumin administered along with standard chemotherapeutic agents, it reduced the chemo-resistance in the cells and also decreased chemo-related adverse drug reactions (Shakibaei et al., 2013). In UC, curcumin maintains the remission (Kumar et al., 2012). Animal studies also observed important role of curcumin as it reduces various inflammatory signs and also inhibited cancer growth. Curcumin along with gemcitabine is effective in pancreatic cancer management (Kunnumakkaraet al., 2007). Curcumin is also potent in inflammatory bowel syndrome, peptic ulcer, and others.

2.5.14 ELETTARIA CARDAMOMUM (ZINGIBERACEAE)

Elettaria is also known as true cardamom or green cardamom. It is native to Indian region and is used widely as spices, having medicinal values. The potent activities of *Elettaria* like antioxidant, anti-inflammatory, and antiproliferative make it an effective agent in colon cancer (Majdalawieh &Carr, 2010) and gastric ulcer (Jamal et al., 2006).

2.5.15 *LIPPIA SIDOIDES* (VERBENACEAE)

Lippia sidoides in Brazil is mainly employed as an antiseptic. It decreased the gastric damage produced by ethanol in mice, thus producing protective effect (Monteiro et al., 2007).

2.5.16 *PANAX GINSENG* (ARALIACEAE)

It is commonly known as ginseng and valued in the traditional medicine since ancient times for anti-aging, antidiabetic, and helpful in protein synthesis. Active components of the compound like ginsenosides Rh2 and Rg3 are responsible for antineoplastic effects (Kim et al., 2004). By inhibiting various pro-inflammatory and pro-apoptotic factors, it prevented ethanol-induced gastric mucosal damage (Huang et al., 2013).

2.5.17 *PHYLLANTHUS AMARUS* (EUPHORBIACEAE)

Phyllanthus amarus is a small plant which is known for its pharmacological properties in India. It has shown valuable pharmacological properties like antiviral, antibacterial, anti-inflammatory, antimicrobial, anticancer, antidiabetic, and others. It has a gastro-protective effect in radiation-induced changes in the mouse intestine as shown in histopathological evaluation and also *P. amarus* showed clastogenic effects by decreasing micronuclei (Harikumar & Kuttan, 2007). *P. amarus* is also effective in neoplasms as it can alter cell signaling mechanisms or by stopping cell cycling or scavenging of carcinogen-induced free radicals (Raphael et al., 2006).

2.5.18 *ROSMARINUS OFFICINALIS* (LAMIACEAE)

Rosmarinus officinalis, mainly referred to as rosemary, grows primarily in the Mediterranean region and acts as an anti-ulcer and anticancerous agent. In colon and pancreatic neoplasm, rosemary extracts increase the level of metabolic-related gene and suppress epigenetic modulator which found to reverse the cancer growth activities (González-Vallinas et al.,

2014). The anti-inflammatory, vasodilation, and antioxidant activities of rosemary provide benefit in gastric ulcer conditions (Amaral et al., 2013).

2.5.19 *SOLANUM TUBEROSUM* (SOLANACEAE)

S. tuberosum, popularly known as potato, is a native to the Andes region and among world's largest food crops produced. But the vital role *of S. tuberosum* in medicine field is also established. Anti-inflammatory properties prevented inflammation in colitis in colonic tissue in mice (Lee et al., 2014). Its fiber components reversed the toxic effect produced by acrylamide in small intestinal region (Dobrowolski et al., 2012). Antioxidant activity of *S. tuberosum* prevents tissue damage in intestinal region of rats (Kudoh et al., 2003).

2.5.20 *WITHANIA SOMNIFERA* (SOLANACEAE)

Withania somnifera is efficacious in IBD, colon cancer, and in ulcer under experimental conditions. In IBD model, anti-inflammatory and antioxidant activities of *Withania* reversed the inflammation. Moreover, the topical use of *W. somnifera* is observed to be as effective as mesalamine (Pawar et al., 2011). This herb has the potential to influence the immune system in colon cancer (Muralikrishnan et al., 2010). The antioxidant activity of this herb helps to restore normal condition in stress-induced gastric ulcer (Bhatnagar et al., 2005).

2.6 MEDICINAL PLANTS IN NEUROLOGICAL DISEASE

The human race has entered into an era where we have understood almost everything and also solved the puzzle of the universe. But brain is still called the enigma and the disease related to it still gives biggest challenges to the neurologists. Numerous therapies have been used to overcome the brain-related diseases and herbal medicine offers several options to modify the progress and symptoms of different neurological disease (Table 2.5).

TABLE 2 5 Medicinal Plants in Neurological Disease.

Name	Family	Action	Disease state
Acanthopanax senticosus	Araliaceae	Reduces apoptosis	SH-SY5Y cells, ethanol-induced apoptosis, human neuroblastoma cell line SK-N-MC
Acorus calamus	Acoraceae	Antispasmodic, carminative, antioxidant	Pentylenetetrazole-induced seizures model in rats, FeCl(3)-induced epileptogenesis
Cannabis sativa	Cannabaceae	Anticonvulsant, antioxidant, anti-inflammatory;	Neurodegeneration
Crinum glaucum	Liliaceae	Interacts with glutaminergic and GABAergic system	Picrotoxin, strychnine, isoniazid, pentylenetetrazol and *N*-methyl-D-aspartate (NMDA)-induced seizures in mice, hexobarbitone-induced sleeping
Ginkgo biloba	Ginkgoaceae	Anti-apoptotic, antioxidative, anti-inflammatory	Rat pheochromocytoma (PC12) cells
Zataria multiflora	Lamiaceae	Anticholinesterase, antioxidant, anti-inflammatory	PTZ-induced mice, amyloid beta-induced AD

2.6.1 *ACANTHOPANAX SENTICOSUS* (ARALIACEAE)

Acanthopanax senticosus decreased the expression of α-synuclein which leads to Parkinson's disease by damaging dopaminergic neurons (Li et al., 2014). In dopaminergic neuronal damage, induced by 1-methyl-4-phenyl-1,2,3,6-tetrahydropyridine hydrochloride (MPTP) in C57BL/6, *A. senticosus* also shows neuroprotective effect (Liu et al., 2012). This herb has found to reduce caspase-3 level, thereby reducing apoptosis in ethanol-induced apoptosis of neuroblastoma cell line SK-N-MC (Jang et al., 2003).

2.6.2 *ACORUS CALAMUS* (ACORACEAE)

Acorus calamus is reported to help in the rejuvenation of nervous system. It is conventionally employed for the treatment of epilepsy, diarrhea, fever, kidney, and liver ailments. Moreover, both rhizomes as well as leaves observed to have vital activities like antispasmodic and carminative (Sharma et al., 2014). *A. calamus* when used along with anticonvulsant drugs provides synergistic effects (Katyal et al., 2012). This herb is also demonstrated to prevent $FeCl_3$-induced epileptogenesis by tempering antioxidant proteins (Hazra et al., 2007).

2.6.3 CANNABIS SATIVA (CANNABACEAE)

Cannabis sativa shows numerous of biologic effects within the nervous system. Many preclinical studies showed its potency in epilepsy (Szaflarski & Bebin, 2014; Devinskyet al., 2014). Beneficial role of the extract has been reported in neuronal injury, neurodegeneration, and psychiatric disease. It is also demonstrated to be effective in Parkinson's disease (Chagas et al., 2014; Lotan et al., 2014). *Cannabis* also showed anticonvulsant effect in rodents (Hill et al., 2013).

2.6.4 CRINUM GLAUCUM (LILIACEAE)

Crinum glaucum is a bulbous herb known for its potent anticonvulsant nature due to which it significantly prolonged the seizures onset period in mice and antagonized *N*-methyl-D-aspartate-induced turning behavior. *C. glaucum* has also reported to act as a hypnotic agent as it increased sleep duration, induced by hexobarbitone, and elevated duration of open-arm exploration in elevated plus maze proved its anxiolytic activity. *C. glaucum* also interacts with glutaminergic and GABAergic systems to show its efficacy on neurological system (Ishola et al., 2013).

2.6.5 GINKGO BILOBA (GINKGOACEAE)

Ginkgo biloba has anti-inflammatory, antioxidant properties and also helps in programmed cell death which provides neuroprotection. Ginkgolide K extracted from *G. biloba* plays a significant role in rat pheochromocytoma (PC12) cells from apoptosis by inhibiting caspase-3, thereby providing a good alternative in preventing cerebral ischemia (Ma et al., 2012).

2.6.6 *ZATARIA MULTIFLORA* (LAMIACEAE)

Zataria multiflora mainly grows in Asia. Extract of *Z. multiflora* was found to prolong the initiation clonic seizures and also reported to prevent pentylenetetrazole-induced tonic convulsions in mice (Mandegary et al., 2013). In Amyloid β-induced Alzheimer's disease rats, the herb reduced severity of the disease state due to its anticholinesterase, antioxidant, and anti-inflammatory activities (Majlessi et al., 2012).

2.7 CONCLUSION

Numerous plants have shown promising pharmacological activities in different states of ailments. With the emergence of new concept of CAM, there comes a flood of employing phytotherapy around the globe. India is said to be the goldmine of documented traditional medicinal plants. There are some important restraints for accepting plant use for medicinal purpose like safety, stability, pesticide-free herbs, single plant use, and others. The future perspective of phytotherapy is to isolate, purify, characterize, and optimize the therapeutically active component of the plants. These optimization strategies are the only means which can evaluate the efficacy claims of these medicinal plants. These standardization processes will finally help physicians and the consumers to come up with better options.

KEYWORDS

- **lifestyle disorders**
- **medicinal plants**
- **bioactive compounds**
- **antioxidant compounds**
- **nutrition**

REFERENCES

Aggarwal, B. B.; Kumar, A.; Bharti, A. C. Anticancer Potential of Curcumin: Preclinical and Clinical Studies. *Anticancer Res.* **2003,** *23*(1A), 363–398.

Ajay, S. S. Hypoglycemic Activity of *Coccinia indica* (Cucurbitaceae) Leaves. *Int. J. PharmTech. Res.* **2009,** *1*(3), 892–893.

Alam, S.; Krupanidhi, K.; Rao, K. R. S. Evaluation of In-Vitro Antioxidant Activity of *Amaranthus tricolor* Linn. *Asian J. Pharmacol. Toxicol.* **2013,** *01*(01), 12–16.

Al-Saeedi FJ. Study of the Cytotoxicity of Asiaticoside on Rats and Tumour Cells. *BMC Cancer* **2014,** *14,* 220.

Amaral, G. P.; de Carvalho, N. R.; Barcelos, R. P.; Dobrachinski, F.; Portella Rde, L.; da Silva, M. H.; Lugokenski, T. H.; Dias, G. R.; da Luz, S. C.; Boligon, A. A.; Athayde, M. L.; Villetti, M. A.; Antunes Soares, F. A.; Fachinetto, R. Protective Action of Ethanolic Extract of *Rosmarinus officinalis* L. in Gastric Ulcer Prevention Induced by Ethanol in Rats. *Food Chem. Toxicol.* **2013,** *55*, 48–55.

Ankita; Prasad, K. Characterization of Dehydrated Functional Fractional Radish Leaf Powder. *Pharm. Lett.* **2015a**, *7*(1), 269–279.

Ankita; Prasad, K. Chemical and Physical Characterization of Dehydrated Functional Fenugreek Leaf Powder. *Asian J. Chem.* **2015b,** *27*(10), 3697–3703.

Ayodhya, S.; Kusum, S.; Anjali, S. Hypoglycaemic Activity of Different Extracts of Various Herbal Plants Singh. *Int. J. Ayurveda Res. Pharm.* **2010,** *1*(1), 212–224.

Balaraman, A. K.; Singh, J.; Dash, S.; Maity, T. K. Antihyperglycemic and Hypolipidemic Effects of *Melothria maderaspatana* and *Coccinia indica* in Streptozotocin Induced Diabetes in Rats. *Saudi Pharm. J.* **2010,** *18*, 173–178.

Bandyopadhyay, U.; Biswas, K.; Sengupta, A.; Moitra, P.; Dutta, P.; Sarkar, D.; Debnath, P.; Ganguly, C. K.; Banerjee, R. K. Clinical Studies on the Effect of Neem (*Azadirachta indica*) Bark Extract on Gastric Secretion and Gastroduodenal Ulcer. *Life Sci.* **2004,** *75*(24), 2867–2878.

Bharani, A.; Ahirwar, L.K.; Jain, N. *Terminalia arjuna* Reverses Impaired Endothelial Function in Chronic Smokers. *Indian Heart J.* **2004,** *56*(2), 123–128.

Bharani, A.; Ganguli, A.; Mathur, L. K.; Jamra, Y.; Raman, P. G. Efficacy of Terminalia arjuna in Chronic Stable Angina: A Double-Blind, Placebo-Controlled, Crossover Study Comparing *Terminalia arjuna* with Isosorbide Mononitrate. *Indian Heart J.* **2002,** *54*(2), 170–175.

Bhatnagar, M.; Sisodia, S. S.; Bhatnagar, R. Antiulcer and Antioxidant Activity of *Asparagus racemosus* Willd. and *Withania somnifera* Dunal in rats. *Ann. N.Y. Acad. Sci.* **2005,** *1056*, 261–278.

Bnouham, M.; Ziyyat, A.; Mekhfi, H.; Tahri, A.; Legssyer, A. Medicinal Plants with Potential Antidiabetic Activity—A Review of Ten Years of Herbal Medicine Research (1990–2000). *Int. J. Diabetes Metab.* **2006,** *14*, 1–25.

Bunpo, P.; Kataoka, K.; Arimochi, H.; Nakayama, H.; Kuwahara, T.; Bando, Y.; Izumi, K.; Vinitketkumnuen, U.; Ohnishi, Y. Inhibitory Effects of *Centella asiatica* on Azoxymethane-Induced Aberrant Crypt Focus Formation and Carcinogenesis in the Intestines of F344 rats. *Food Chem. Toxicol.* **2004,** *42*(12), 1987–1997.

Chagas, M. H.; Eckeli, A. L.; Zuardi, A. W.; Pena-Pereira, M. A.; Sobreira-Neto, M. A.; Sobreira, E. T.; Camilo, M. R.; Bergamaschi, M. M.; Schenck, C. H.; Hallak, J. E.;

Tumas, V.; Crippa, J. A. Cannabidiol Can Improve Complex Sleep-Related Behaviours Associated with Rapid Eye Movement Sleep Behaviour Disorder in Parkinson's Disease Patients: A Case Series. *J. Clin. Pharm. Ther.* **2014,** *39*(5), 564–566.

Chao, C. Y.; Lii, C. K.; Hsu, Y. T.; Lu, C. Y.; Liu, K. L.; Li, C. C.; Chen, H. W. Induction of Heme Oxygenase-1 and Inhibition of TPA-Induced Matrix Metalloproteinase-9 Expression by Andrographolide in MCF-7 Human Breast Cancer Cells. *Carcinogenesis.* **2013,** *34*(8), 1843–1851.

Chattopadhyay, R. R. Possible Mechanism of Antihyperglycemic Effect of *Azadirachta indica* Leaf Extract. Part IV. *Gen Pharmacol.* **1996,** *27*(3), 431–434.

Chattopadhyay, S.; Bisaria, V. S.; Panda, A. K.; Srivastava, A. K. Cytotoxicity of In Vitro Produced Podophyllotoxin from *Podophyllum hexandrum* on Human Cancer Cell Line. *Nat Prod Res.* **2004,** *18*(1), 51–57.

Chauhan, A.; Sharma, P. K.; Srivastava, P.; Kumar, N.; Duehe, R. Plants Having Potential Antidiabetic Activity: A Review. *Der Pharm. Lett.* **2010,** *2*(3), 369–387.

Chen, G. W.; Chung, J. G.; Hsieh, C. L.; Lin, J. G. Effects of the Garlic Components Diallyl Sulfide and Diallyl Disulfide on Arylamine *N*-acetyltransferase Activity in Human Colon Tumour Cells. *Food Chem. Toxicol.* **1998,** *36*(9–10), 761–770.

Chen, Q. Y.; Zheng, Y.; Jiao, D. M.; Chen, F. Y.; Hu, H. Z.; Wu, Y. Q.; Song, J.; Yan, J.; Wu, L. J.; Lv, G. Y. Curcumin Inhibits Lung Cancer Cell migration and Invasion through Rac1-dependent Signaling Pathway. *J. Nutr. Biochem.* **2014,** *25*(2), 177–185.

Chung, J. G.; Lu, H. F.; Yeh, C. C.; Cheng, K. C.; Lin, S. S.; Lee, J. H. Inhibition of *N*-Acetyltransferase Activity and Gene Expression in Human Colon Cancer Cell Lines by Diallyl Sulfide. *Food Chem. Toxicol.* **2004,** *42*(2), 195–202.

Dallak, M.; Bashir, N.; Abbas, M.; Elessa, R.; Haidara, M.; Khalil, M.; et al. Concomitant Down Regulation of Glycolytic Enzymes, Upregulation of Gluconeogenic Enzymes and Potential Hepatonephroprotective Effects Following the Chronic Administration of the Hypoglycemic, Insulinotropic *Citrullus colocynthis* Pulp Extract. *Am. J. Biochem. Biotechnol.* **2009,** *5*(4), 153–161.

Deng, J.; Lv, X. T.; Wu, Q.; Huang, X. N. Ginsenoside Rg(1) Inhibits Rat Left Ventricular Hypertrophy Induced by Abdominal Aorta Coarctation: Involvement of Calcineurin and Mitogen-activated Protein Kinase Signaling. *Eur. J. Pharmacol.* **2009,** *608*(1–3), 42–47.

Devinsky, O.; Cilio, M. R.; Cross, H.; Fernandez-Ruiz, J.; French, J.; Hill, C.; Katz, R.; Di Marzo, V.; Jutras-Aswad, D.; Notcutt, W. G.; Martinez-Orgado, J.; Robson, P. J.; Rohrback, B. G.; Thiele, E.; Whalley, B.; Friedman, D. Cannabidiol: Pharmacology and Potential Therapeutic Role in Epilepsy and Other Neuropsychiatric Disorders. *Epilepsia* **2014,** *55*(6), 791–802.

di Paola, R.; Esposito, E.; Mazzon, E.; Caminiti, R.; Toso, R. D.; Pressi, G.; Cozzocrea, S. 3,5-Dicaffeoyl-4-malonylquinic Acid Reduced Oxidative Stress and Inflammation in a Experimental Model of Inflammatory Bowel Disease. *Free Radical Res.* **2010,** *44*(1), 74–89.

Dobrowolski, P.; Huet, P.; Karlsson, P.; Eriksson, S.; Tomaszewska, E.; Gawron, A.; Pierzynowski, S. G. Potato Fiber Protects the Small Intestinal Wall against the Toxic Influence of Acrylamide. *Nutrition* **2012,** *28*(4), 428–435.

Dwivedi, S. *Terminalia arjuna* Wight & Arn.—A Useful Drug for Cardiovascular Disorders. *J. Ethnopharmacol.* **2007,** *114*(2), 114–129. Epub 2007 Aug 10.

Eamlamnam, K.; Patumraj, S.; Visedopas, N.; Thong-Ngam, D. Effects of *Aloe vera* and Sucralfate on Gastric Microcirculatory Changes, Cytokine Levels and Gastric Ulcer Healing in Rats. *World J. Gastroenterol.* **2006,** *12*(13), 2034–2039.

Fu, J. H.; Zheng, Y. Q.; Li, P.; Li, X. Z.; Shang, X. H.; Liu, J. X. Hawthorn Leaves Flavonoids Decreases Inflammation Related to Acute Myocardial Ischemia/Reperfusion in Anesthetized Dogs. *Chin. J. Integr. Med.* **2013,** *19*(8), 582–588.

Gangar, S. C.; Sandhir, R.; Koul, A. Anti-clastogenic Activity of *Azadirachta indica* Against Benzo(*a*)pyrene in Murine Forestomach Tumorigenesis Bioassay. *Acta Polym. Pharm.* **2010,** *67*(4), 381–390.

Goel, H. C.; Prasad, J.; Sharma, A.; Singh, B. Antitumour and Radioprotective Action of *Podophyllum hexandrum*. *Indian J. Exp. Biol.* **1998,** *36*(6), 583–587.

González-Vallinas, M.; Molina, S.; Vicente, G.; Zarza, V.; Martín-Hernández, R.; García-Risco, M. R.; Fornari, T.; Reglero, G.; Ramírez de Molina, A. Expression of MicroRNA-15b and the Glycosyltransferase GCNT3 Correlates with Antitumor Efficacy of Rosemary Diterpenes in Colon and Pancreatic Cancer. *PLoS ONE* **2014,** *9*(6), e98556.

Gopal, V. B., Subhash L. B.; Parag P. K.; Girish N. Z. Anti-Nociceptive and Anti-Inflammatory Activity of Hydroalcoholic Extract of Leaves of *Amaranthus tricolor* L. *Pharm. Lett.* **2013,** *5*(3), 48–55.

Gray, A. M.; Flatt, P. R. Antihyperglycemic Actions of *Eucalyptus globulus* (Eucalyptus) Are Associated with Pancreatic and Extra-Pancreatic Effects in Mice. *J. Nutr.* **1998,** *128*(12), 2319–2323.

Gull, A.; Prasad, K.; Kumar, P. Effect of Millet Flours and Carrot Pomace on Cooking Qualities, Color and Texture of Developed Pasta, *LWT—Food Sci. Technol.* **2015.** DOI:10.1016/j.lwt.2015.03.008.

Guo, J. S.; Cheng, C. L.; Koo, M. W. Inhibitory Effects of *Centella asiatica* Water Extract and Asiaticoside on Inducible Nitric Oxide Synthase During Gastric Ulcer Healing in Rats. *Planta Med.* **2004,** *70*(12), 1150–1154.

Haq, R.; Prasad, K. Nutritional and Processing Aspects of Carrot (*Daucus carota*)—A Review. *South Asian J. Food Technol. Environ.* **2015,** *1*(1), 1–14.

Harikumar, K. B.; Kuttan, G.; Kuttan, R. Inhibition of Viral Carcinogenesis by *Phyllanthus amarus*. *Integr. Cancer Ther.* **2009,** *8*(3), 254–260.

Harikumar, K. B.; Kuttan, R. An Extract of *Phyllanthus amarus* Protects Mouse Chromosomes and Intestine from Radiation Induced Damages. *J. Radiat. Res.* **2007,** *48*(6), 469–476.

Hartojo, W.; Silvers, A. L.; Thomas, D. G.; Seder, C. W.; Lin, L.; Rao, H.; Wang, Z.; Greenson, J. K.; Giordano, T. J.; Orringer, M. B.; Rehemtulla, A.; Bhojani, M. S.; Beer, D. G.; Chang, A. C. Curcumin Promotes Apoptosis, Increases Chemosensitivity, and Inhibits Nuclear Factor kappaB in Esophageal Adenocarcinoma. *Transl. Oncol.* **2010,** *3*(2), 99–108.

Hazra, R.; Ray, K.; Guha, D. Inhibitory Role of *Acorus calamus* in Ferric Chloride-Induced Epileptogenesis in Rat. *Hum. Exp. Toxicol.* **2007,** *26*(12), 947–953.

Hill, T. D.; Cascio, M. G.; Romano, B.; Duncan, M.; Pertwee, R. G.; Williams, C. M.; Whalley, B. J.; Hill, A. J. Cannabidivarin-Rich Cannabis Extracts are Anticonvulsant in Mouse and Rat via a CB1 Receptor-Independent Mechanism. *Br. J. Pharmacol.* **2013,** *170*(3), 679–692.

Ho, J. H.; Hong, C. Y. Cardiovascular Protection of Magnolol: Cell-Type Specificity and Dose-Related Effects. *J. Biomed. Sci.* **2012,** Jul 31;19:70.

Huang, C. C.; Chen, Y. M.; Wang, D. C.; Chiu, C. C.; Lin, W. T.; Huang, C. Y.; Hsu, M. C. Cytoprotective Effect of American Ginseng in a Rat Ethanol Gastric Ulcer Model. *Molecules* **2013,** *19*(1), 316–326.

Husain, I.; Chander, R.; Saxena, J. K.; Mahdi, A. A.; Mahdi, F. Antidyslipidemic Effect of *Ocimum sanctum* Leaf Extract in Streptozotocin Induced Diabetic Rats. *Indian J. Clin. Biochem.* **2015,** *30*(1), 72–77.

Ilavarasan, J. R.; Monideen, S.; Vijayalakshmi, M. Antiulcer Activity of *Aegle marmelos* Linn. *Anc. Sci. Life* **2002,** *21*(4), 256–259.

Ishola, I. O.; Olayemi, S. O.; Idowu, A. R. Anticonvulsant, Anxiolytic and Hypnotic Effects of Aqueous Bulb Extract of *Crinum glaucum* A. chev (Amaryllidaceae): Role of GABAergic and Nitrergic Systems. *Pak. J. Biol. Sci.* **2013,** *16*(15), 701–710.

Jagetia, G. C.; Rao, S. K. Evaluation of the Antineoplastic Activity of Guduchi (*Tinospora cordifolia*) in Ehrlich Ascites Carcinoma Bearing Mice. *Biol. Pharm. Bull.* **2006,** *29*(3), 460–466.

Jamal, A.; Javed, K.; Aslam, M.; Jafri, M. A. Gastroprotective Effect of Cardamom, *Elettaria cardamomum* Maton. Fruits in Rats. *J. Ethnopharmacol.* **2006,** *103*(2), 149–153. Epub 2005 Nov 17.

Jang, M. H.; Shin, M. C.; Kim, Y. J.; Kim, C. J.; Chung, J. H.; Seo, J. C.; Kim, E. H.; Kim, K. Y.; Lee, C. Y.; Kim, K. M. Protective Effect of *Acanthopanax senticosus* Against Ethanol-Induced Apoptosis of Human Neuroblastoma Cell Line SK-N-MC. *Am. J. Chin. Med.* **2003,** *31*(3), 379–388.

Kalthur, G.; Mutalik, S.; Pathirissery, U. D. Effect of Withaferin A on the Development and Decay of Thermotolerance in B16F1 Melanoma: A Preliminary Study. *Integr. Cancer Ther.* **2009,** *8*(1), 93–97.

Katyal, J.; Sarangal, V.; Gupta, Y. K. Interaction of Hydroalcoholic Extract of *Acorus calamus* Linn. with Sodium Valproate and Carbamazepine. *Indian J. Exp. Biol.* **2012,** *50*(1), 51–55.

Khaliq, F.; Parveen, A.; Singh, S.; Hussain, M. E.; Fahim, M. *Terminalia arjuna* Improves Cardiovascular Autonomic Neuropathy in Streptozotocin-Induced Diabetic Rats. *Cardiovasc. Toxicol.* **2013,** *13*(1), 68–76.

Khatua, T. N.; Adela, R.; Banerjee, S. K. Garlic and Cardioprotection: Insights into the Molecular Mechanisms. *Can. J. Physiol. Pharmacol.* **2013,** *91*(6), 448–458.

Kim, D. S.; Kim, S. J.; Kim, M. C.; Jeon, Y. D.; Um, J. Y.; Hong, S. H. The Therapeutic Effect of Chelidonic Acid on Ulcerative Colitis. *Biol. Pharm. Bull.* **2012,** *35*(5), 666–671.

Kim, H. S.; Lee, E. H.; Ko, S. R.; Choi, K. J.; Park, J. H.; Im, D. S. Effects of Ginsenosides Rg3 and Rh2 on the Proliferation of Prostate Cancer Cells. *Arch. Pharm. Res.* **2004,** *27*(4), 429–435. Ko, B. S.; Choi, S. B.; Park, S. K.; Jang, J. S.; Kim, Y. E.; Park, S. Insulin Sensitizing and Insulinotropic Action of Berberine from *Cortidis rhizoma*. *Biol. Pharm. Bull.* **2005,** *28*(8), 1431–1437.

Kshirsagar, R. P.; Darade, S. S.; Takale, V. Effect of *Alangium salvifolium* (Alangiaceae) on Dexamethasone Induced Insulin Resistance in Rats. *J. Pharm. Res.* **2010,** *3*(11), 2714–2716.

Kudoh, K.; Matsumoto, M.; Onodera, S.; Takeda, Y.; Ando, K.; Shiomi, N. Antioxidative Activity and Protective Effect Against Ethanol-Induced Gastric Mucosal Damage of a Potato Protein Hydrolysate. *J. Nutr. Sci. Vitaminol. (Tokyo)* **2003,** *49*(6), 451–455.

Kumar, G. S.; Shetty, A. K.; Salimath, P. V. Modulatory Effect of Fenugreek Seed Mucilage and Spent Turmeric on Intestinal and Renal Disaccharidases in Streptozotocin Induced Diabetic Rats. *Plant Foods Hum. Nutr.* **2005,** *60*(2), 87–91.

Kumar, S.; Ahuja, V.; Sankar, M. J.; Kumar, A.; Moss, A. C. Curcumin for Maintenance of Remission in Ulcerative Colitis. *Cochrane Database Syst. Rev.* **2012,** *10*, CD008424.

Kumar, V.; Mahdi, F.; Khanna, A. K.; Singh, R.; Chander, R.; Saxena, J. K.; Mahdi, A. A.; Singh, R. K. Antidyslipidemic and Antioxidant Activities of *Hibiscus rosa sinensis* Root Extract in Alloxan Induced Diabetic Rats. *Indian J. Clin. Biochem.* **2013,** *28*(1), 46–50.

Kunnumakkara, A. B.; Guha, S.; Krishnan, S.; Diagaradjane, P.; Gelovani, J.; Aggarwal, B. B. Curcumin Potentiates Antitumor Activity of Gemcitabine in an Orthotopic Model of Pancreatic Cancer Through Suppression of Proliferation, Angiogenesis, and Inhibition of Nuclear Factor-kappaB-regulated Gene Products. *Cancer Res.* **2007,** *67*(8), 3853–3861.

Lai, Y. H.; Yu, S. L.; Chen, H. Y, Wang, C. C.; Chen, H. W.; Chen, J. J. The HLJ1-Targeting Drug Screening Identified Chinese Herb Andrographolide that can Suppress Tumour Growth and Invasion in Non-small-cell Lung Cancer. *Carcinogenesis* **2013,** *34*(5), 1069–1080.

Lee, S. J.; Shin, J. S.; Choi, H. E.; Lee, K. G.; Cho, Y. W.; An, H. J.; Jang, D. S.; Jeong, J. C.; Kwon, O. K.; Nam, J. H.; Lee, K. T. Chloroform Fraction of *Solanum tuberosum* L. cv Jayoung Epidermis Suppresses LPS-induced Inflammatory Responses in Macrophages and DSS-Induced Colitis in Mice. *Food Chem. Toxicol.* **2014,** *63*, 53–61.

Leyon, P. V.; Kuttan, G. Effect of *Withania somnifera* on B16F-10 Melanoma Induced Metastasis in Mice. *Phytother. Res.* **2004,** *18*(2), 118–122.

Li, X. Z.; Zhang, S. N.; Wang, K. X.; Liu, H. Y.; Yang, Z. M.; Liu, S. M.; Lu, F. Neuroprotective Effects of Extract of *Acanthopanax senticosus* Harms on SH-SY5Y Cells Overexpressing Wild-Type or A53T Mutant α-Synuclein. *Phytomedicine* **2014,** *21*(5), 704–711.

Liang, C. J.; Lee, C. W.; Sung, H. C.; Chen, Y. H.; Wang, S. H.; Wu, P. J.; Chiang, Y. C.; Tsai, J. S.; Wu, C. C.; Li, C. Y.; Chen, Y. L. Magnolol Reduced TNF-α-Induced Vascular Cell Adhesion Molecule-1 Expression in Endothelial Cells via JNK/p38 and NF-κB Signalling Pathways. *Am. J. Chin. Med.* **2014,** *42*(3), 619–637.

Liu, B.; Asare-Anane, H.; Al-Romaiyan, A.; Huang, G.; Amiel, S. A.; Jones, P. M.; Persaud, S. J. Characterisation of the Insulinotropic Activity of an Aqueous Extract of *Gymnema sylvestre* in Mouse Beta-Cells and Human Islets of Langerhans. *Cell Physiol. Biochem.* **2009,** *23*(1–3), 125–132.

Liu, S. M.; Li, X. Z.; Huo, Y.; Lu, F. Protective Effect of Extract of *Acanthopanax senticosus* Harms on Dopaminergic Neurons in Parkinson's Disease Mice. *Phytomedicine* **2012,** *19*(7), 631–638.

Liu, Y. L.; Zhou, Y.; Sun, L.; Wen, J. T.; Teng, S. J.; Yang, L.; Du, D. S. Protective Effects of *Gingko biloba* Extract 761 on Myocardial Infarction via Improving the Viability of Implanted Mesenchymal Stem Cells in the Rat Heart. *Mol. Med. Rep.* **2014,** *9*(4), 1112–1120.

Lotan, I.; Treves, T. A.; Roditi, Y.; Djaldetti, R. Cannabis (medical marijuana) Treatment for Motor and Non-Motor Symptoms of Parkinson Disease: An Open-Label Observational Study. *Clin. Neuropharmacol.* **2014,** *37*(2), 41–44.

Lu, J. S.; Liu, Y. Q.; Li, M.; Li, B. S.; Xu, Y. Protective Effects and Its Mechanisms of Total Alkaloids from *Rhizoma coptis chinensis* on *Helicobacter pylori* LPS Induced Gastric Lesion in Rats. *Zhongguo Zhong Yao Za Zhi* **2007,** *32*(13), 1333–1336.

Lu, M.; Sun, L.; Zhou, J.; Zhao, Y.; Deng, X. Dihydroartemisinin-Induced Apoptosis is Associated with Inhibition of Sarco/Endoplasmic Reticulum Calcium ATPase Activity in Colorectal Cancer. *Cell Biochem. Biophys.* **2015,** *73*(1), 137-145.

Ma, S.; Liu, H.; Jiao, H.; Wang, L.; Chen, L.; Liang, J.; Zhao, M.; Zhang, X. Neuroprotective Effect of Ginkgolide K on Glutamate-Induced Cytotoxicity in PC 12 Cells via Inhibition of ROS Generation and Ca(2+) Influx. *Neurotoxicology* **2012,** *33*(1), 59–69.

Mahmoudzadeh-Sagheb, H.; Heidari, Z.; Bokaeian, M.; Moudi, B. Antidiabetic Effects of *Eucalyptus globulus* on Pancreatic Islets: A Stereological Study. *Folia Morphol. (Warsz).* **2010,** *69*(2), 112–118.

Majdalawieh, A. F.; Carr, R. I. In Vitro Investigation of the Potential Immunomodulatory and Anti-cancer Activities of Black Pepper (*Piper nigrum*) and Cardamom (*Elettaria cardamomum*). *J. Med. Food* **2010,** *13*(2), 371–381.

Majlessi, N.; Choopani, S.; Kamalinejad, M.; Azizi, Z. Amelioration of Amyloid β-Induced Cognitive Deficits by *Zataria multiflora* Boiss. Essential Oil in a Rat Model of Alzheimer's Disease. *CNS Neurosci. Ther.* **2012,** *18*(4), 295–301.

Maliyakkal, N.; Appadath Beeran, A.; Balaji, S. A.; Udupa, N.; Ranganath Pai, S.; Rangarajan, A. Effects of *Withania somnifera* and *Tinospora cordifolia* Extracts on the Side Population Phenotype of Human Epithelial Cancer Cells: Toward Targeting Multidrug Resistance in Cancer. *Integr. Cancer Ther.* **2015,** *14*(2), 156–171.

Mall, G. K.; Mishra, P. K.; Prakash, V. Antidiabetic and Hypolipidemic Activity of *Gymnema sylvestre* in Alloxan Induced Diabetic Rats. *Glob. J. Biotech. Biochem.* **2009,** *4*(1), 37–42.

Mandegary, A.; Sharififar, F.; Abdar, M. Anticonvulsant Effect of the Essential Oil and Methanolic Extracts of *Zataria multiflora* Boiss. *Cent. Nerv. Syst. Agents Med. Chem.* **2013,** *13*(2), 93–97.

Mangathayaru, K.; Kuruvilla, S.; Balakrishna, K.; Venkhatesh, J. Modulatory Effect of *Inula racemosa* Hook. F. (Asteraceae) on Experimental Atherosclerosis in Guinea-Pigs. *J. Pharm. Pharmacol.* **2009,** *61*(8), 1111–1118.

Mikaili, P.; Maadirad, S.; Moloudizargari, M.; Aghajanshakeri, S.; Sarahroodi, S. Therapeutic uses and Pharmacological Properties of Garlic, Shallot, and their Biologically Active Compounds. *Iran. J. Basic Med. Sci.* **2013,** *16*(10), 1031–1048.

Modak, M.; Dixit, P.; Londhe, J.; Ghaskadbi, S.; Paul, A.; Devasagayam, T. Indian Herbs and Herbal Drugs Used for the Treatment of Diabetes. *J. Clin. Biochem. Nutr.* **2007,** *40*(3), 163–173.

Monteiro, M. V.; de Melo Leite, A. K.; Bertini, L. M.; de Morais, S. M.; Nunes-Pinheiro, D. C. Topical Anti-Inflammatory, Gastroprotective and Antioxidant Effects of the Essential Oil of *Lippia sidoides* Cham. Leaves. *J. Ethnopharmacol.* **2007,** *111*(2), 378–382.

Montopoli, M.; Bertin, R.; Chen, Z.; Bolcato, J.; Caparrotta, L.; Froldi, G. *Croton lechleri* Sap and Isolated Alkaloid Taspine Exhibit Inhibition against Human Melanoma SK23 and Colon Cancer HT29 Cell Lines. *J. Ethnopharmacol.* **2012,** *144*(3), 747–753.

Muralikrishnan, G.; Dinda, A. K.; Shakeel, F. Immunomodulatory Effects of *Withania somnifera* on Azoxymethane Induced Experimental Colon Cancer in Mice. *Immunol. Invest.* **2010,** *39*(7), 688–698.

Nadova, S.; Miadokova, E.; Alfoldiova, L.; Kopaskova, M.; Hasplova, K.; Hudecova, A.; Vaculcikova, D.; Gregan, F.; Cipak, L. Potential Antioxidant Activity, Cytotoxic and Apoptosis-Inducing Effects of *Chelidonium majus* L. Extract on Leukemia Cells. *Neurol. Endocrinol. Lett.* **2008,** *29*(5), 649–652.

Ojha, S.; Bharti, S.; Sharma, A. K.; Rani, N.; Bhatia, J.; Kumari, S.; Arya, D. S. Effect of *Inula racemosa* Root Extract on Cardiac Function and Oxidative Stress against Isoproterenol-Induced Myocardial Infarction. *Indian J. Biochem. Biophys.* **2011,** *48*(1), 22–28.

Park, M. Y.; Kwon, H. J.; Sung, M. K. Dietary Aloin, Aloesin, or Aloe-gel Exerts Anti-inflammatory Activity in a Rat Colitis Model. *Life Sci.* **2011,** *88*(11–12), 486–492.

Patel, D. K.; Prasad, S. K.; Kumar, R.; Hemalatha, S. An Overview on Antidiabetic Medicinal Plants having Insulin Mimetic Property. *Asian Pac. J. Trop. Biomed.* **2012,** *2*(4), 320–330.

Pawar, P.; Gilda, S.; Sharma, S.; Jagtap, S.; Paradkar, A.; Mahadik, K.; Ranjekar, P.; Harsulkar, A. Rectal Gel Application of *Withania somnifera* Root Extract Expounds Anti-Inflammatory and Muco-Restorative Activity in TNBS-Induced Inflammatory Bowel Disease. *BMC Complement. Altern. Med.* **2011,** *11*, 34.

Peer, P. A.; Trivedi, P. C.; Nigade, P. B.; Ghaisas, M. M.; Deshpande, A. D. Cardioprotective Effect of *Azadirachta indica* A. Juss. on isoprenaline induced Myocardial Infarction in Rats. *Int. J. Cardiol.* **2008,** *126*(1), 123–126.

Pichu, S.; Krishnamoorthy, S.; Shishkov, A.; Zhang, B.; McCue, P.; Ponnappa, B. C. Knockdown of Ki-67 by Dicer-Substrate Small Interfering RNA Sensitizes Bladder Cancer Cells to Curcumin-Induced Tumor Inhibition. *PLoS ONE.* **2012,** *7*(11), e48567.

Prasad, K. K.; Debi, U.; Prasad, K.; Sinha, S. K.; Nain, C. K.; Singh, K. Obesity—Relation to Gut Microecology: An Update for Gastroenterologists. *Med. Progr.* **2009,** 345–353.

Prasad, K.; Janve, B.; Sharma, R. K.; Prasad, K. K. Compositional Characterization of Traditional Medicinal Plants: Chemometric Approach. *Arch. Appl. Sci. Res.* **2010,** *2*(5), 1–10.

Puri, A.; Saxena, R.; Saxena, R. P.; Saxena, K. C.; Srivastava, V.; Tandon, J. S. Immunostimulant Agents from *Andrographis paniculata. J. Nat. Prod.* **1993,** *56*(7), 995–999.

Puri, D. The Insulinotropic Activity of a Nepalese Medicinal Plant *Biophytum sensitivum*: Preliminary Experimental Study. *J. Ethnopharmacol.* **2001,** *78*(1), 89–93.

Rajeshkumar, N. V.; Joy, K. L.; Kuttan, G.; Ramsewak, R. S.; Nair, M. G.; Kuttan, R. Antitumour and Anticarcinogenic Activity of *Phyllanthus amarus* Extract. *J. Ethnopharmacol.* **2002,** *81*(1), 17–22.

Raphael, K. R.; Sabu, M.; Kumar, K. H.; Kuttan, R. Inhibition of *N*-Methyl *N'*-nitro-*N*-Nitrosoguanidine (MNNG) Induced Gastric Carcinogenesis by *Phyllanthus amarus* Extract. *Asian Pac. J. Cancer Prev.* **2006,** *7*(2), 299–302.

Ravi, K.; Ramachandran, B.; Subramanian, S. Effect of *Eugenia jambolana* Seed Kernel on Antioxidant Defence System in Streptozotocin-Induced Diabetes in Rats. *Life Sci.* **2004,** *75*(22), 2717–2731.

Ribes, G.; Sauvaire, Y.; Da Costa, C.; Baccou. J. C.; Loubatieres-Mariani, M. M. Antidiabetic Effects of Subfractions from Fenugreek Seeds in Diabetic Dogs. *Proc. Soc. Exp. Biol. Med.* **1986,** *182*(2), 159–166.

Rozza, A. L.; de Mello Moraes, T.; Kushima, H.; Nunes, D. S.; Hiruma-Lima, C. A.; Pellizzon, C. H. Involvement of Glutathione, Sulfhydryl Compounds, Nitric Oxide, Vasoactive Intestinal Peptide, and Heat-Shock Protein-70 in the Gastroprotective Mechanism of *Croton cajucara* Benth. (Euphorbiaceae) Essential Oil. *J. Med. Food.* **2011,** *14*(9), 1011–1017.

Sandborn, W. J.; Targan, S. R.; Byers, V. S.; Rutty, D. A.; Mu, H.; Zhang, X.; Tang, T. *Andrographis paniculata* Extract (HMPL-004) for Active Ulcerative Colitis. *Am. J. Gastroenterol.* **2013,** *108*(1), 90–98.

Sengupta, A.; Ghosh, S.; Das, S. Modulatory Influence of Garlic and Tomato on Cyclooxygenase-2 Activity, Cell Proliferation and Apoptosis during Azoxymethane Induced Colon Carcinogenesis in Rat. *Cancer Lett.* **2004,** *208*(2), 127–136.

Sengupta, A.; Ghosh, S.; Das, S. Tomato and Garlic Can Modulate Azoxymethane-Induced Colon Carcinogenesis in Rats. *Eur. J. Cancer Prev.* **2003,** *12*(3), 195–200.

Shakibaei, M.; Mobasheri, A.; Lueders, C.; Busch, F.; Shayan, P.; Goel, A. Curcumin Enhances the Effect of Chemotherapy against Colorectal Cancer Cells by Inhibition of NF-κB and Src Protein Kinase Signaling Pathways. *PLoS ONE* **2013,** *8*(2), e57218.

Shao, Z. M.; Shen, Z. Z.; Liu, C. H.; Sartippour, M. R.; Go, V. L.; Heber, D.; Nguyen, M. Curcumin Exerts Multiple Suppressive Effects on Human Breast Carcinoma Cells. *Int. J. Cancer*. **2002,** *98*(2), 234–240.

Sharma, V.; Singh, I.; Chaudhary, P. *Acorus calamus* (The Healing Plant): A Review on its Medicinal Potential, Micropropagation and Conservation. *Nat. Prod. Res.* **2014,** *28*(18), 1454–1466.

Singh, L. W. Traditional Medicinal Plants of Manipur as Anti-diabetics. *J. Med. Plant Res.* **2011,** *5*(5), 677–687.

Sivam, G. P.; Lampe, J. W.; Ulness, B.; Swanzy, S. R.; Potter, J. D. *Helicobacter pylori*—In Vitro Susceptibility to Garlic (*Allium sativum*) Extract. *Nutr. Cancer* **1997,** *27*(2), 118–121.

Ströfer, M.; Jelkmann, W.; Depping, R. Curcumin Decreases Survival of Hep3B Liver and MCF-7 Breast Cancer Cells: The Role of HIF. *Strahlenther Onkol.* **2011,** *187*(7), 393–400.

Sun, Y. F.; Song, C. K.; Viernstein, H.; Unger, F.; Liang, Z. S. Apoptosis of Human Breast Cancer Cells Induced by Microencapsulated Betulinic Acid from Sour Jujube Fruits through the Mitochondria Transduction Pathway. *Food Chem.* **2013,** *138*(2–3), 1998–2007.

Surh, Y. J.; Chun, K. S.; Cha, H. H.; Han, S. S.; Keum, Y. S.; Park, K. K.; Lee, S. S. Molecular Mechanisms Underlying Chemopreventive Activities of Anti-inflammatory Phytochemicals: Down-regulation of COX-2 and iNOS Through Suppression of NF-kappa B Activation. *Mutat. Res.* **2001,** 480–481, 243–268.

Szaflarski, J. P.; Bebin, E. M. Cannabis, Cannabidiol, and Epilepsy—From Receptors to Clinical Response. *Epilepsy Behav.* **2014,** *41*, 277–282.

Thirumalai, T.; Therasa, S. V.; Elumalai, E. K.; David, E. Hypoglycemic Effect of *Brassica juncea* (seeds) on Streptozotocin Induced Diabetic Male Albino Rat. *Asian Pac. J. Trop. Biomed.* **2011,** *1*(4), 323–325.

Tian, F.; Song, M.; Xu, P. R.; Liu, H. T.; Xue, L. X. Curcumin Promotes Apoptosis of Esophageal Squamous Carcinoma Cell Lines Through Inhibition of NF-kappaB Signaling Pathway. *Ai Zheng* **2008,** *27*(6), 566–570.

Tjong, Y.; Ip, S.; Lao, L.; Fong, H. H.; Sung, J. J.; Berman, B.; Che, C. Analgesic Effect of *Coptis chinensis* Rhizomes (*Coptidis rhizoma*) Extract on Rat Model of Irritable Bowel Syndrome. *J. Ethnopharmacol.* **2011,** *135*(3), 754–761.

Tripathi, Y. B.; Tripathi, P.; Upadhyay, B. N. Assessment of the Adrenergic Beta-blocking Activity of Inula Racemosa. *J. Ethnopharmacol.* **1988,** *23*(1), 3–9.

Tsai, C. C.; Chan, P.; Chen, L. J.; Chang, C. K.; Liu, Z.; Lin, J. W. Merit of Ginseng in the Treatment of Heart Failure in Type 1-like Diabetic Rats. *Biomed. Res. Int.* **2014,** *2014*, 484161.

Venkatesh, S.; Thilagavathi, J.; Shyam Sundar, D. Anti-Diabetic Activity of Flowers of *Hibiscus rosasinensis*. *Fitoterapia*. **2008,** *79*(2), 79–81. Epub 2007 Aug 11.

Vijayan, N. A.; Thiruchenduran, M.; Devaraj, S. N. Anti-Inflammatory and Anti-Apoptotic Effects of *Crataegus oxyacantha* on Isoproterenol-Induced Myocardial Damage. *Mol. Cell Biochem.* **2012,** *367*(1–2), 1–8.

Wasman, S. Q.; Mahmood, A. A.; Chua, L. S.; Alshawsh, M. A.; Hamdan, S. Antioxidant and Gastroprotective Activities of *Andrographis paniculata* (*Hempedu bumi*) in Sprague–Dawley Rats. *Indian J. Exp. Biol.* **2011,** *49*(10), 767–772.

Yadav, V. R.; Prasad, S.; Sung, B.; Gelovani, J. G.; Guha, S.; Krishnan, S.; Aggarwal, B. B. Boswellic Acid Inhibits Growth and Metastasis of Human Colorectal Cancer in Orthotopic Mouse Model by Downregulating Inflammatory, Proliferative, Invasive and Angiogenic Biomarkers. *Int. J. Cancer*. **2012,** *130*(9), 2176–2184.

Yang, Z.; Grinchuk, V.; Ip, S. P.; Che, C. T.; Fong, H. H.; Lao, L.; Wu, J. C.; Sung, J. J.; Berman, B.; Shea-Donohue, T.; Zhao, A. Anti-inflammatory Activities of a Chinese Herbal Formula IBS-20 In Vitro and In Vivo. *Evid. Based Complement Alternat. Med.* **2012b,** 491496, 1–12.

Yang, Z.; Ding, J.; Yang, C.; Gao. Y.; Li, X.; Chen, X.; Peng, Y.; Fang, J.; Xiao, S. Immunomodulatory and Anti-Inflammatory Properties of Artesunate in Experimental Colitis. *Curr. Med. Chem.* **2012a,** *19*(26), 4541–4551.

Yunos, N. M.; Mutalip, S. S.; Jauri, M. H.; Yu, J. Q.; Huq, F. Anti-proliferative and Pro-Apoptotic Effects from Sequenced Combinations of Andrographolide and Cisplatin on Ovarian Cancer Cell Lines. *Anticancer Res.* **2013,** *33*(10), 4365–4371.

Zhou, J.; Ong, C. N.; Hur, G. M.; Shen, H. M. Inhibition of the JAK-STAT3 Pathway by Andrographolide Enhances Chemosensitivity of Cancer Cells to Doxorubicin. *Biochem. Pharmacol.* **2010,** *79*(9), 1242–1250.

CHAPTER 3

SECONDARY METABOLITES OF BASIL AND THEIR POTENTIAL ROLES

KAMLESH PRASAD[1*], NEEHA V. S.[1], VASUDHA BANSAL[2], MOHAMMED WASIM SIDDIQUI[3], and ISABELLA MONTENEGRO BRASIL[4]

[1]*Department of Food Engineering and Technology, SLIET, Longowal 148106, Punjab, India*

[2]*Departmentof Civil & Environmental Engineering, Hanyang University, 22 Wangsimni-Ro, Seoul 133791, South Korea*

[3]*Department of Food Science and Postharvest Technology, Bihar Agricultural University, Sabour, Bhagalpur 813210, Bihar, India*

[4]*Department of Food Technology, Federal University of Ceará, Av. Mister Hall 2297 Bl. 858, Campus do Pici, CEP 60455-760, Fortaleza, CE, Brazil*

**Corresponding author, E-mail: dr_k_prasad@rediffmail.com*

CONTENTS

ABSTRACT

Basil (*Ocimum sanctum* L.), considered as "Queen of herbs" in India and "King of herbs" in Greek, is considered as a miracle healing solace to the physical, mental, and spiritual state of human being. It is acknowledged as a principal adaptogen as it adapts and copes with varied range of stresses and restores disturbed physiological and psychological functions to a normal healthy state. Scientific research gave substantial evidence that basil is capable in reducing the incidence of non-communicable diseases (NCD). Cardiovascular disease (CVDs), cancer, chronic obstructive pulmonary diseases, and diabetes are the four leading NCDs devastating the population worldwide. Ayurveda termed this as "Raj rog" which is nothing but lifestyle disorders. These lifestyle disorders increased the prevalence of NCD leading to a drastic shift in the health burden from communicable to NCD. Basil assures to decelerate this health burden as its administration is going to improve the perseverance and tolerance to various internal (diseases, aging, stress, and free radicals accumulation) and external factors (allergies, inflammation, pollution, noise, and radiation) faced by the human race in their daily life. To the possible extent, this chapter briefly provides the overview of basil and highlighting some of its secondary metabolites with their role. Literature survey also revealed some exciting benefits about basil in other sectors. As this herb produces ozone, thus, it may help in preventing environment-related issues. In textile industry, basil extracts are used to make ecofriendly antibacterial cloths, and in corrosion sector, basil oil or leaf extract is coated on the tin and aluminum metals to reduce corrosion. Till date, this abundantly available and easily cultivated herb is enormously used in pharmaceutical and cosmetic sector but very limited work is reported on the aspect of food application with scientific approach. Food is an indispensible part of human life, and food industries are tremendously involved in making innovative food product, which can improve the quality of food and thereby the living standard. Basil acts as an excellent antioxidant, gelling agent, thickening agent, stabilizing agent, and increases the shelf life of food items. Therefore, lots of scope exists in the food sector toward exploring basil-incorporated food. This chapter, thus, opens basil's door to various other sectors like food, environment, corrosion, and others, and ultimately will encourage the mass cultivation of basil.

3.1 INTRODUCTION

The holy basil (*Ocimum sanctum* L.), known as "Tulsi" in India (Kadam et al., 2012), is derived from Sanskrit which means matchless one. In Indian literature, basil is praised as the "Queen of herbs" and within Ayurveda it is termed as "the incomparable one." Sumit and Geetika (2012) and Kadam et al. (2011) considered it as "elixir of life" as it promotes longevity. Ancient scriptures (Rig Veda, Padma Purana, and Tulsi Kavacham) written between 5000BC and 1200AD also praise the medicinal greatness of Tulsi (Singh & Gilca, 2008). Indians, therefore, acknowledge it as one of the most esteemed herbs for the spiritual seekers of the world (Joshi, 2007). The term basil is derived from Greek word "basilica" which means royal plant. This herb was believed to be spotted by St. Constantine and his mother St. Helen while discovering the Holy Cross. Therefore, the ancient Greeks called basil as the "king of herbs" (Mondal et al., 2009; Tilebeni, 2011; Verma & Kothiyal, 2012; Mathew & Sankar, 2013).

3.2 ORIGIN AND DISTRIBUTION

Basil is supposed to be originated in the Middle East. However, it is reported that basil is brought to Europe from India through the Middle East route during the 16th century and subsequently to America in the 17th century (Verma & Kothiyal, 2012). Also it is native to the areas in Asia and Africa (Miller & Miller, 2003) and was traded through the medieval spice routes in Asia, India, Africa, and Mediterranean regions. It is also profusely cultivated in Arabian countries, Australia, Malaysia, and West Africa (Mondal et al., 2009) and now can be seen in the entire world (Tilebeni, 2011). Table 3.1 represents worldwide production of genus *Ocimum* and Table 3.2 shows its name in different languages. Apart from these countries, universities like Saskatchewan, North Carolina, Hawaii, Purdue, and Minnesota extension are reported to be involved in the production of various basil varieties. Although in India, almost in every home basil is cultivated, the report of mass production on basil is scarcely reported.

TABLE 3.1 Worldwide Production Areas of Genus *Ocimum*.

Climatic condition	Country of cultivation
Warm climate	India, Pakistan, the Comores, Madagascar, Haiti, Uatemala, Reunion, Thailand, Indonesia, Russia (Georgia, East-Caucasus), and South Africa
Mediterranean climate	Egypt, Morocco, France, Israel, Bulgaria, USA (Arizona, California, New Mexico), Italy, Greece, and Turkey
Temperate climate	Hungary, Poland, Germany, Balkan countries, and Slovakia

Source: Joseph and Nair (2013b).

TABLE 3.2 Name of *Ocimum* in Different Languages.

Country	Name
Hindi	Tulsi
Spanish	Alba Laca
French	Basilic
German	Basilienkraut
Arabic	Raihan
Dutch	Basilicum
Italian	Basilico
Portuguese	Manjericao
Russian	Basilik
Japanese	Meboki
Chinese	Lo-Le
Thai	Horapa, Horapha
English	Sweet Basil

Source: Anon (2014a).

3.3 BOTANICAL DESCRIPTION AND MORPHOLOGY

Basil is an aromatic herb of genus *Ocimum* from Lamiaceae family (Table 3.3). It is an erect, soft, hairy, annual to perennial herb, having a height of 30–90 cm. The leaves of basil are transitional between elliptic and oblong shape, ovate, lanceolate, acuminate, and the leaf margin is either smooth or has sharp saw-like teeth (serrated); both sides of the leaves are covered with erect hairs (pubescent) and dotted minute glands. The flowers are crimson or purplish in racemes which are close whorled

(Vishwabhan et al., 2011; Mathew & Sankar, 2013). In India, green-colored basil leaves are known as Sri or Rama Tulsi and purple leaves are known as Krishna or Shyama Tulsi (*Ocimum tenuiflorum* or *O. sanctum*) (Prakash & Gupta, 2005; Ravi et al., 2012; Joseph & Nair, 2013b). Basil has more than 150 genotypes (Hussain et al., 2008). Some known important species of genus *Ocimum* are *O. sanctum* (Holy basil, Tulsi), *Ocimum basilicum* (Sweet basil, Ban tulsi) *Ocimum gratissimum* (Clove basil, Vana tulsi), *Ocimum kilimandscharicum* (Camphor basil, Kapur tulsi), *Ocimum americanum* (Hoary basil, Dulal tulsi), *Ocimum basilicum thyrsiflora* (Queen or shyam basil), and *Ocimum basilicum purpurascens* (Red rubin basil). All these varieties are cultivated in various parts of the world and have potent medicinal properties (Das & Vasudevan, 2006).

TABLE 3.3 Taxonomic Ranks.

Kingdom	Plantae
Division	Magnoliophyta
Class	Magnoliopsida
Order	Lamiales
Family	Lamiaceae
Genus	*Ocimum*
Species	*Sanctum*
Binomial name	*Ocimum sanctum* L.

Source: Singh et al. (2012), Joseph and Nair (2013a).

Sweet basil (*O. basilicum*) is an erect, almost glabrous herb, 30–90 cm high. Leaves are ovate, lanceolate, acuminate, toothed or entirely glabrous on both surfaces, and have white or pale purple flowers in simple or much branched racemes.

Vana tulsi (*O. gratissimum*) bears pale yellow flowers and is tall (1–2.5 m high) and branched. Leaves are ovate, coarsely crenate, gland dotted, and pubescent on both surfaces; flowers are same as sweet basil, however moderately close whorled, rugose, brown, with glandular depression but not mucilaginous.

Kapur tulsi (*O. kilimandscharicum*) leaves are ovate or oblong, acute narrow at base, deeply serrated, pubescent on either sides as *Ocimum sanctum*; flowers in 4–6 flowered whorls on long villose racemes; nutlets ovoid to ovoid oblong, black to brown.

Dulal tulsi (*O. americanum*) is an erect, sweet, pubescent herb, with a height of 30–60 cm. Leaves are ellipticlanceolate, entire or faintly toothed, almost glabrous, and dotted gland with small flowers (Prakash & Gupta, 2005; Singh et al., 2012; Verma & Kothiyal, 2012; Joshi, 2013). Table 3.4 shows basils' scientific name, common name, and Indian name. Figure 3.1 shows different types of basils as available on net.

TABLE 3.4 Basil Varieties.

Scientific name	Common name	Indian name
O. tenuiflorum or *O. sanctum*	Holy basil	Sri or Rama Tulsi
O. basilicum L.	Sweet basil	Ban Tulsi
O. gratissimum L.	Clove basil	Vana Tulsi
O. kilimandscharicum L.	Camphor Basil	Kapur Tulsi
O. americanum L.	Hoary Basil	Dulal Tulsi
O. basilicum thyrsiflora	Queen of Siam basil	–
O. basilicum purpurascens	Red rubin basil	Krishna or Shyama Tulsi

Source: Anon. (2014b,d).

FIGURE 3.1 Pictorial views of basil.

3.4 CULTIVATION

Cultivation and production of basil depends upon the climate. In tropical climate, it is a perennial herb, while in temperate regions, an annual herb, and in these areas, basil can be sown directly from seed or can be transplanted. Only a few varieties of basil grow well under competitive situations, most are highly frost sensitive. Therefore, in colder climates and during winter, it can be productively cultivated as indoor or pot plant (Tilebeni, 2011). Regarding storage of basil, Hamasaki et al. (1994) reported not to cultivate basil below 5°C for long duration and avoid overwatering to prevent the incidence of "damping off" disease in basil. Plucking flowers during the growth phase will increase the yield of basil leaves. Four to 5 days are required for seed germination and if it is stored in dry conditions then it remains viable for years. Efficient production of basil can be achieved by cultivating with 8 in. difference between plants and between rows and in partial or full sun.

3.5 PRIMARY METABOLITES AND CHEMICAL COMPOSITION

Plant-made nutritive metabolites that are directly involved in growth and development, respiration and photosynthesis, and hormone and protein synthesis are known as primary metabolites. Hounsome et al. (2008) recognized carbohydrate, amino acids, vitamins, organic acids, and fatty acids as some of the primary metabolites present in basil. Table 3.5 shows chemical composition of basil leaves. Prasad et al. (2010) compared and classified some important medicinal plants using principal component analysis and hierarchical cluster analysis. On comparison, they found that Jamun and Basil get clustered as having similar high energy and protein levels. Their study also revealed that tulsi is loaded with ashes and fat mainly. Nair et al. (2012) also found that the leaves of basil (*O. sanctum*) showed about 15.21% (w/w) of total ash as compared to the percentage ash value of leaves of Neem (*Azadirachta indica*) and Karanj (*Millettia pinnata*), which lied between 6% and 13% w/w. This shows that basil is a good source of mineral.

TABLE 3.5 Proximate Analysis of *Ocimum sanctum*.

Crude protein (%)	1.28 ± 0.16
Crude fat (%)	9.03 ± 0.60
Total ash (%)	7.97 ± 0.65
Crude fiber (%)	10.48 ± 1.20
Acid insoluble ash (%)	0.70 ± 0.13
Moisture (%)	6.72 ± 0.62

Source: Prasad et al. (2010).

3.6 SECONDARY METABOLITES

The non-nutritive plant chemicals that are produced by the plant for defense and self-protection from adverse conditions are termed as secondary metabolites. Researchers have proved that these secondary metabolites are also beneficial to human and are indirectly useful in the growth, protection, and development (Jain & Srivastava, 2013). According to British Nutrition Foundation, phenolic and polyphenolic compounds (~8000 compounds), terpenoids (~25,000 compounds), alkaloids (~12,000 compounds), and sulfur-containing compounds are the four major classifications of secondary metabolites (Hounsome et al., 2008). Basil is piled with major secondary metabolites such as phenols, flavonoids, alkaloids, terpenoids, aldehydes, steroids, glycosides, saponins, tannins, and essential oils (Mathew & Sankar, 2013). Eugenol is the major constituent of essential oil found in the extract of different parts of basil (Klimankova et al., 2008; Veillet et al., 2010; Vishwabhan et al., 2011; Singh et al., 2012). The metabolites of basil's varieties of tulsi (*O. sanctum*) are fatty acids (palmitic, stearic, oleic, linoleic, and linolenic acid) obtained from seed oil (also termed as fixed oil). The oil is of greenish-yellow color with good drying properties (Mondal et al., 2009). The seeds contain mucilage (hexouronic acid, pentose, and ash) which on hydrolysis yields xylose and an acid polysaccharide. The leaves of *O. sanctum* contain 0.7% volatile oil comprising about 71% eugenol and 20% methyl eugenol. The oil also contains carvacrol and sesquiterpene hydrocarbon caryophyllene. Fresh leaves and stem of *O. sanctum* extract yielded some phenolic compounds (antioxidants) such as cirsilineol, circimaritin, isothymusin, apigenin, and rosameric acid with appreciable quantities of eugenol. Two flavonoids, orientin, and vicenin from aqueous extract of leaf have been reported.

Ursolic acid apigenin, luteolin, apigenin-7-*O*-glucuronide, luteolin-7-*O*-glucuronide, orientin, and molludistin have also been isolated from the leaf extract. It contains a number of sesquiterpenes and monoterpenes bornyl acetate, β-elemene, neral, α- and β-pinenes, camphene, campesterol, cholesterol, stigmasterol, and β-sitosterol.

Sweet Tulsi (Ocimum basilicum): The essential oils of the aerial parts of *Ocimum basilicum* have 74.19% of the total oil. Major constituents of oil are linalool (64.35%), 1,8-cineole (12.28%), eugenol (3.21%), germacrene D (2.07%), alpha-terpineol (1.64%), and rho-cymene (1.03%). The components are linalool, (*Z*)-cinnamic acid methyl ester, cyclohexene, α-cadinol, 2,4-diisopropenyl-1-methyl-1-vinylcyclohexane, 3,5-pyridine-dicarboxylic acid, 2,6-dimethyl diethyl ester, β-cubebene, guaia-1,11-diene, cadinene (E)-cinnamic acid methyl ester, and β-guaiene. Gas chromatography and mass spectroscopy (GC–MS) were used to identify the essential oils (linalool (54.95%), methylchavikol (11.98%), methylcinnamat (7.24%), and linolen (0.14%).

Vana Tulsi (Ocimum gratissimum): The characteristic and composition of the volatile oil of plants growing in different areas vary. Two types of oil may be distinguished, one contained thymol as chief constituent and the other eugenol. Acidic xylan composed of D-xylose (48%), L-arabinose (16%), D-galactose (16%), and D-galactouronic acid (20%) has been reported to be isolated from the capsular, mucilaginous polysaccharide complex of the seeds.

Kapur Tulsi (Ocimum kilimandscharicum): The yields of oil and camphor vary according to the locality, season of harvest, and the plant material distilled. Leaves contain the maximum proportion of camphor and oil, followed by flowers and stem, which contain only minute quantities. The distilled oil is light yellow in color, with a strong odor of camphor. The oil constituted D-camphor, D-limonene, terpineol, and unidentified sesquiterpenes and sesquiterpene alcohol eugenol is also reported. The essential oil contained 62% 1.8-cineole, indicating the occurrence of a new chemo for this species. In addition to 1.8-cineole, 16 oxygen-containing compounds and 14 monoterpene hydrocarbons were identified, of which limonene and betapinene were the major ones.

Dulal Tulsi (Ocimum americanum): Plant material after hydrodistillation gave oil with 0.7% yield. Nine were monoterpene hydrocarbons (3, 1.6%), 4 were oxygenated monoterpenes (35.4%), 12 sesquiterpene hydrocarbons (23, 44%), 7 sesquiterpene alcohols (7.7%), and a few unidentified compounds (1.8%). Major constituents of oil included α-pinene (8.3%), sabinene (8.0%), limonene (7.8%), camphor (26.7%), and selinene (10.9%) (Verma & Kothiyal, 2012).

3.7 PHYTOCHEMICALS

The phytochemical compounds present in the methanolic extracts of basil leaf powder were identified using GC–MS. The active compounds with their peak area (%), retention time (RT), and corresponding health benefits are presented in Table 3.6. Major compounds present in the basil leaf powder with structure are presented in Figure 3.2 and they are eugenol (11.66%), stigmast-5-En-3-ol-(3-Beta)-(β-Sitosterol) (11.16%),

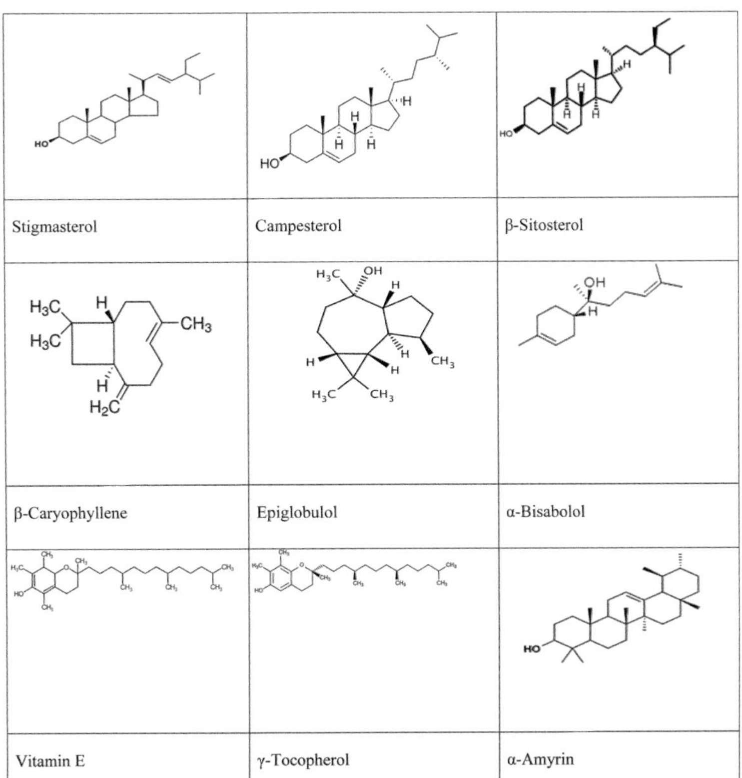

FIGURE 3.2 Phytochemicals with structures.

TABLE 3.6 Phytochemical Compounds of Basil Leaf Powder and Health Benefits.

S. No.	Compound name	RT (min)	Peak area (%)	Health benefits	References
1	Eugenol	15.133	11.66	Anti-inflammatory, antioxidant, antibacterial, cardio-protection, vasodilating effect, antibiotic activity, anticancer, antimycotic properties	Damiani et al. (2003), Ahmad et al. (2010), Afzali et al. (2014), Kong et al. (2014)
2	Methyleugenol	16.268	0.93	Antibacterial properties, relaxant, and antispasmodic actions on the ileum, antimycotic properties	Ahmad et al. (2010), Cerkezkayabekir et al. (2010)
3	β-Sitosterol	51.855	11.16	Cardio-protection, hypocholesterolemic activity, anticancer, modulating the immune system, prevents rheumatoid arthritis, antimicrobial activity	Tasan et al. (2006), Awad et al. (2007), Marangoni and Poli (2010), Sen et al. (2012), Saeidnia et al. (2014)
4	Stigmasterol	50.399	7.29	Anti-inflammatory effects inhibited both acute inflammation and chronic inflammation, hypocholesterolemic activity, bacteriostatic activity, prevents ovarian cancer	Woyengo et al. (2009), Hashem et al. (2011)
5	Campesterol	49.668	2.09	Hypocholesterolemic activity, anticancer properties (prostate, lung, and breast cancers)	Choi et al. (2007)
6	β-Caryophyllene	16.676	0.93	Reduced production of prostaglandin E2 (PGE2), anti-inflammatory, antimicrobial, anticarcinogenic, antibiotic, antioxidant, and local anesthetic properties	Kuwahata et al. (2012), Legault et al. (2013), Zhu et al. (2013)
7	Epiglobulol	18.176	0.87	Antiseptic and cytotoxic behavior	Jain et al. (2012)
8	α-Bisabolol	22.928	0.81	Anti-inflammatory activities, antimycotic, and anti-bacterial properties	Miguel (2010), Saad et al. (2013)

TABLE 3.6 *(Continued)*

S. No.	Compound name	RT (min)	Peak area (%)	Health benefits	References
9	Squalene	42.929	6.31	Anti-inflammatory, antimicrobial, antiviral, cytotoxic, and cardiovascular effects, hypocholesterolemic activity, antioxidant activity, anticancer, cardio-protection, radio-protection	Gregory and Kelly (1999), Amarowicz (2009), Silva et al. (2012), Guneş (2013)
10	α-Amyrin	54.096	0.56	Alpha-amyrin, together with beta-amyrin, acts as an antinociceptive agent, has anti-inflammatory, anti-cancer, antitumor, and hepatoprotective actions	Wagh et al. (2012)
11	Phytol	31.367	4.46	Anticancer	Oyugi et al. (2011)
12	Vitamin E	47.61	0.34	Antioxidant activity	Singh et al. (2007)
13	γ-Tocopherol	46.19	0.39	Antioxidant activity and prevents thrombosis. γ-Tocopherols inhibit lipid peroxidation damage and trap mutagenic electrophiles more efficiently than α-tocopherols	Singh et al. (2007)
14	Palmitic acid	28.684	5.5	LDL cholesterol raising properties	Williams (2000)
15	Oleic acid	24.643	0.85	Hypocholesterolemic, anticancer properties, and provides cardio protection	Win (2005)
16	Steric acid	32.365	0.8	Hypocholesterolemic activity	Williams (2000)

stigmasta-5,22-Dien-3-ol (7.29%) (stigmasterol), squalene (6.31%), *n*-hexadecanoic acid (palmitic acid) (5.5%), phytol (4.46%), campesterol (2.09%), and methyleugenol (0.93%). Other compounds are also listed in Table 3.6. Figure 3.3 shows the gas chromatogram of basil leaf powder extract. The found result of basil analysis was well in accordance with the studies conducted by Lee et al. (2005), Hussain et al. (2008), Klimankova et al. (2008), Mondal et al. (2009), Verma and Kothiyal (2012). They also stated that basil constitutes eugenol, methyleugenol, oleic acid, palmitic acid, steric acid β-sitosterol, stigmasterol, campesterol, tocopherol, β-caryophyllene, and bisabolol as some of the main compounds. Our studies indicated the absence of some phytochemical compounds in basil leaf powder. This might be because of pretreatment applied to basil leaf powder, variation among species, or cultivation in different geographic locations. Table 3.7 shows the class-wise category of phytochemical compounds identified in basil leaf powder.

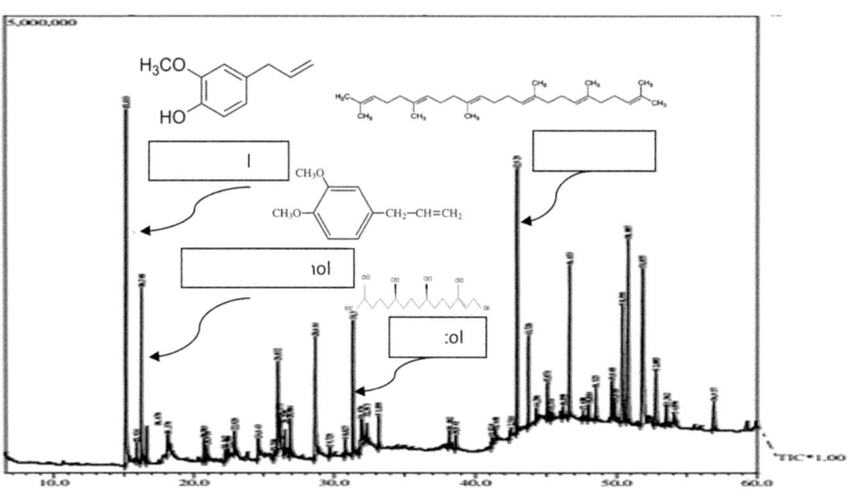

basil leaf powder by GC–MS.

FIGURE 3.3 Chromatogram of basil leaf powder by GC–MS

TABLE 3.7 Class-wise Category of Phytochemical Compounds.

S. No.	Class	Compound name	Peak area (%)
1	Phenylpropenes	Eugenol	11.66
		Methyleugenol	0.93
2	Phytosterol	β-Sitosterol	11.16
		Stigmasterol	7.29
		Campesterol	2.09
3	Sesquiterpene	β-Caryophyllene	0.93
		Epiglobulol	0.87
		α-Bisabolol	0.81
4	Triterpene	Squalene	6.31
		α-Amyrin	0.56
5	Diterpene	Phytol	4.46
6	Antioxidant	Vitamin E	0.34
		γ-Tocopherol	0.39
7	Fatty acid	Palmitic acid	5.5
		Oleic acid	0.85
		Steric acid	0.8

3.8 PHENYLPROPENES

Lignin and lignin biosynthetic pathways give intermediate products like coumaryl, coniferyl, and sinapyl alcohol. Phenylpropenes, also known as allylbenzenes, are derived from these intermediate products. They are one of the main plant secondary metabolites characterized by a benzene ring bearing a propenyl side chain (Louie et al., 2007).

Eugenol is a phenylpropene and the major proportion is present in clove oil. Eugenol is therefore the chief component of clove oil (Afzali et al., 2014). Spices like nutmeg and cinnamon also have good amount of eugenol. Leaves of bay tree and basil also possess appreciable amount of eugenol (Jaganathan & Supriyanto, 2012). Eugenol synthase converts coniferyl acetate to eugenol (Louie et al., 2007). Juliani and Simon (2002) extracted the essential oil from basil through hydrodistillation and estimated the amount of eugenol on retention index basis as

1358. Eugenol provides protection from both Gram-positive and Gram-negative microorganisms (Afzali et al., 2014). Antibacterial activity is due to the hydrophobicity of eugenol that separates lipids of cell membrane and mitochondria of bacteria by changing its structure so as to increase the penetrability of the cell membrane. Blocking properties of eugenol toward proton-motive stream, electron stream, and active transport causes coagulation of cell content, thus providing antibiotic activity (Kong et al., 2014). This compound also promises to reduce the leading NCDs, that is, eugenol has cardio-protection, anticancer, and vasodilating activity (Damiani et al., 2003; Kong et al., 2014). The antioxidant mechanism of eugenol is by giving hydrogen atoms thus reduces free radical generation. Eugenol can reduce the possibility of apoptosis of B-cell lymphoma-2, Cylcooxygenase-2, and interleukin-1β, and increase the treatment efficacy of gemcitabine. It thus inhibits programmed cell death and tumor cell invasion (Kong et al., 2014).

Methyl eugenol (ME) is also a phenylpropene and it is the methyl ether of eugenol. It is found in more than 450 plant species (like basil, fennel, lemongrass, nutmeg, pimento, star anise, and tarragon). In the food industry, it is used as a flavoring agent specifically in bakery goods, beverages, and confectioneries and for improving the aroma of cosmetic products (Groh et al., 2013; Haq et al., 2014). Apart from the flavoring and aromatic properties, Cerkezkayabekir et al. (2010) reported that methyleugenol also has some biological activities like antibacterial properties, relaxant, and antispasmodic actions on the ileum. Ahmad et al. (2010) reported that both the mentioned phenylpropenes have antimycotic properties. In the *basilicum* basil, Miele et al. (2001) identified the presence of methyleugenol, which was found to be around 0.93%.

3.9 PHYTOSTEROL

Phytosterols also known as plant sterols are steroid compounds abundantly present in the first three kingdoms, that is, Animalia, Plantae, and Fungi. Structurally, it is similar to cholesterol and when consumed competes with cholesterol and displaces it from intestinal micelles, thus reducing the cholesterol adsorption in the body (Saeidnia et al., 2014). Experiments showed that phytosterols possess several biological qualities which can greatly reduce the rate of NCD. Some of the qualities are hypocholesterolemic activity, protection against cancer especially the colon, breast, and prostate (Tasan et al., 2006; Marangoni & Poli, 2010).

β-Sitosterol, one of the constituents of phytosterol capable in curing heart disease, has hypocholesterolemia, modulates the immune system, prevents cancer, protects from rheumatoid arthritis, tuberculosis, alopecia, and benign prostatic hypertrophy (Awad et al., 2007; Saeidnia et al., 2014). Sen et al. (2012) experimented and found out that β-sitosterol can inhibit the growth of microorganisms. The presence of β-sitosterol was analyzed by Rahman et al. (2009) in the dehydrated basil leaves. In the investigation, β-sitosterol was found with a peak area of 11.16%.

Other phytosterols revealed were stigmasterol (7.29%) and campesterol (2.09%). Stigmasterol reduces inflammation, cholesterol, and bacteria (Hashem et al., 2011). Woyengo et al. (2009) reported that when stigmasterol is taken in higher dosage, that is, >23 mg per day can lower down the risk of ovarian cancer. Campesterol is abundant in seeds, nuts, cereals, beans, legumes, and vegetable oils. It is similar in structure to cholesterol and similarly metabolized by intestinal bacteria. Due to the structural similarity, campesterol competes with cholesterol and thus have lowering action. Campesterol also exhibits chemo-preventive effects against prostate, lung, and breast cancers (Choi et al., 2007).

3.10 TERPENE

Terpenes are secondary metabolites derived from five carbon isoprene units and are present in head-to-tail arrangement. It can also be configured with varying degrees of unsaturation, functional groups, ring closures, and oxidation, giving rise to a rich diversity of structural classes (Zhang et al., 2008). Sesquiterpene are a class of terpenes that consist of three isoprene units. It is formed by the combination of geranyl pyrophosphate and isopentenyl pyrophosphate. All sesquiterpenes significantly reduced production of prostaglandin E2 (Zhu et al., 2013).

β-Caryophyllene is a sesquiterpene and the common constituent of the essential oil of numerous plants such as clove, oregano, thyme, black pepper, and cinnamon. It is known to possess anti-inflammatory, antimicrobial, anticarcinogenic, antibiotic, antioxidant, and local anesthetic properties (Legault et al., 2013; Kuwahata et al., 2012; Lee et al., 2005). Klimankova et al. (2008) reported the presence of sesquiterpenes in basil. The present analysis of the dried leaf extract revealed the presence of β-caryophyllene with a peak area 0.93%. Other sesquiterpenes revealed in the analysis were epiglobulol (0.87%) and α-bisabolol (0.81%). Epiglobulol, one of the sesquiterpenes, has antiseptic and cytotoxic behavior (Jain et al., 2012). Natural α-bisabolol is a monocyclic unsaturated sesquiterpene alcohol. It has anti-inflammatory activities, antimycotic and antibacterial properties (Miguel, 2010; Saad et al., 2013).

Triterpene are a class of terpenes that consist of six isoprene units. It is formed by the combination of farnesyl pyrophosphate and farnesyl pyrophosphate. They are present in all organisms, especially in plants. The triterpene acids exhibit unique biological and pharmacological activities, including anti-inflammatory, antimicrobial, antiviral, cytotoxic, and cardiovascular effects (Silva et al., 2012). Squalene is a triterpene compound rich in shark liver oil. However, it is widely distributed in nature reasonably in olive oil, palm oil, wheat germ oil, amaranth oil, and rice bran oil (Gregory & Kelly, 1999). Analysis of the dried basil leaf methanolic extract also revealed the presence of squalene (peak area 6.31%). Gregory and Kelly (1999) reported that dietary supplementation with 1 g of squalene daily for 9 weeks increases very low-density lipoprotein (VLDL), intermediate density lipoprotein (IDL), and low-density lipoprotein (LDL) cholesterol concentrations in the serum by 12, 34, 28, and 12%, respectively. Squalene feeding also produced a significant improvement in fecal mass containing cholesterol, its non-polar derivatives, and bile acids, suggesting that although cholesterol synthesis probably increased by as much as 50%, fecal elimination was also upregulated, resulting in no net effect on serum cholesterol concentrations. In vitro experimental evidence indicated squalene as a highly effective oxygen-scavenging agent. It was reported that the antioxidant activity is comparable to butylated hydroxyl toluene (BHT). Squalene can also suppress growth of tumor cells, partially prevent the development of chemically induced cancer, and cause reduction of some already existing tumors. Furthermore, Amarowicz (2009) reported that squalene exerted a significant antioxidant activity in mild UV-mediated polyunsaturated fatty acid (PUFA) oxidation. Also combined administration of squalene and PUFA concentration observed a significant lowering of lipid peroxidation. Squalene altogether has anticancer properties, radioprotection, cardio-protection, antioxidant, and hypocholesterolemic effect. Squalene also protects the skin from aging and also from UV-associated skin disorders (Guneş, 2013). α-Amyrin belongs to the class of pentacyclic triterpenoids (Wagh et al., 2012). This compound is found to be present in methanolic extract of dehydrated tulsi powder (0.56%). α-Amyrin together with β-amyrin acts as an antinociceptive agent, has anti-inflammatory, anticancer, and hepatoprotective actions (Wagh et al., 2012). Diterpene, a type of terpene, is an organic compound composed of four isoprene units. Phytol is an acyclic C20 diterpene alcohol which functions as a preventive agent against epoxide-induced breast cancer carcinogenesis (Oyugi et al., 2011). Sathianarayanan et al. (2010) found the presence of phytol in dried methanolic extract of tulsi leaf as 3.9%, whereas 4.46% was found in the carried investigation.

3.11 ANTIOXIDANT

Antioxidants are molecules that prevent oxidation of vital molecules present in the body and stop free radical production, thus preventing from damaging the cells (Hamid et al., 2010). Vitamin E is a lipid-soluble antioxidant and is present in four forms α, β, γ, and δ tocopherols and four tocotrienols (α, β, γ, and δ-tocotrienols). α-Tocopherols and γ-tocopherols are responsible for the antioxidant activity of vitamin E. Daily consumption of small amount of γ-tocopherols with usual dietary intake from mixed food sources may provide protection from oxidation damage and prevent thrombosis. γ-Tocopherols inhibit lipid peroxidation damage and trap mutagenic electrophiles more efficiently than α-tocopherols (Singh et al., 2007). Vitamin E and γ-tocopherol were found in the methanolic extract to the extent of 0.34% and 0.39% dehydrated basil powder, respectively.

3.12 FATTY ACID

Fatty acids are the main component of oils and fats having long hydrocarbon chain and one or more carboxylic groups. The most common being palmitic (C16:0), stearic (C18:0), oleic (C18:1), and linoleic (C18:2). Oleic acid (omega-9) is a monounsaturated fatty acid found in animal and vegetable oils, such as olive oil (contains 55–80%), almonds, avocados, cashews, hazelnuts, macadamia nuts, peanuts, pistachio nuts, and sesame oil. In the food industry, oleic acid is used to make synthetic butters and cheeses. It is also used to flavor baked goods, candy, and ice cream. Oleic acid is an effective hypocholesterolemic agent. It also possesses anticancer properties and provides cardio protection. High administration of oleic acid can lower blood levels of cholesterol and reduce the risk of heart problems (Win, 2005). Dehydrated basil powder has 0.85 % peak area of oleic acid (9-octadecenoic acid). Stearic acid (peak area 0.8%) is an 18-carbon saturated fatty acid denoted as *n*-octadecanoic acid. Stearic acid is a fatty acid in nature, commonly found in many animal and vegetable fats, but is usually higher in animal fat than vegetable fat (Loften et al., 2014). Williams (2000) stated that stearic acid does not increase total cholesterol. The present analysis has revealed the presence of stearic acid. Palmitic acid (5.5%) is a 16-carbon-saturated fatty acid denoted as *n*-hexadecanoic acid. Palmitic acid is the most common saturated fatty acid found in plants, animals, and many microorganisms. Major sources of C16:0 are palm oil, palm kernel oil, coconut oil, and milk fat (Loften et al., 2014). Williams (2000) stated that lauric, myristic, and palmitic acids have LDL cholesterol raising properties. GC–MS analysis revealed the presence of palmitic acid with a peak area of 5.5% in basil leaf powder.

Other compounds found are 1-octadecanol (0.55%), 2-methyloctacosane (3.39%), nonacosane (0.99%), diethyl phthalate (0.74%), 1,2-benezenedicarboxylic acid (0.41%), and methyl (3-oxo-2-pentylcyclopentyl) acetate (0.35%).

3.13 MEDICINAL PROPERTIES

The world health scenario shows that there is a drastic shift in the occurrence of communicable to NCD (Neeha & Priyamvadah, 2013). According to the projections by the World Health Organization (WHO), deaths caused by communicable disease will approximately halve between 2004 and 2030 in low-income countries, while death attributed to NCD-related diseases will nearly be doubled. NCD is seriously going to affect the global economic status. It is predicted that by 2015, CVD and diabetics alone are expected to reduce the GDP by 5% leading to a big shift in health burden (Mattke et al., 2011; Mahmood et al., 2013). The WHO factsheet published during 2013 reported the four main types of NCDs as CVDs, cancers, chronic respiratory diseases, and diabetes. All these diseases are driven by forces that include aging, tobacco use, physical inactivity, use of alcohol, unhealthy diet, rapid unplanned urbanization, and the globalization of unhealthy lifestyles (Neeha & Priyamvadah, 2013). As per Ayurveda, these symptoms of disease mentioned above reflect "Raj rog" which is nothing but lifestyle disorders (Neeha & Priyamvadah, 2013; Jain & Srivastava, 2013). The predominant example of globalization of unhealthy lifestyles is unhealthy diets and environmental issues; it may show up in individuals as hypertension, hyperglycemia, high blood cholesterol, and obesity (Anon., 2014c,e).

The health scenario of India also shows same trend (Scheutz, 2013). CVD constituted 52% of mortality followed by chronic respiratory diseases, cancers, diabetes, and injuries. It has been expected that these diseases will lose India's 237 billion dollars by 2005–2015 (Anon., 2014c). Kumar and Kaushik (2012) report that CVD is expected to cause more than 50% of deaths by 2020. Even adolescence group are being suffered by CVD and this will double by 2025. Unhealthy environment condition, improper diet, poor infrastructure, and unpleasant working environment are also provoking hypertension, another major problem affecting the nation in a faster phase (Nongkynrih et al., 2004). Considering obesity, India is following the trend of other developing countries that are steadily becoming more obese. The morbid obesity is affecting 5% of the country's population. According to the National Family Health Survey (NFHS) conducted in 2006, 10% of India's population was having more body fat than what is considered healthy. Obesity causes several chronic diseases, including type-2 diabetes, hypertension, CVDs, cancers and various psychosocial problems (Gothankar, 2011; Neeha & Priyamvadah, 2013). Basil assures to reduce the health burden by compacting stress (Mondal et al., 2009), inflammation (Vetal et al., 2013), radiation damage (Mondal et al., 2009), aging factors (Kwee & Niemeyer, 2011), obesity (Vetal et al., 2013), cancer (Kwee & Niemeyer, 2011; Vetal et al., 2013), heart disease, and diabetics (Mondal et al., 2009). Basil enhances stamina and endurance; increases the body's efficient use of oxygen; modulates the immune system (Mondal et al., 2009); stimulates digestion (Kwee & Niemeyer, 2011); has antibiotic (Suppakul et al., 2008), antiviral (Hussain et al., 2008), and antifungal properties (Ghanjaoui et al., 2011; Karagozlu et al., 2011); enhances the efficacy of many other therapeutic treatments; and provides ample antioxidants and other nutrients (Verma & Kothiyal, 2012). Overall, basil is acknowledged as a principal adaptogen as it adapts and copes with a wide range of stresses, and restores disturbed physiological and psychological functions to a normal healthy state (Geetha et al., 2004; Prasad et al., 2010; Nair et al., 2012; Shafqatullah et al., 2013).

Basil is comprised of many potent health benefits. Some of the reported health benefits are detailed below.

3.13.1 ANTIDIABETIC

Basil has potential to control diabetes mellitus (Mondal et al., 2009). To support this, Vishwabhan et al. (2011) analyzed and observed the stimulatory effect of *O. sanctum* on physiological pathways of insulin secretion. Ravi et al. (2012) and Kumar et al. (2012) reported that *O. tenuiflorum* variety of basil has hypoglycemic activity. Result showed that basil's aldehyde reductase activity may help in reducing the complications faced by diabetic patients such as cataract, retinopathy, and others (Singh et al., 2012).

3.13.2 CANCER PROTECTION

Carcinogen-metabolizing enzymes like Cytochrome P 450, Cytochrome b5, aryl hydrocarbon hydroxylase, and glutathione S-transferase are important in detoxification of carcinogens and mutagens. It has been experimentally observed that the alcoholic extract of *O. sanctum* leaves has a modulatory influence on carcinogen-metabolizing enzymes (Kwee & Niemeyer, 2011; Kumar et al., 2012; Vetal et al., 2013). Vishwabhan et al. (2011) experimented and stated that *O. sanctum* can significantly reduce tumor cell size and increase the lifespan of mice having Sarcoma-180 solid tumors. Kumar et al. (2013) reported that the ethanolic extract of *Ocimum* can prevent lung carcinoma. The seed oil of *O. sanctum* was also analyzed by Prakash and Gupta (2005) and was found to possess chemo-preventive properties. The alcoholic extract of *O. sanctum* leaves decreased the number of tumors (Rastogi et al., 2007).

3.13.3 CARDIO PROTECTION

The essential fatty acids like linoleic and linolenic acids present in the *O. tenuiflorum* leaves exhibited cardio tonic and cardiac stimulant effect (Ravi et al., 2012). The linoleic and linolenic acid produce series 1 and 3 (PGE1 and PGE3) prostaglandins and inhibit the formation of series 2 prostaglandins (PGE2). Consumption of basil regularly for a long time can protect from isoproterenol-induced myocardial necrosis through enhancement of endogenous antioxidant (Singh et al., 2012).

3.13.4 ANTI-OBESITY

O. sanctum reduced high serum level of total cholesterol, triglyceride, LDL cholesterol, and phospholipids, and significantly increased the high-density lipoprotein cholesterol and total fecal sterol contents (Das & Vasudevan, 2006). Vishwabhan et al. (2011) reported that linolenic acid and linoleic acid contained in *O. sanctum*-fixed oil were possibly responsible for cholesterol lowering activity.

3.13.5 ANTIOXIDANT ACTIVITY

O. sanctum has significant ability to scavenge highly reactive free radicals (Archana & Namasivayam, 2000; Geetha et al., 2004; Samson et al., 2007). In a study, it was found that *O. sanctum* can protect rat heart from chronic restraint stress-induced changes through its central effect (Sood et al., 2006). The results of the study indicated the protective nature of the plant material on the brain tissues against the detrimental effect of noise stress. Kelm et al. (2000) and Singh et al. (2012) reported that the antioxidant property is due to the presence of phenolic compounds like cirsilineol, cirsimaritin, isothymusin, apigenin, and rosmarinic acid, and appreciable quantities of basil eugenol.

3.13.6 NEURO-PROTECTION

The ethanolic extract of *O. sanctum* leaves treats the neurodegenerative diseases. The extract contained rosmarinic acid and ursolic acid, and therefore, it is predicted that the protective effect may be due to the amelioration of excitotoxicity and oxidative stress. There was an increase in the activity of antioxidant enzymes with reduction in lipid peroxidation and nitrite concentration. This led to an improvement in the behavioral activities (Mondal et al., 2009; Rajagopal et al., 2013).

3.13.7 ADAPTOGEN

An adaptogen is an agent that reduces the intensity and negativity of stress caused by tension, emotional difficulties, poor lifestyle habits, disease and infection, pollution, and other factors. Tulsi is the most effective adaptogen known. The immune-stimulant capacity of basil may be responsible for the adaptogenic action of plant. The alcoholic extract of basil increased the physical endurance of swimming, prevented stress-induced ulcers, and milk-induced leukocytosis, respectively in rats and mice, indicating induction of non-specifically increased resistance against a variety of stress-induced biological changes by basil in animals (Singh et al., 2012). Kumar et al. (2013) reported that increased stress in the body leads to grave consequences and in this regard basil is proven valuable as the phytochemical like α-linolenic acid of the plant provides antioxidant as well as anti-infective and immune-enhancing properties.

3.13.8 RADIATION PROTECTION

Basil reduces toxicity of radio therapies and prevents radiation-induced cancer which is exhibited in animal models, human cell fibrosarcoma (in vivo), and human cervical cancer cell line (in vitro) (Kumar et al., 2013).

3.13.9 ANTI-INFLAMMATORY ACTIVITY

Linolenic acid present in *O. sanctum*-fixed oil has the capacity to block both the cyclooxygenase and lipoxygenase pathways of arachidonate metabolism and could be responsible for the anti-inflammatory activity of the oil (Singh et al., 1996; Das & Vasudevan, 2006). The anti-inflammatory effects of basil are comparable to drugs like ibuprofen, naproxen, and aspirin (Sumit & Geetika, 2012).

3.13.10 OTHER HEALTH BENEFITS

The oral administration of *O. sanctum* leaves extract induced a protective resistance against *Salmonella typhimurium* infection (Goel et al., 2010). *O. sanctum*-fixed oil may be considered to be a drug of natural origin which possesses both anti-inflammatory and antiulcer activity (Singh & Majumdar, 1999). *O. sanctum* has antifungal activity (Amber et al., 2010), and the leaf extracts may be a useful source for dermatophytic infections (Balakumar et al., 2011). *O. sanctum* also has antibacterial, antimicrobial, antifungal, antiulcer, anticataleptic, anticataract, anticoagulant and anticonvulsant activities, antiarthritic, and antithyroid activity (Karagozlu et al., 2011; Kadian & Parle, 2012; Kumar et al., 2012).

3.13.10 BASIL'S AIR PURIFICATION AND OZONE PROTECTION

Modern research states that basil produces ozone (Tewari et al., 2012), thus purifies the atmosphere and protects from global warming. However, this purification ability of basil was stated long back in Padma Purana, Uttara Khanda. According to the ancient scriptures, wherever the aroma of basil is carried by the wind, it purifies the atmosphere and frees all animals from all baser tendencies. Basil plant kept at doorstep of the house or in its vicinity keeps the atmosphere pure, supplies oxygen in greater quantities, ensures the health of its occupants, and keeps it free from poisonous insects. It possesses bacteriostatic properties. The emanations from basil plants are in fact fatal to mosquitoes. Snakes also cannot tolerate the aroma of basil and keep away from it. Thus, basil helps not only in curing physical ailments but also in promoting purity, sanctity, and spiritual progress (Daruwalla, 2013). Basil extracts are used in textile industries to make eco-friendly antibacterial cloths (Sathianarayanan et al., 2010), and in corrosion prevention sector, basil oil or leaf extract is coated on the tin and aluminum can to reduce corrosion (Halambek et al., 2013; Kumpawat et al., 2013). Table 3.8 shows some of the medicinal properties and health benefits of basil.

TABLE 3.8 Basil: Medicinal Properties and Health Benefits.

Sr. No.	Health benefits	Mode of action	Secondary metabolites involved in the action	References
1	Antidiabetic	• Aids in insulin secretion • Hypoglycemic activity • Aldose reductase activity	β-Sitosterol	Mondal et al. (2009), Vishwabhan et al. (2011), Ravi et al. (2012), Kumar et al. (2012), Singh et al. (2012)
2	Cancer protection	• Modulate carcinogen metabolizing enzymes • Reduce the number and cell size of tumors • Inhibited lung carcinoma	Eugenol, stigmasterol, campesterol, β-caryophyllene, epiglobulol, squalene, phytol, vitamin E, oleic acid	Prakash and Gupta (2005), Rastogi et al. (2007), Kwee and Niemeyer (2011), Vishwabhan et al. (2011), Kumar et al. (2012), Kumar et al. (2013), Vetal et al. (2013)
3	Cardioprotection	• Exhibited cardio tonic and cardiac stimulant effect • Inhibited the formation of series 2 prostaglandins (PGE2) • Protection against isoproterenol-induced myocardial necrosis	Linoleic and linolenic acids, eugenol, β-sitosterol, stigmasterol, campesterol, squalene, palmitic acid, oleic acid, steric acid	Ravi et al. (2012), Singh et al. (2012)
4	Anti-obesity	• Depressed high serum levels of total cholesterol, triglyceride, LDL-C, and phospholipids • Increased HDL-cholesterol and total fecal sterol content	Linolenic acid and linoleic acid, steric acid, oleic acid, palmitic acid, squalene, campesterol, stigmasterol, β-sitosterol	Das and Vasudevan (2006), Vishwabhan et al. (2011), Mohan et al. (2011)
5	Antioxidant property	• Scavenge highly reactive free radicals • Protects heart from chronic restraint stress • Protected brain tissues from noise stress	α-Linolenic acid, cirsilineol, cirsimaritin, isothymusin, apigenin and rosmarinic acid, and appreciable quantities of eugenol	Archana and Namasivayam (2000), Kelm et al. (2000), Geetha et al. (2004), Sood et al. (2006), Samson et al. (2007), Singh et al. (2012), Kumar et al. (2013)

TABLE 3.8 *(Continued)*

Sr. No.	Health benefits	Mode of action	Secondary metabolites involved in the action	References
6	Neuro-protection	• Treat neurodegenerative diseases • Reduction in the level of lipid peroxidation and nitrite concentration leading to improvement in behavioral activities	Rosmarinic acid and ursolic acid	Mondal et al. (2009), Rajagopal et al. (2013)
7	Adaptogen	• Anti-infective and immune-enhancing properties • Increased physical endurance • Prevented stress-induced ulcers and milk-induced leukocytosis	α-Linolenic acid, eugenol, methyleugenol, β-sitosterol, β-caryophyllene	Awad et al. (2007), Ahmad et al. (2010), Cerkezkayabekir et al. (2010), Singh et al. (2012) , Kumar et al. (2013), Afzali et al. (2014), Kong et al. (2014), Saeidnia et al. (2014)
8	Radiation protection	• Reduces toxicity caused by radiotherapies • Prevents radiation-induced cancer	Squalene	Kumar et al. (2013)
9	Anti-inflammatory	• Blocks the cyclooxygenase and lipoxygenase pathways of arachidonate metabolism • Anti-inflammatory effects of basil are comparable to ibuprofen, naproxen, and aspirin	Linolenic acid, squalene, α-bisabolol, β-caryophyllene, stigmasterol, eugenol	Singh et al. (1996), Das and Vasudevan (2006), Sumit and Geetika (2012)
10	Others health benefits			
	Antibacterial, antimicrobial, antifungal, antiulcer, anticataleptic, anticataract, anticoagulant, and anticonvulsant activities, antiarthritic and antithyroid activities			Singh and Majumdar (1999), Goel et al. (2010), Karagozlu et al. (2011), Kadian and Parle (2012), Kumar et al. (2012)

3.14 BASIL IN FOOD

Till date this economic, abundantly available, and easily cultivated herb is enormously used in medicines and cosmetic sector but very limited work is reported on the aspect of food application with scientific approach. Food is an indispensible part of human life and food industry are tremendously involved in making innovative food product which can improve the quality of food and thereby the living standard. Therefore, lots of scope exists toward exploring basil-incorporated food. Basil acts as an excellent antioxidant, gelling agent, thickening agent, stabilizing agent, and preserves the food items. Table 3.9 shows some of the limited application of basil in food industry.

3.14.1 SHELF LIFE IMPROVEMENT

The water insoluble fraction of tulsi (*O. sanctum* L.) improves the oxidative stability of ghee (butter fat). This improvement in ghee quality was because of the presence of phenolic compounds present in Tulsi leaves providing good antioxidant property (Bandyopadhyay et al., 2006). Volatile substance considered as the potential precursors of antioxidants can increase the total antioxidant capacity. Politeo et al. (2007) found that essential oil of basil also contained the four common volatile compounds, that is, eugenol, chavicol, linalool, and α-terpineol. The antioxidant capacity of basil was compared with essential oil and BHT. 2,2′-diphenyl-1-picrylhydrazyl radical scavenging method showed superior free radical scavenging capacity of basil while ferric ion reducing antioxidant power analysis showed that basil was slightly less effective than essential oil and BHT. Even then, this confirms the potential use of basil in the food industry to increase the shelf life of foodstuffs. Lipid peroxidation leads to the unpleasant tastes, off flavors, change in color, rheological properties, solubility, and potential formation of toxic compounds. Juntachote et al. (2006) inhibited the lipid peroxidation in cooked pork meat by blending dried holy basil powder and galangal powder in cooked pork meat.

TABLE 3.9 Basil: Application in Food Industry.

Sr. No.	Benefits		Secondary metabolites	References
1	Shelf life improvement	• Extends the oxidative stability of ghee (butterfat).	Phenolic substances, eugenol, chavicol, linalool, and α-terpineol	Juntachote et al. (2006), Politeo et al. (2007)
		• Inhibited the lipid peroxidation in cooked pork meat		
2	Food-quality improvement	• Improved the quality broiler diet. The addition of tulsi leaf powder reduced serum LDL-cholesterol in broilers.	Cholesterol biosynthesis, increased fecal bile secretion and stimulation of receptor-mediated catabolism of LDL	Lanjewar et al. (2009), Amirdivani and Baba (2011)
		• The presence of herbal extracts altered the yogurt bacteria fermentation of milk leading to enhanced acidification of yogurts, formation of bioactive peptides, antioxidant activities, antimicrobial activity, and anticancer properties	Enhanced the metabolic activity of yogurt bacteria (enhanced acidification), increased total phenolic content and DPPH activity (α-tocopherol, β-carotene, and ferulic acid), higher anti-angiotensis-1 converting enzyme (ACE)	
3	Primary packaging material	• Edible film made by adding basil improved the water barrier property	Linalool and methylchavikol	Suppakul et al. (2008), Bonilla et al. (2011)
		• Basil-incorporated LDPE-based film enhanced the quality and safety of cheese		
4	Thickening, gelling, and stabilizing agent	• Basil seed gum acts as a hydrocolloid and will impart better structural quality	Glucomannan, xylan, and glucan	Rafe et al. (2013), Osano et al. (2014)
		• Basil seed gum imparts smooth gelling property and can be applied in creamy food products		

3.14.2 FOOD-QUALITY IMPROVEMENT

Even in the absence of risk factors (like high blood pressure, cigarette, smoking, obesity, and diabetes) obesity is associated with coronary heart disease. This is due to increased consumption of cholesterol-rich food. The addition of tulsi leaf powder reduced serum LDL cholesterol in broiler diet (Lanjewar et al., 2009). Yogurt, a fermented milk product, categorized under functional food is an excellent medium to deliver probiotics to the consumers. Amirdivani and Baba (2011) investigated the effects of mixing dill, peppermint, and basil with milk on fresh and refrigerated yogurt. The herbs were dried, grounded, and kept in clean, dry, and airtight dark glass bottles at room temperature. Water extract for the three herbs was then prepared by soaking each herb in distilled water (1:10 ratio) for 12 h at 70°C, followed by centrifugation (2000 rpm, 10 min at 4°C). The obtained supernatant was then stored at 4°C in refrigerator. Then the extracts were used for the preparation of herbal-yogurts. The presence of herbal extracts altered the yogurt bacteria and fermentation of milk leading to enhanced acidification of yogurts, formation of bioactive peptides, enhanced antioxidant, antimicrobial, and anticancer properties (Amirdivani & Baba, 2011).

3.14.3 PRIMARY PACKAGING MATERIAL

Edible films and coatings are prepared from biopolymers and are able to protect food products in extending their shelf life. Basil essential oil was used by Bonilla et al. (2011) to make edible film. Their study revealed that addition of basil essential oil improved the water barrier property of edible film. Suppakul et al. (2008) studied the feasibility of low-density polyethylene-based films containing principal constituents of basil and found out that the study enhanced the quality and safety of cheese.

3.14.4 THICKENER, STABILIZER, AND GELLING AGENT

Basil seed gum (BSG) is an innovative and cost-effective hydrocolloid, extracted from seeds of basil herb. It contains water-soluble polysaccharides like glucomannan, xylan, and glucan. The influence of BSG, as a novel hydrocolloid, on the rheological properties and microstructure of gels, made from heat-denatured β-lacto globulin (BLG), was examined based on gelling profiles and by observing the interaction between BSG and BLG. The SEM results showed that BLG has a globular structure in comparison with BSG which has fibril and globular structure and the mean diameter of its globular structure is higher than that of BLG. When the proportion of BSG was increased than BLG, fine microstructure was obtained with better water retention capacity. This increase in proportion improved the gel structure. In addition, microstructural study showed that BLG to BSG mixture is a bicontinuous network in which BSG is dispersed in BLG continuous phase. Hence, the application of BSG as a thickening and gelling agent in the food industry due to its particular behavior and convenience of extraction, opens the door to many unique products with better structural quality (Rafe et al., 2013). Osano et al. (2014) also emphasized that since basil seeds are available in abundance, there is great potential to utilize the gum of basil seeds as a functional ingredient in the food industry as a thickening and stabilizing agent.

3.15 SUGGESTED PROCESS IMPROVEMENTS

Antioxidants have been widely used as food additives to avoid food degradation and regular consumption of basil leaf prevents from the occurrence of lifestyle-related disorders and aging. However, there are concerns about the use of synthetic antioxidants, such as butylated hydroxyl anisole and BHT, because of their instability and their possible activity as promoters of carcinogenesis. Sgherri et al. (2010) aimed to study the antioxidant and nutraceutical properties of basil (*O. basilicum*) grown in normal and in hydroponics condition. The antioxidant activities of aqueous and lipid extracts of basil leaves were evaluated both by spectrophotometric and electron paramagnetic resonance detection. It was observed that hydroponic cultivation significantly improved antioxidant activity. This greenhouse cultivation maximized the yield, antioxidant activity, and allowed a constant supply of the material throughout the year.

Eugenol is one of the active phytochemical compounds of basil. To maximize the yield of the active component, Ghosh et al. (2013) performed supercritical carbon dioxide (SC-CO_2) extraction with Krishna tulsi and comparatively evaluated against conventional extractions. The comparative study revealed that SC-CO_2 extraction gave a better yield of eugenol and phenolic content.

In another study, to characterize basil (*O. basilicum* L.), traditional sun drying method was compared with oven drying (50°C) (Zcan et al., 2005). Oven drying was found beneficial as it increased the drying rates and reduced the drying time. Optimum condition for oven drying is at 50°C to 17.31% moisture content after 15h and sun drying to 23.79% moisture content after 28 h. It was also observed that the mineral content of oven-dried herbs was higher than sun-dried herbs. In another study, basil leaves were dried using conventional hot air (50, 60, and 70°C) and low-pressure superheated steam (LPSS) dryers. Extract of fresh and dried leaves were obtained by simultaneous distillation–extraction technique and the aroma compounds were identified and quantified using capillary GC–MS and GC, respectively. Result showed that LPSS-dried product retained the original aroma profile while air-dried showed significant variation in the relative proportion of aroma compounds. Barbieri et al. (2004) thus concluded that from the product quality and production-cost point of view, the LPSS drying technique is better than conventional air-drying. LPSS assures to preserve the basil aroma compounds. To determine the highest essential oil yield, Pirbalouti et al. (2013) characterized the two varieties of basil (purple and green). Initially the leaves were dried under different drying conditions like sunlight (40°C), shade (60°C), microwave oven (500 W) and freeze-drying. Samples dried through each method were then used to extract the essential oil by hydrodistillation and the analysis was performed by GC and GC–MS. It was observed that highest essential oil yields were obtained from shade-dried method in both the varieties followed by freeze-drying method. Basil's (*O. basilicum* L.) leaves are very perishable and rapidly decline in quality after harvesting leading to serious commercial losses. Sweet basil, a plant of tropical origin is most susceptible to chilling and when stored at a temperature below 12°C causes chilling injury. Senescence of green leaves increases chlorophyll and protein degradation, and accumulates amino acids in detached leaves. Hassan and Mahfouz (2010) investigated the effect of 1-MCP treatment on the shelf life and postharvest quality of basil leaves. Result showed that single treatment of 1-MCP improved shelf life by inhibiting ethylene action and also maintained the higher volatile oil of sweet basil. Moreover, this treatment significantly reduced the degradation of protein and chlorophyll of harvested leaves. Table 3.10 shows suggested process improvements of basil.

TABLE 3.10 Basil: Suggested Process Improvements.

Sr. No.	Techniques		Benefits	Reference
1	Cultivation	Hydroponics cultivation	Maximized the yield (vitamin E, lipoic acid, total phenols, and rosmarinic acid), antioxidant activity, and allowed a constant supply of the material throughout the year	Sgherri et al. (2010)
2	Extraction	Supercritical carbon dioxide (SC-CO_2) extraction	Maximized the yield of the active component eugenol and phenolic contents	Ghosh et al. (2013)
3	Drying	Oven drying	Oven drying method (at 50°C and moisture content 17.31%) is recommended in the case of better mineral content retention	Zcan et al. (2005)
		Low-pressure superheated steam (LPSS) dryers	LPSS for aroma retention	Barbieri et al. (2004)
		Sunlight drying	Sunlight drying for highest essential oil yield	Pirbalouti et al. (2013)
4	Treatment	1-Methylcyclopropene (MCP) treatment on basil	Extended the shelf life of basil and maintained higher volatile oil percentage of basil leaves	Hassan and Mahfouz (2010)

3.16 CONCLUSION

Basil, an abundantly available and easily cultivated herb, has proved as a miracle solace to the physical, mental, and spiritual state of human being. The GC–MS analysis of basil powder identified various secondary and primary metabolites like phenylpropenes (eugenol, methyleugenol), phytosterol (β-sitosterol, stigmasterol, campesterol), sesquiterpene (β-caryophyllene, epiglobulol, α-bisabolol), triterpene (squalene, α-amyrin), diterpene (phytol), antioxidant (vitamin E, γ-tocopherol), and fatty acids (palmitic acid, oleic acid and steric acid). These compounds proved to have following health benefits: anti-inflammatory, antioxidant, antibacterial, antibiotic, antimycotic, antispasmodic, anticancer, antidiabetic, antistress, cardio-protection, radiation protection, immune-modulating, hypocholesterolemic, and vasodialating activity. Knowing the potentiality of basil, medical and cosmetic sectors are enormously engaged in extracting its benefits. Though food is an indispensable part of human life and basil's incorporation is going to change the phase of food industry, very limited work is reported on the aspect of food application with scientific approach. Therefore, lots of scope exists toward exploring basil-incorporated food as it acts as an excellent antioxidant, gelling agent, thickening agent, stabilizing agent, and moreover preserves the food items. Thus, this chapter opens basil's door to various other sectors like food, environment, corrosion, and others, and ultimately will encourage the mass cultivation of basil for the goodwill to mankind worldwide.

KEYWORDS

- **Tulsi**
- ***Ocimum sanctum***
- **basil**
- **herb**
- **secondary metabolites**
- **health**
- **food**

REFERENCES

Afzali, D.; Zarei, S.; Fathirad, F.; Mostafavi, A. Gold Nanoparticles Modified Carbon Paste Electrode for Differential Pulse Voltammetric Determination of Eugenol. *Mater. Sci. Eng.* **2014,** *43*, 97–101.

Ahmad, A.; Khan, A.; Khan, L. A.; Manzoor, N. In Vitro Synergy of Eugenol and Methyleugenol with Fluconazole against Clinical Candida Isolates. *J. Med. Microbiol.* **2010,** *59*, 1178–1184.

Amarowicz, R. Editorial—Squalene: A Natural Antioxidant?. *Eur. J. Lipid Sci. Technol.* **2009,** *111*, 411–412.

Amber, K.; Aijaz, A.; Immaculata, X.; Lugman, K. A.; Nikhat, M. Anticandidal Effect of *Ocimum sanctum* Essential Oil and Its Synergy with Fluconazole and Ketoconazole. *Phytomedicine* **2010,** 17(12), 921–925.

Amirdivani, S.; Baba, A. S. Changes in Yogurt Fermentation Characteristics, and Antioxidant Potential and In Vitro Inhibition of Angiotensin-1 Converting Enzyme Upon the Inclusion of Peppermint, Dill and Basil. *LWT—Food Sci. Technol.* **2011,** *44*, 1458–1464.

Anon. www.indianspices.com/html/s062ibas.htm, 2014a.

Anon. www.extension.psu.edu/plants/gardening/herbs/basil-green-ruffles, 2014b.

Anon.www.ncd.in/, 2014c.

Anon.www.wekipedia.org, 2014d.

Anon.www.who.int/mediacentre/factsheets/fs355/en/, 2014e.

Archana, R.; Namasivayam, A. Effect of *Ocimum sanctum* on Noise Induced Changes in Neutrophil Functions. *J. Ethnopharmacol.* **2000,** *73*(1,2), 81–85.

Awad A. B.; Chinnam, M.; Fink, C, S.; Bradford, P. G. β-Sitosterol Activates FAS Signaling in Human Breast Cancer Cells. *Phytomedicine* **2007,** *14*, 747–754.

Balakumar, S.; Rajan, S.; Thirunalasundari, T.; Jeeva, S. Antifungal Activity of *Ocimum sanctum* Linn. (Lamiaceae) on Clinically Isolated Dermatophytic Fungi. *Asian Pac. J. Trop. Med.* **2011,** *4*(8), 654–657.

Bandyopadhyay, M.; Chakraborty, R.; Raychaudhuri, U. A Process for Preparing a Natural Antioxidant Enriched Dairy Product (Sandesh). *LWT—Food Sci. Technol.* **2006,** *40*, 842–851.

Barbieri, S.; Elustondo, M.; Urbicain, M. Retention of Aroma Compounds in Basil Dried with Low Pressure Superheated Steam. *J. Food Eng.* **2004,** *65*, 109–115.

Bonilla, J.; Vargas, M.; Atares, L.; Chiralt, A. Physical Properties of Chitosan-Basil Essential Oil Edible Films as Affected by Oil Content and Homogenization Conditions. *Procedia Food Sci.* **2011,** *1*, 50–56.

Cerkezkayabekir, A.; Kizilay, G.; Ertan, F. Ultra Structural Changes in Rat Liver by Methyleugenol and Evaluation of Some Biochemical Parameters. *Turkey J. Biol.* **2010,** *34*, 439–445.

Choi, J. M.; Lee, E. O.; Lee, H. J.; Kim, K. H.; Ahn, K. S.; Shim, B. S.; Kim, N.; Song, M. C.; Biek, N. I.; Kim, S. H. Identification of Campesterol from *Chrysanthemum coronarium* L.: Its Antiangiogenic Activities. *Phytother. Res.* **2007,** *21*, 954–959.

Damiani, E. N.; Rossoni, L. V.; Vassallo, D. V. Vasorelaxant Effects of Eugenol on Rat Thoracic Aorta. *Vasc. Pharmacol.* (2003), *40*, 59–66.

Daruwalla, B. *The Scintillating and Illuminous Power of Jupiter*. 2013, p 1.

Das, S. K.; Vasudevan, D. M. Tulsi: The Indian Holy Power Plant. *Nat. Prod. Radiance*. (2006), *5*(4), 279–283.

Geetha, Kedlaya, R.; Vasudevan, D. M. Inhibition of Lipid Peroxidation by Botanical Extracts of *Ocimum sanctum*: In Vivo and In Vitro Studies. *Life Sci.* **2004,** *76*(1,19), 21–28.

Ghanjaoui, M. E.; Cervera, M. L.; Rhazi, M. E.; Guardia, M. D. L. Validated Fast Procedure for Trace Element Determination in Basil Powder. *Food Chem*. **2011,** *125*, 1309–1313.

Ghosh, S.; Chatterjee, D.; Das, S.; Bhattacharjee, P. Supercritical Carbon Dioxide Extraction of Eugenol-rich Fraction from *Ocimum sanctum* Linn. and a Comparative Evaluation with Other Extraction Techniques: Process Optimization and Phytochemical Characterization. *Ind. Crops Prod.* **2013,** *47*, 78–85.

Goel, A.; Kumar, S.; Bhatia, A. K. Effect of *Ocimum sanctum* on the Development of Protective Immunity against *Salmonella typhimurium* Infection through Cytokines. *Asian Pac. J. Trop. Med.* **2010,** *60*(10), 682–686. DOI:10.1016/S1995-7645(10)60 165-4.

Gothankar, J. S. Prevalence of Obesity and Its Associated co-morbidities amongst Adults—A Review. *Nat. J. Comm. Med.* **2011,** *2*(2), 221–224.

Gregory, S.; Kelly, N. D. Squalene and Its Potential Clinical Uses. *Altern. Med. Rev*. **1999,** *4*(1).

Groh, I. A. M.; Chen, C.; Lüske, C.; Cartus, A. T.; Esselen, M. Plant Polyphenols and Oxidative Metabolites of the Herbal Alkenylbenzene Methyleugenol Suppress Histone Deacetylase Activity in Human Colon Carcinoma Cells. *J. Nutr. Metab*. **2013,** *2013*, 10 pages, Article ID 821082.

Guneş, F. E. Medical Use of Squalene as a Natural Antioxidant. *J. Marmara Univ. Inst. Health Sci.* **2013,** *3*(4), 220–228.

Halambek, J.; Zutinic, A.; Berkovic, K. *Ocimum basilicum* L. Oil as Corrosion Inhibitor for Aluminium in Hydrochloric Acid Solution. *Int. J. Electrochem. Sci.* **2013,** *8*, 11201–11214.

Hamasaki, R. T.; Valenzuela, H. R.; Tsuda, D. M.; Uchida, J. Y. Fresh Basil Production Guidelines for Hawai'r. *Res. Ext. Ser*. **1994,** 1–9.

Hamid, A. A.; Aiyelaagbe, O. O.; Usman, L. A.; Ameen, O. M.; Lawal, A. Antioxidants: Its Medicinal and Pharmacological Applications. *Afr. J. Pure Appl. Chem*. **2010,** *4*(8), 142–151.

Haq. I.; Vreysen M. J. B.; Caceres C.; Shelly, T. E.; Hendrichs, J. Methyl Eugenol Aromatherapy Enhances the Mating Competitiveness of Male *Bactrocera carambolae* Drew & Hancock (Diptera: Tephritidae). *J. Insect Physiol*. **2014,** *68*, 1–6.

Hashem, H. A.; Bassuony, F. M.; Hassanein, R. A.; Baraka, D. M.; Khalil, R. R. Stigmasterol Seed Treatment Alleviates the Drastic Effect of NaCl and Improves Quality and Yield in Flax Plants. *Austr. J. Crop Sci*. 2011, *5*(13), 1858–1867.

Hassan, F. A. S.; Mahfouz, S. A. Effect of 1-Methylcyclopropene (1-MCP) Treatment on Sweet Basil Leaf Senescence and Ethylene Production during Shelf-Life. *Postharv. Biol. Technol*. **2010,** *55*, 61–65.

Hounsome, N.; Hounsome, B.; Tomos, D.; Edwards-Jones, G. Plant Metabolites and Nutritional Quality of Vegetables. *J. Food Sci*. **2008,** *73*(4), 48–65.

Hussain, A. I.; Anwar, F.; Sherazi, S. T. H.; Przybylski, R. Chemical Composition, Antioxidant and Antimicrobial Activities of Basil (*Ocimum basilicum*) Essential Oils Depends on Seasonal Variations. *Food Chem*. **2008,** *108*, 986–995.

Jaganathan, S. K.; Supriyanto, E. Antiproliferative and Molecular Mechanism of Eugenol-induced Apoptosis in Cancer Cells. *Molecules* **2012,** *17*, 6290–6304.
Jain, R.; Srivastava, B. K. Role of Diet and Dietetics Principle as Mentioned in Ayurveda in Prevention of Lifestyle Disorders. *Int. J. Ayurveda Herb. Med.* **2013,** *3*(1), 1120–1128.
Jain, S C.; Pancholi, B.; Jain, R. Antimicrobial, Free Radical Scavenging Activities and Chemical Composition of *Peltophorum pterocarpum* baker ex K. Heyne stem extract. *Der Pharma Chem.* **2012,** *4*(5), 2073–2079.
Joseph, B.; Nair, V. M. Ethanopharmacological and Phytochemical Aspects of *Ocimum sanctum* Linn—The Elixir of Life. *Br. J. Pharm. Res.* **2013a,** *3*(2), 273–292.
Joseph, B.; Nair, V. M. *Ocimum Sanctum* L. (Holy Basil): Pharmacology behind Its Anti-cancerous Effect. *Int. J. Pharm. Biol. Sci.* **2013b,** *4*(2), 556–575.
Joshi, D. K. The Sacred Basil (Tulsi). *Orissa Rev.* **2007,** 4–6.
Joshi, R. K. Chemical Composition of the Essential Oil of Camphor Basil (*Ocimum kilimandscharicum* Guerke). *Glob. J. Med. Plant Res.* **2013,** *1*(2), 207–209.
Juliani, H. R.; Simon, J. E. Antioxidant Activity of Basil. In *Trends in New Crops and New Uses*;. Janick, J., Whipkey, A., Eds.; ASHS Press: Alexandria, VA, 2002.
Juntachote, T.; Berghofer, E.; Siebenhandl, S.; Bauer, F. The Antioxidant Properties of Holy Basil and Galangal in Cooked Ground Pork. *Meat Sci.* **2006,** *72*, 446–456.
Kadam, D. M.; Goyal, R. K.; Gupta, M. K. Mathematical Modeling of Convective Thin Layer Drying. *J. Med. Plants Res.* **2011,** *5*(19), 4721–4730.
Kadam, P. V.; Yadav, K. N.; Jagdale, S. K.; Shivatare, R. S.; Bhilwade, S. K.; Patil, M. J. Evaluation of *Ocimum sanctum* and *Ocimum basilicum* Mucilage—As a Pharmaceutical Excipient. *J. Chem. Pharm. Res.* **2012,** *4*(4), 1950–1955.
Kadian, R.; Parle, M. Therapeutic Potential and Phytopharmacology of Tulsi. *Int. J. Pharmacy Life Sci.* **2012,** *3*(7), 1858–1867.
Karagozlu, N.; Ergonul, B.; Ozcan, D. Determination of Antimicrobial Effect of Mint and Basil Essential Oils on Survival of *E. coli* O157:H7 and *S. typhimurium* in Fresh-cut Lettuce and Purslane. *Food Control.* **2011,** *22*, 1851–1855.
Kelm, M. A.; Nair, M. G.; Strasburg, G. M.; Dewitt, D. L. Antioxidant and Cyclooxygenase Inhibitory Phenolic Compounds from *Ocimum sanctum* Linn. *Phytomedicine* **2000,** *7*(1), 7–13.
Klimankova, E.; Holadova, K.; Hajslova, J.; Cajka T.; Poustka, J.; Koudela, M. Aroma Profiles of Five Basil (*Ocimum basilicum* L.) Cultivars Grown under Conventional and Organic Conditions. *Food Chem.* **2008,** *107*, 464–472.
Kong, X.; Xiwang, L. I. U.; Jianyong, L. I.; Yang, Y. Advances in Pharmacological Research of Eugenol. *Curr. Opin. Compl. Altern. Med.* **2014,** *1*(1), 8–11.
Kumar, A.; Rahul A.; Chakraborty S.; Tiwari R.; Latheef, S. K.; Dhama, K. *Ocimum sanctum* (Tulsi): A Miracle Herb and Boon to Medical Science—A Review. *Int. J. Agron. Plant Prod.* **2013,** *4*(7), 1580–1589.
Kumar, P. K.; Kumar, M. R.; Kavitha, K.; Singh, J.; Khan, R. Pharmacological Actions of *Ocimum sanctum*—Review Article. *Int. J. Adv. Pharmacy, Biol. Chem.* **2012,** *1*(3), 408–414.
Kumar, S.; Kaushik, A. Editorial: Non-communicable Diseases: A Challenge. *Indian J. Comm. Health* **2012,** *24*(4), 252–254.

Kumpawat, N.; Chaturvedi, A.; Upadhyay, R. K. Comparative Study of Corrosion Inhibition Efficiency of Naturally Occurring Eco-Friendly Varieties of Holy Basil (Tulsi) for Tin in HNO_3 Solution. *Iran. J. Mater. Sci. Eng.* **2013,** *10*(4), 43–48.

Kuwahata, H.; Katsuyama, S.; Komatsu, T.; Nakamura, H.; Corasaniti, M. T.; Bagetta, G.; Sakurada, S.; Sakurada, T. Local Peripheral Effects of β-Caryophyllene through CB2 Receptors in Neuropathic Pain in Mice. *Pharmacol. Pharmacy* **2012,** *2*, 397–403.

Kwee, E. M.; Niemeyer, E. D. Variations in Phenolic Composition and Antioxidant Properties among 15 Basil (*Ocimum basilicum* L.) Cultivars. *Food Chem.* **2011,** *128*, 1044–1050.

Lanjewar, R. D.; Zanzad, A. A.; Ramteke, B. N.; Lalmuanpuii, Taksande, P. E.; Patankar, R. B. Incorporation of Tulsi (*Ocimum sanctum*) Leaf Powder in Diet of Broiler for Quality Meat Production. *Vet. World* **2009,** *2*(9), 340–342.

Lee, S. J.; Umano, K.; Shibamoto, T.; Lee, K. G. Identification of Volatile Components in Basil (*Ocimum basilicum* L.) and Thyme Leaves (*Thymus vulgaris* L.) and Their Antioxidant Properties. *Food Chem.* **2005,** *91*, 132–137.

Legault, J.; Cote, P. A.; Ouellet, S., Simard, S.; Pichette, A. Iso-caryophyllene Cytotoxicity Induced by Lipid Peroxidation and Membrane Permeabilization in L-929 Cells. *J. Appl. Pharm. Sci.* **2013,** *3*(8), 25–31.

Loften, J. R.; Linn J. G.; Drackley, J. K.; Jenkins, T. C.; Soderholm C. G.; Kertz, A. F. Invited Review: Palmitic and Stearic Acid Metabolism in Lactating Dairy Cows. *J. Dairy Sci.* **2014,** *97*, 4661–4674.

Louie, G. V.; Baiga, T. J., Bowman M. E.; Koeduka, T.; Taylor, J. H.; Spassova, S. M.; Pichersky, E.; Noel, J. P. Structure and Reaction Mechanism of Basil Eugenol Synthase. *PLoS ONE* **2007,** *10*, 1–12.

Mahmood, S. A. I.; Ali, S.; Islam, R. Shifting from Infectious Diseases to Non-communicable Diseases: A Double Burden of Diseases in Bangladesh. *J. Public Health Epidemiol.* **2013,** *5*(11), 424–434.

Marangoni, F.; Poli, A. Phytosterols and Cardiovascular Health. *Pharm. Res.* **2010,** *61*, 193–199.

Mathew, R.; Sankar, D. Plant Cell Culture Technology and its Entrée into the World of *Ocimum. Int. J. Pharmacy Pharm. Sci.* **2013,** *5*(2), 6–13.

Mattke, S.; Haims, M. C.; Guedehoussou, N. A.; Hunter, E. M. G. L.; Klautzer, L.; Mengistu, T. *Improving Access to Medicines for Non-communicable Diseases in the Developing World*, 2011, pp. 1–86.

Miele, M.; Dondero, R.; Ciarallo, G.; Mazzei, M. Methyleugenol in *Ocimum basilicum* L. Cv. Genovese Gigante. *J. Agric. Food Chem.* **2001,** *49*, 517–521.

Miguel, M. G. Antioxidant and Anti-inflammatory Activities of Essential Oils: A Short Review. *Molecules* **2010,** *15*, 9252–9287.

Miller, R.; Miller, S. *Tulsi Queen of Herbs: India's Holy Basil*, 2003, pp. 1–6.

Mondal, S.; Mirdha, B. R.; Mahapatra, S. C. The Science Behind Sacredness of Tulsi (*Ocimum sanctum* Linn.). *Indian J. Physiol. Pharmacol.* **2009,** *53*(4), 291–306.

Nair, L. D.; Santosh, S. K.; Arun, A.; Deepak, M. A Comparative Study on Proximate Analysis Conducted on Medicinal Plants of Chhattisgarh, CG, India. *Res. J. Chem. Sci.* **2012,** *2*(9), 18–21.

Neeha, V. S.; Priyamvadah, K. Nutrigenomics Research: A Review. *J. Food Sci. Technol.* **2013,** *50*(3), 415–428.

Nongkynrih, B.; Patro, B. K.; Pandav, C. S. Review Article on Current Status of Communicable and Non-communicable Diseases in India. *J. Assoc. Phys. India* **2004,** *52*, 118–123.

Osano, J. P.; Hosseini-Parvar, S. H.; Merino, L. M.; Golding, M. Emulsifying Properties of a Novel Polysaccharide Extracted from Basil Seed (*Ocimum basilicum* L.): Effect of Polysaccharide and Protein Content. *Food Hydrocolloids* **2014,** *37*, 40–48.

Oyugi, D. A.; Ayorinde, F. O.; Gugssa, A.; Allen, A.; Izevbigie, E. B.; Eribo, B.; Anderson, W. A. Biological Activity and Mass Spectrometric Analysis of *Vernonia amygdalina* Fractions. *J. Biosci. Technol.* **2011,** *2*(3), 287–304.

Pirbalouti, A. G.; Mahdad, E.; Craker, L. Effects of Drying Methods on Qualitative and Quantitative Properties of Essential Oils of Two Basil Landraces. *Food Chem.* **2013,** *141*, 2440–2449.

Politeo, O.; Jukic, M.; Milos, M. Chemical Composition and Antioxidant Capacity of Free Volatile Aglycones from Basil (*Ocimum basilicum* L.) Compared with Its Essential Oil. *Food Chem.* **2007,** *101*, 379–385.

Prakash, P.; Gupta, N. Therapeutic uses of *Ocimum sanctum* L. (Tulsi) with a Note on Eugenol and its Pharmacological Actions: A Short Review. *Indian J. Physiol. Pharmacol.* **2005,** *49*(2), 125–131.

Prasad, K.; Janve, B.; Sharma, R. K.; Prasad, K. K. Compositional Characterization of Traditional Medicinal Plants: Chemo-metric Approach. *Arch. Appl. Sci. Res.* **2010,** *2*(5), 1–10.

Rafe, A.; Razavi, S. M. A.; Farhoosh, R. Rheology and Microstructure of Basil Seed Gum and β-Lactoglobulin Mixed Gels. *Food Hydrocolloids.* **2013,** *30*, 134–142.

Rahman, S. M. M.; Mukta, Z. A.; Hossain, M. A. Isolation and Characterization of β-Sitosterol-D-glycoside from Petroleum Extract of the Leaves of *Ocimum sanctum* L. *Asian J. Food Agric. Ind.* **2009,** *2*(1), 39–43.

Rajagopal, S. S.; Lakshminarayanan, G.; Rajesh, R.; Dharmalingam, S. R.; Ramamurthy, S.; Chidambaram, K.; Shanmugham, S. Neuroprotective Potential of *Ocimum sanctum* (L.) Leaf Extract in Monosodium Glutamate Induced Excitotoxicity. *Afr. J. Pharmacy Pharmacol.* **2013,** *7*(27), 1894–1906.

Rastogi, S.; Shukla, Y.; Paul, B. N.; Chowdhuri, D. K.; Khanna, S. K.; Das, M. Protective Effect of *Ocimum sanctum* on 3-methylcholanthrene, 7,12-dimethylbenz(*a*)anthracene and Aflatoxin B1 Induced Skin Tumorigenesis in Mice. *Toxicol. Appl. Pharmacol.* **2007,** *224*(3,1), 228–240.

Ravi, P.; Elumalai, A.; Eswaraiah, M. C.; Kasarla, R. A Review on Krishna Tulsi, *Ocimum tenuiflorum* L. *Int. J. Res. Ayurveda Pharmacy* **2012,** *3*(2), 291–293.

Saad, A. H.; Ahmed, S. N.; Mohamed, E. B. Effect of Chamomile oil as a Percutaneous Absorption Enhancer. *Int. Res. J. Pharmacy.* **2013,** *4*(9), 66–71.

Saeidnia, S.; Manayi, A.; Gohari, A. R.; Abdollahi, M. The Story of Beta-sitosterol—A Review. *Eur. J. Med. Plants* **2014,** 4(5).

Samson, J.; Sheeladevi, R.; Ravindran, R. Oxidative Stress in Brain and Antioxidant Activity of *Ocimum sanctum* in Noise Exposure. *Neurotoxicology* **2007,** *28*(3), 679–685.

Sathianarayanan, M. P.; Bhat, N. V.; Kokate, S. S.; Walunj, V. E. Antibacterial Finish for Cotton Fabric from Herbal Products. *Indian J. Fibre Textile Res.* **2010,** *35*, 50–58.

Scheutz, A. M. *A Report on India's Healthcare System—Overview and Quality Improvements*; 2013; pp 1–48.

Sen, A.; Dhavan, P.; Shukla, K. K.; Singh, S.; Tejovathi, G. Analysis of IR, NMR and Antimicrobial Activity of β-Sitosterol Isolated from *Momordica charantia*. *Sci. Secure J.* **2012,** *1*, 9–13.

Sgherri, C.; Cecconami, S.; Pinzino C.; Navari-Izzo, F.; Izzo R. Levels of Antioxidants and Nutraceuticals in Basil Grown in Hydroponics and Soil. *Food Chem.* **2010,** *123*, 416–422.

Shafqatullah, Khurram, M.; Asadullah, Khaliqurrehman and Khan, F. A. Comparative Analyses of *Ocimum sanctum* Stem and Leaves for Phytochemicals and Inorganic Constituents. *Middle-East J. Sci. Res.* **2013,** *13*(2), 236–240.

Silva, M. L.; David, J. P.; Silva, L. C. R. C.; Santos, R. A. F.; David, J. M.; Lima, L. S.; Reis, P. S.; Fontana, R. Bioactive Oleanane, Lupane and Ursane Triterpene Acid Derivatives. *Molecules* **2012,** *17*, 12197–12205.

Singh, N.; Gilca, M. Tulsi and A Potential Protector against Air Travel Health Problems. *Nat. Prod. Radiance* **2008,** *7*(1), 54–57.

Singh, N.; Verma, P.; Pandey, B. R.; Bhalla, M. Therapeutic Potential of *Ocimum sanctum* in Prevention and Treatment of Cancer and Exposure to Radiation: An Overview. *Int. J. Pharm. Sci. Drug Res.* **2012,** *4*(2), 97–104.

Singh, S.; Majumdar, D. K. Evaluation of the Gastric Antiulcer Activity of Fixed Oil of *Ocimum sanctum* (Holy Basil). *J. Ethnopharmacol.* **1999,** *65*(1,1), 13–19.

Singh, S.; Majumdar, D. K.; Rehan, H. M. S. Evaluation of Anti-inflammatory Potential of Fixed Oil of *Ocimumu sanctum* (Holybasil) and Its Possible Mechanism of Action. *J. Ethnopharmacol.* **1996,** *54*(1), 19–26.

Singh. I.; Turner. A. H.; Sinclair. A. J.; Li. D.; Hawley. J. A. Effects of Gamma-Tocopherol Supplementation on Thrombotic Risk Factors. *Asian Pac. J. Clin. Nutr.* **2007,** *16*(3), 422–428.

Sood, S.; Narang, D.; Thomas, M. K.; Gupta, Y. K.; Maulik, S. K. Effect of *Ocimum sanctum* L. on Cardiac Changes in Rats Subjected to Chronic Restraint Stress. *J. Ethnopharmacol.* **2006,** *108*(3,6), 423–427.

Sumit, B.; Geetika, A. Therapeutic Benefits of Holy Basil (Tulsi) in General and Oral Medicine: A Review. *Int. J. Res. Ayurveda Pharmacy* **2012,** *3*(6), 761–764.

Suppakul, P.; Sonneveld, K.; Bigger, S. W.; Miltz, J. Efficacy of Polyethylene-based Antimicrobial Films Containing Principal Constituents of Basil. *LWT—Food Sci. Technol.* **2008,** *41*, 779–788.

Tasan, M.; Bilgin, B.; Gecgel. U.; Demirei, A. S. Phytosterol as Functional Food Ingredients. *J. Tekirdag Agric. Fac.* **2006,** *3*(2): 153–159.

Tewari, D.; Pandey, H. K.; Sah, A. N.; Meena, H. S.; Manchanda, A. Pharmacognostical and Biochemical Investigation of *Ocimum kilimandscharicum* Plants Available in Western Himalayan Region. *Asian J. Plant Sci. Res.* **2012,** *2*(4), 446–451.

Tilebeni, H. G. Review to Basil Medicinal Plant. *Int. J. Agron. Plant Prod.* **2011,** *2*(1), 5–9.

Veillet, S.; Tomao, V.; Chemat, F. Ultrasound Assisted Maceration: An Original Procedure for Direct Aromatization of Olive Oil with Basil. *Food Chem.* **2010,** *123*, 905–911.

Verma, S.; Kothiyal, P. Pharmacological Activities of Different Species of Tulsi. *Int. J. Biopharm. Phytochem. Res.* **2012,** *1*(1), 21–39.

Vetal, M. D.; Lade, V. G.; Rathod, V. K. Extraction of Ursolic Acid from *Ocimum sanctum* by Ultrasound: Process Intensification and Kinetics Studies. *Chem. Eng. Process.* **2013,** *69*, 24–30.

Vishwabhan, S.; Birendra, V. K.; Vishal, S. A Review on Ethnomedical Uses of *Ocimum sanctum* (Tulsi). *Int. Res. J. Pharmacy* **2011,** *2*(10), 1–3.

Wagh, S. J.; Gujar, J. G.; Gaikar, V. G. Experimental and Modeling Studies on Extraction of Amyrins from Latex of Mandar (*Calotropis gigantea*). *Indian J. Chem. Technol.* **2012,** *19*, 427–433.

Williams, C. M. Dietary Fatty Acids and Human Health. *Ann. Zootech.* **2000,** *49*, 165–180.

Win, D. T. Oleic acid—The Anti-breast Cancer Component in Olive Oil. *Assumpt. Univ. J. Technol.* **2005,** *9*(2), 75–78.

Woyengo, T. A.; Ramprasath, V. R.; Jones, P. J. H. Anticancer Effects of Phytosterols. *Eur. J. Clin. Nutr.* **2009,** *63*, 813–820.

Zcan, M. O.; Arslan, D.; Unver, A. Effect of Drying Methods on the Mineral Content of Basil (*Ocimum basilicum* L.). *J. Food Eng.* **2005,** *69*, 375–379.

Zhang, H.; Qiu, M.; Chen, Y.; Chen, J.; Sun, Y.; Wang, C.; Fong, H. H. S. Terpenes: Flavors, Fragrances, Pharmaca, Pheromones, 1st ed. Wiley-VCH Verlag GmbH & Co. KGaA: Germany, 2008.

Zhu, X. X.; Yang, L.; Li, Y. J.; Zhang, D.; Chen, Y.; Kostecka, P.; Kmonickova, E.; Zidek, Z. Effects of Sesquiterpene, Flavonoids and Coumarin Types of Compounds from *Artemisia annua* L. on Production of Mediators of Angiogenesis. *Pharm. Rep.* **2013,** *65*, 410–420.

CHAPTER 4

SECONDARY METABOLITES IN SPICES AND MEDICINAL PLANTS: AN OVERVIEW

AMIT BARAN SHARANGI

Department of Spices and Plantation Crops, Faculty of Horticulture, Bidhan Chandra Krishi Viswavidyalaya, Agricultural University, Nadia, Mohanpur 741252, West Bengal, India

E-mail: dr_absharangi@yahoo.co.in

CONTENTS

ABSTRACT

So far, over 4000 phytochemicals have been recognized and about 150 phytochemicals have been studied in detail. Several of them have a key role in protecting plants from herbivores and microbes whereas some others attract pollinators and seed-dispersing animals, act as allelopathic agents, UV protectants, and signal molecules toward formation of nitrogen-fixing root nodules in legumes. Secondary metabolites are also having multifarious uses as dye, fiber, glue, oil, wax, flavoring agent, drug and perfume, and, most importantly, they are viewed as components of potential natural drugs, antibiotics, insecticides, and herbicides. They are believed to enhance human immunity too. Increasing clinical evidences of secondary metabolites having potential health benefits have registered a steady growth in their consumption pattern at the rate of 5–10% per year.

4.1 INTRODUCTION

Use of plants for curing ailments particularly for human beings and that of plant-derived medicines predates written human history. Ethnobotany, the study of plants, is a recognized branch of science to identify and discover future medicines in an efficient way (Elumalai & Eswariah, 2002). As part of their regular metabolic activities, all plants produce some chemical compounds known as phytochemicals. They may be categorized into (1) primary metabolites comprising carbohydrates, lipids, and proteins, which are found in all plants having essential roles associated with photosynthesis, respiration, growth, and development; and (2) secondary metabolites comprising compounds found in a smaller range of plants, serving a more specific function and accumulating in high concentrations in some species. These chemicals were not being addressed properly for long and are attracting attention only recently.

So far, over 4000 such phytochemicals have been recognized and about 150 phytochemicals have been studied in detail (American Cancer Society, 2000). Several of them have a key role in protecting plants from herbivores and microbes whereas some others attract pollinators and seed-dispersing animals, act as allelopathic agents, UV protectants, and signal molecules toward formation of nitrogen-fixing root nodules in legumes.

Secondary metabolites are also having multifarious uses as dye, fiber, glue, oil, wax, flavoring agent, drug and perfume, and, most importantly, they are viewed as components of potential natural drugs, antibiotics, insecticides, and herbicides (Croteau et al., 2000). They are believed to enhance human immunity too (Atoui et al., 2005). Increasing clinical evidences of secondary metabolites having potential health benefits have registered a steady growth in their consumption pattern at the rate of 5–10% per year (Tham et al., 1998).

Herbs and spices have a documented history of medicinal uses (Tapsell et al., 2006). They could be called the first "functional food," however, have grossly become forgotten for being overtaken by the modern westernized diet. Some of the herbs and spices include black pepper (*Piper nigrum*), celery leaf (*Apium graveolens*), clove (*Syzygium aromaticum*), coriander (*Coriandrum sativum*), cumin (*Cuminum cyminum*), curry leaf (*Murraya koenigii*), garlic (*Allium sativum*), ginger (*Zingiber officinale*), nutmeg (*Myristica fragrans*), onion (*Allium cepa*), chilli (*Capsicum frutescens*), red pepper (*Capsicum annum*), star anise (*Illicium verum*), turmeric (*Curcuma longa*), tamarind (*Tamarindus indica*), and others.

Dietary spices have profound influence on gastrointestinal, cardiovascular, reproductive, and nervous systems of our body resulting in diverse metabolic and physiologic actions (Kochhar, 2008). Many of them are routinely used to be proclaimed as antimicrobial agents and posses several medicinal values as well. There are many ways to hunt for new bioactive chemicals in higher plants (Farnsworth & Loub, 1983), since they have an immense potential to produce new drugs that benefits mankind. Systemic screening may result in the exploration and exploitation of novel effective products (Janovska et al., 2003).

Plants can synthesize surprisingly wide variety of phytochemicals (Str. 4.1) to contribute to its unique defense chemistry. Interestingly most of them are derived from a few biochemical motifs, for example, alkaloids (a class of chemical compounds containing naturally occurring organic nitrogen-containing bases), flavonoids (polyphenolic compounds comprising 15 carbons, with 2 aromatic rings connected by a linear 3-carbon chain), polyphenols (also known as phenolics, compounds contain phenol rings), anthocyanins, glucosinolates, cyanogenic glycosides (a molecule in which a sugar is bound to a non-carbohydrate moiety), terpenes (consisting of a branched chain of 5 carbon atoms), coumarins, tannins (a polyphenolic compound binding and precipitating proteins and various other organic compounds including amino acids and alkaloids), saponins (include compounds that are glycosylated steroids, triterpenoids, and steroid alkaloids), steroids, lignans, essential oils (EOs), and fragrances. Many culinary spices (e.g., clove, garlic, ginger, onion, mustard, pepper, and turmeric) have their bioactive constituents characterized (Achinewhu et al., 1995). Bioactive compounds bestow protection against cardiovascular and cancer diseases. Bioactive compounds confer protection against cardiovascular and cancer diseases.

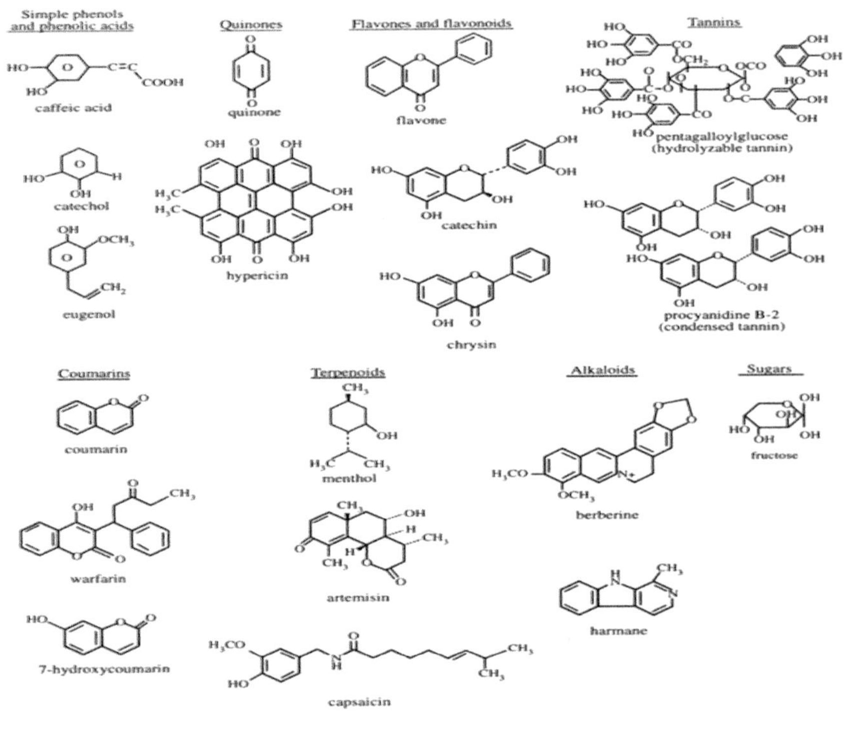

STRUCTURE 4.1 Chemical structures of some common phytochemicals.

Seed spices like coriander, cumin, fennel, fenugreek, black cumin, dill, ajwain, and others, are rich in natural phytochemicals having complementary and overlapping actions including antioxidant effects, anticancer, antidiabetic, antimicrobial activity, hypolipidemic effect, insecticidal use along with other uses as in menstrual disorders, in indigestion, in hypertension, in modulation of detoxification enzymes, in stimulation of immune system, in reduction of inflammation, in modulation of steroid metabolism, and in improvement of several other human disorders (Rathore et al., 2013). Wang (2014) explored the effects of secondary compounds produced by tea, black pepper, onion, and ginger against bacteria. She discovered the relationship between the secondary compound and the inhibition capacity of the Gram-positive and Gram-negative bacteria. Antibacterial compounds in order of effectiveness were as follows: green tea–water solution > black pepper–acetone solution > onion–water solution > ginger–water solution. Spices impart flavor, aroma, color, and taste to food preparations and often mask undesirable odors. The aroma and the taste in food are contributed by volatile oils and oleoresins, respectively. Volatile oils give the aroma, and oleoresins impact the taste.

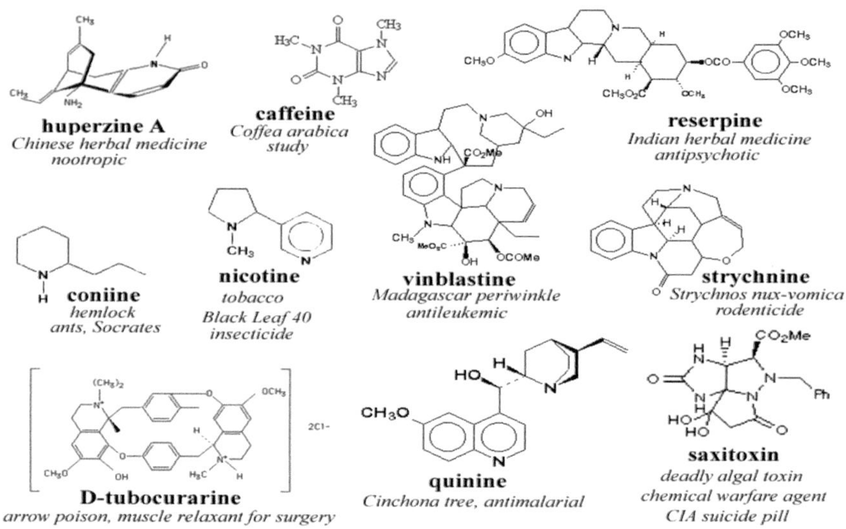

STRUCTURE 4.2 Basic structures of some alkaloids.

Medicinal plants essentially contain complex mixtures of bioactive compounds including the popular alkaloids (Str. 4.2). They affect multiple targets and, in general, show low toxicity. Some of the active secondary metabolites can have advantages in treating chronic diseases. Scognamiglio et al. (2015) analyzed seven aromatic Mediterranean plant species used in traditional cuisine with NMR-based metabolic approach.

Seasonal metabolic changes on primary as well as secondary metabolites had been detected and quantified. Flavonoids (apigenin, quercetin, and kaempferol derivatives) (Str. 4.3) and phenylpropanoid derivatives (e.g., chlorogenic and rosmarinic acid) were the main identified polyphenols.

But why these plants are so special? This may possibly be explained with the biological properties ascribed to these plant species as a result of being rich in these metabolites. In aromatic plants, part of their therapeutic effects comes from their EOs.

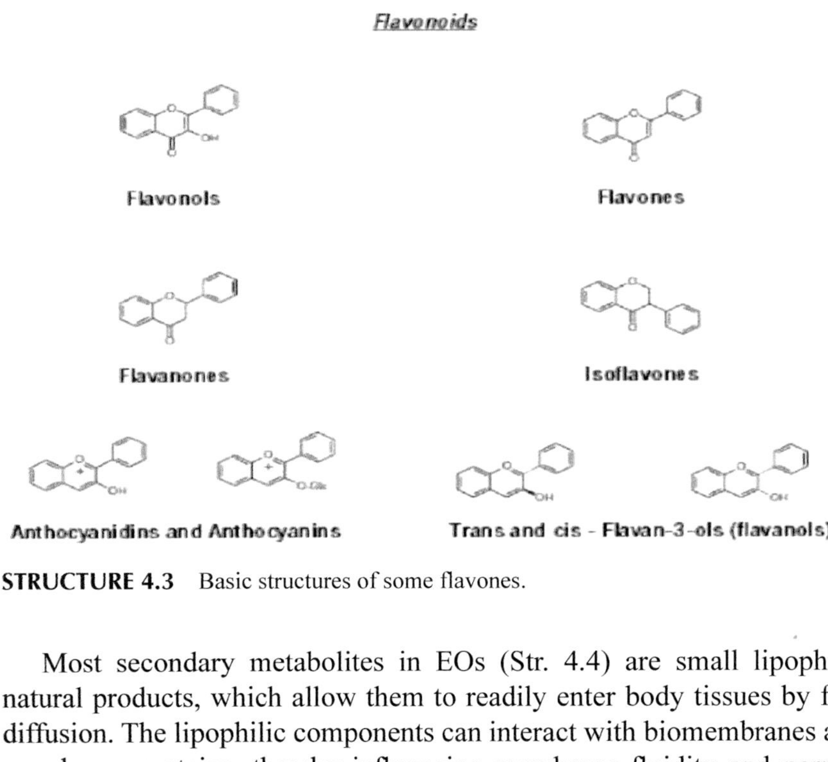

STRUCTURE 4.3 Basic structures of some flavones.

Most secondary metabolites in EOs (Str. 4.4) are small lipophilic natural products, which allow them to readily enter body tissues by free diffusion. The lipophilic components can interact with biomembranes and membrane proteins, thereby influencing membrane fluidity and permeability. This phenomenon explains the antibacterial, antifungal, antiviral, and cytotoxic nature of most of the components of EOs (Sharopov et al., 2015). Aromatic compounds produce varied range of flavorants, having wide applications in food industry to impart flavor and increase the appeal of products. These compounds are classified by functional groups, for example, alcohols, aldehydes, amines, esters, ethers, ketones, thiols, and others. In spices, the volatile oils constitute these components (Menon,

2000). Several active ingredients of spices, namely, capsaicin (chilli), piperine (black pepper), curcumin (turmeric), eugenic acid (clove), ferulic acid (turmeric), and myristic acid (mace, amla) have been reported to influence lipid metabolism predominantly by mobilization of fatty acids (Srinivasan & Satyanarayana, 1987).

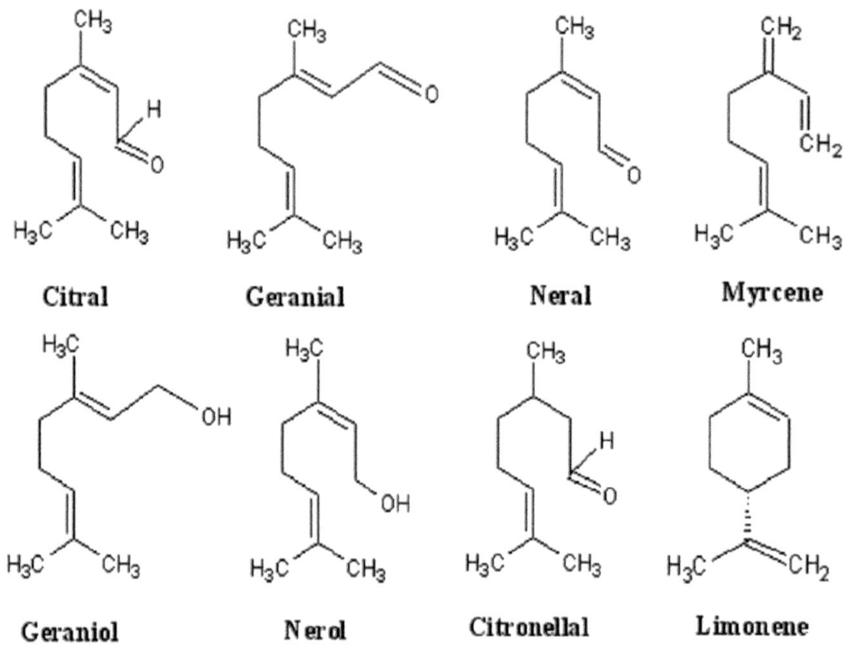

STRUCTURE 4.4 Basic structures of some essential oil.

Some important bioactive phytochemicals often found in certain herbs and medicinal plants in the form of secondary metabolites have been described in Table 4.1.

Drought stress leads to stomatal closure resulting in a highly over-reduced metabolic status. The enhancement of reductive power ($NADPH^{+}H^{+}$), in turn, causes an increase in the synthesis of all highly reduced secondary metabolites such as isoprenoids, phenols, or alkaloids. Intentional drought stress during the growth of spice and medicinal plants could enhance the concentration of natural products (Al-Gabbiesh et al., 2015). Favarin et al. (2013) described the anti-inflammatory activity of various medicinal plant extracts of *Ginkgo biloba* and *Punica granatum*,

(a) Nutmeg & mace
(b) Black pepper
(c) Clove
(d) Garlic
(e) Ginger
(f) Turmeric
(g) Onion
(h) Coriande
(i) Cumin

FIGURE 4.1 Photographs (a–i) of some common spices rich in secondary metabolites.

and such secondary metabolites like epigallocatechin-3-gallate and ellagic acid. They highlighted the therapeutic potential of these plant-derived compounds for treatment of acute lung injury. The following paragraphs will describe the significant research works done on secondary metabolites present in several commonly used and popular spice crops like black pepper, coriander, celery leaf, clove, cumin, curry leaf, fenugreek, garlic, ginger, nutmeg, onion, red pepper (chilli), star anise, tamarind, and turmeric (Fig. 4.1 a–i).

TABLE 4.1 Bioactive Phytochemicals in Herbs and Medicinal Plants (after Saxena et al., 2013).

Secondary metabolites	Biological function
Alkaloids, terpenoids, volatile flavor compounds, amines	Neuropharmacological agents, antioxidants, cancer chemoprevention
Carotenoids, polyphenols, curcumine, flavonoids	Inhibitors of tumor, inhibited development of lung cancer, antimetastatic activity
Cellulose, hemicellulose, gums, mucilages, pectins, lignins	Water-holding capacity, delay in nutrient absorption, binding toxins, and bile acids
Polyphenolic compounds, flavonoids, carotenoids, tocopherols, ascorbic acid	Oxygen free radical quenching, inhibition of lipid peroxidation
Reductive acids, tocopherols, phenols, indoles, aromatic isothiocyanates, coumarins, flavones, carotenoids, retinoids, cyanates, phytosterols	Inhibitors of procarcinogen activation, inducers of drug binding of carcinogens, inhibitors of tumorogenesis
Terpenoids, alkaloids, phenolics	Inhibitors of microorganisms, reduce the risk of fungal infection

4.2 BLACK PEPPER *(PIPER NIGRUM)*

Black pepper is considered as the "King of spices" throughout the world. Secondary metabolites of *P. nigrum* can be used as antiapoptotic, antibacterial, antidepressant, antidiarrheal, anti-inflammatory, antifungal, antimutagenic, antimetastatic, antioxidative, antipyretic, antispasmodic, antispermatogenic, antitumor, antithyroid, as ciprofloxacin potentiator, hepatoprotective, insecticidal, against cold extremities, gastric ailments, and intermittent fever, and others (Ahmad et al., 2012).

Phytochemical investigations on *P. nigrum* roots indicated presence of several alkaloids, namely, piperine (Str. 4.5), pellitorine, and aristolactam AII and 3,4-methylenedioxy benzoic acid in the roots. The ethyl acetate extract of the roots of *P. nigrum* was found with encouraging results against the larvae of *Aedes* (Wen, 2009). The black pepper oil contains β-and α-pinenes, δ-limonene, and β-caryophyllene as major components. Pepper has long been recognized as a carminative. This property is supposed to be due to the stimulating effect of gastric acid secretion by piperine (Ononiwu et al., 2002). The important compounds in the fresh pepper are *trans*-linalool oxide and α-terpineol. Black pepper was found to be an impressive antioxidant (Karthikeyan & Rani, 2003; Vijayakumar et al., 2004; DíSouza et al., 2004) and anti-inflammatory effects (Pratibha et al., 2004).

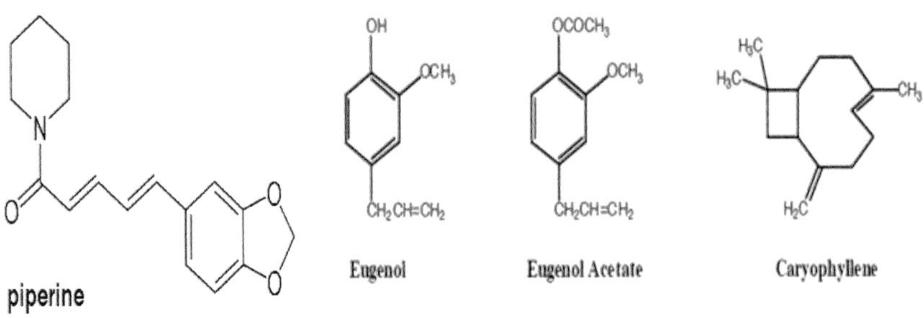

STRUCTURE 4.5 Secondary metabolites present in black pepper, clove, and curry leaf.

4.3 CELERY LEAF *(APIUM GRAVEOLENS)*

Celery, of the family Apiaceae, a biennial herb, is cultivated and consumed worldwide. It is low in calorie and rich in carotenoids, flavonoids, volatile oils, and fiber (Sowbhagya, 2014). Polyacetylenes present in celery leaves have positive impact on human cancer cells with a clear reduction in tumor formation as found with a mammalian in vivo model (Christensen & Brandt, 2006). Limonene (40.5%), β-selinene (16.3%), cis-ocimene (12.5%), and β-caryophyllene (10.5%) are major volatile oil constituents present in celery leaf (Ehiabhi et al., 2003).

4.4 CLOVE *(SYZYGIUM AROMATICUM)*

Cloves are actually the immature dried buds of the clove tree. There are three major constituents of clove oil, namely, eugenol (70–80%), β-caryophyllene (5–12%) and eugenyl acetate (15%), together comprising 99% of the oil (Str. 4.5). Cloves also contain flavonoids, galloyltannins, phenolic acids, and triterpenes. The minor constituents like methyl amyl ketone, methyl salicylate, and others, are responsible for the characteristic pleasant odor of cloves. The buds had their use in folk medicine especially as diuretic, odontalgic, stomachic, tonicardiac, aromatic, carminative, and stimulant. It has now been proved to have very good antioxidant properties and it could also be used in the treatment of ulcers, coronary diseases, blood pressure as herbal medicine (Behera et al., 2013; Mishra & Sharma, 2014).

4.5 CORIANDER *(CORIANDRUM SATIVUM)*

Coriander (*C. sativum*) seeds containing phytochemicals like linalool, carvone, geraniol, limonene, borneol, camphor, and others, have significant healing properties. In certain parts of Europe, coriander has long been referred to as antidiabetic plant. In some parts of India, it has traditionally been used for its anti-inflammatory properties. It is antioxidant (Saxena et al., 2013), hypoglycemic (Selvan, 2003; Ertas et al., 2005) and inhibits aflatoxin, bacteria (Baratta et al., 1998), swellings, diarrhea, anemia, indigestion, mouth ulcer, small pox, menstrual, and skin disorders. Coriander is believed as a natural aphrodisiac and previously it was extensively used in certain preparations, combined with other herbs (Kumar et al., 1977).

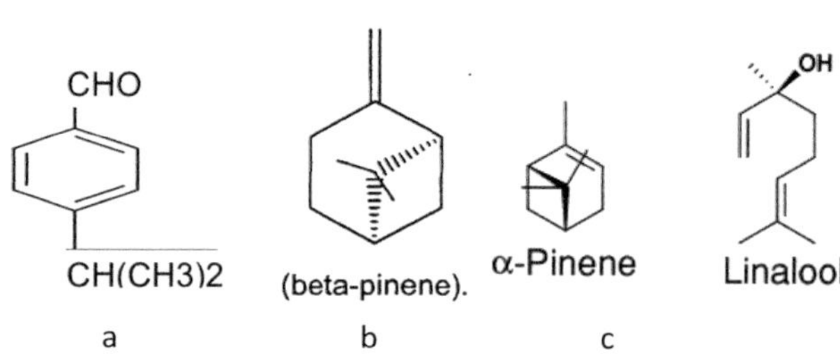

STRUCTURE 4.6 Chemical structures of (a) cuminaldehyde, (b) β-pinene, (c) α-pinene, and (d) linalool.

4.6 CUMIN *(CUMINUM CYMINUM)*

Among the seed spices, cumin fruits have a characteristic bitter flavor and strong aroma due to their EO content. Of this, cuminaldehyde (40–65%) is the primary aroma compound for which bitterness is ascribed to it (Str. 4.6c). Cuminaldehyde is also responsible for the antimicrobial and antimutagenic properties. Rahimi et al. (2012) identified 22 compounds in cumin EO through GC and GC–MS methods of analysis including α-terpinene-7-al, cumin aldehyde, α-terpinene-7-al, *p*-cymene, and β-pinene, respectively. The characteristic flavor of cumin is due to the presence of monoterpenes such as α-pinene and cis-β-farnesene (Str. 4.6). Cumin is used against dyspepsia, jaundice, diarrhea and having anti-inflammatory, diuretic, carminative, and antispasmodic properties (Mohammad Pour et al., 2012).

4.7 CURRY LEAF *(MURRAYA KOENIGII)*

It is a perennial shrub or small tree commonly cultivated for its leaves in India, Sri Lanka, and other Asian countries. The curry leaf plant is highly prized for its characteristic aroma and medicinal potentialities. A good number of leaf EO constituents and alkaloids, namely, carbazole, murrayacine, and koenigine have been extracted from this plant (Mallavarapu et al., 1999). Many oxygenated mono and sesquiterpenes are present in it, for example, cis-ocimene (34.1%), β-caryophyllene (9.5%), α-pinene (19.1%), D-terpenene (6.7%), and β-phellandrene which is responsible for the intense odor associated with the stalk and flower parts of curry leaves (Onayade & Adebajo, 2000). Both *M. koenigii* and *Brassica juncea* showed significant hypoglycemic action in experimental rats (Khan et al., 1996).

In a study, high levels of phenolic acids (especially gallic acid) and flavonoids (especially myricetin, epicatechin, and quercetin) in curry leaf exhibited a significant anticancer activity. Moreover, contents of phenolic acids and flavonoids and the related antioxidant activity of curry leaf extracts can be successfully utilized for further pharmaceutical applications (Ghasemzadeh et al., 2014). Sivakumar and Meera (2013) showed that curry leaf is capable of directly quenching free radicals to cease the radical chain reaction, can act as reducing agent, and facilitates chelating of transition metals to suppress the initiation of radical formation.

4.8 FENUGREEK *(TRIGONELLA FOENUM-GRAECUM)*

Fenugreek is an annual leguminous herb grown extensively in many Asian, Middle Eastern, and European countries. It contains alkaloids, mainly trigonelline (0.2–0.36%) choline (0.5%), gentianine, and carpaine, flavonoids, such as apigenin, luteolin, orientin, quercetin, vitexin, isovitexin, saponins (0.6–1.7%, and volatile oils (*n*-alkanes and sesquiterpenes) (Krishnaswamy, 2008) (Str. 4.7). The importance of diosgenin in the synthesis of oral contraceptives and sex hormones has been documented (Chapagain & Wiesman, 2005). It is an excellent antioxidant and used against diabetes, cancer, inflammation, obesity, hypertension, and other disorders (Rathore et al., 2013).

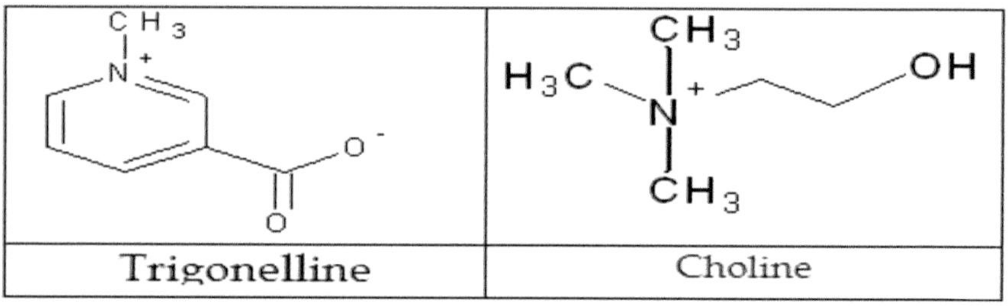

STRUCTURE 4.7 Chemical structures of trigonelline and choline.

4.9 GARLIC *(ALLIUM SATIVUM)*

Garlic has been used all throughout the recorded history for culinary and medicinal applications (Karuppiah & Rajaram, 2012). The therapeutic effects of garlic as hypolipidemic, antithrombotic, antihypertensive, antihyperglycemic, antihypercholesterolemic and immunomodulatory is well known. Photochemical analyses of the ethanolic extracts by Arekemase et al. (2013) showed the presence of many secondary metabolites such as saponins, tannins, alkaloid steroids, and glycosides in garlic (Str. 4.8). The major bioactive components responsible for the health benefits of garlic are assumed to be allylic sulfur compounds. The use of herbs and spices to displace fats and salt in the food may reduce cardiovascular risk, and that specific herb or spice is more convincingly, the garlic. Consumption of garlic or its oil has been associated with a reduction in total cholesterol, low-density lipoprotein (LDL) cholesterol, triglyceride levels, and thereby the cardiovascular risk. Allicin and ajoene, the major sulfur components of garlic, were shown to inhibit inducible nitric oxide syntheses by reducing the protein and mRNA and thus to promote vasodialation.

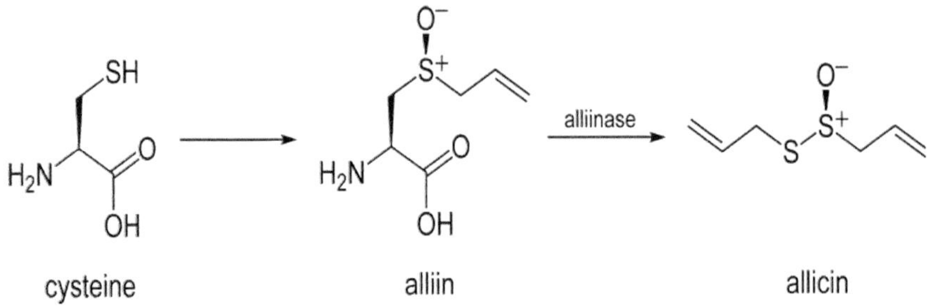

STRUCTURE 4.8 Chemical structures of alliin and allicin.

4.10 GINGER *(ZINGIBER OFFICINALE)*

Ginger is the most widely used herb especially in Asia and contains amazing ranges of health-promoting bioactive constituents (Rozanida et al., 2005). It is a mixture of over several hundred known constituents, including gingerols, shagaols, β-carotene, caffeic acid, curcumin, salicylate, and capsaicin. Ginger owes its characteristic organoleptic properties to two classes of constituents. The steam-volatile oil consisting of sesquiterpene hydrocarbons, monoterpene hydrocarbons, and oxygenated monoterpenes is mainly responsible for the aroma of ginger (Purseglove et al., 1981) while its pungency is due to the non-steam-volatile components also known as the gingerols. The major sesquiterpene hydrocarbon constituent of ginger oil is α-zingiberene. Certain ginger oil possesses a particular "lemony" aroma, due to its high content of the isomers, neral and geranial often collectively referred to as citral (Wohlmuth et al., 2006). The concentration of secondary metabolites of ginger (Str. 4.9) can be influenced by environmental conditions such as light intensity, temperature, insects, biotic and abiotic factors, which can often alter the concentration of the active constituents and can be harmful to consumers (Ghasemzadeh et al., 2010). The primary and secondary metabolite synthesis and antioxidant activities of ginger can be enhanced through controlled environment and CO2 enrichment (Ghasemzadeh & Jaafar, 2012).

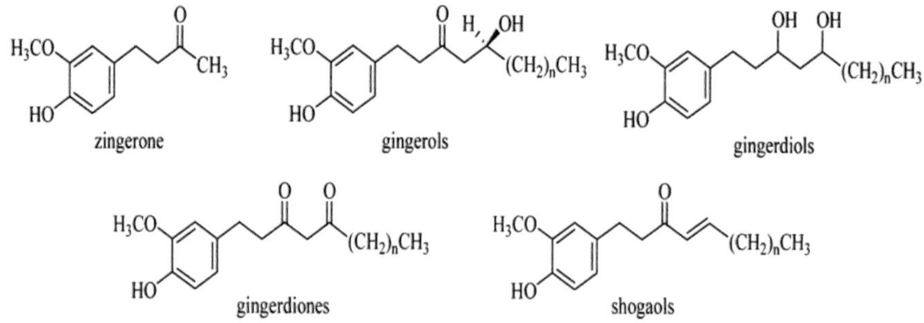

STRUCTURE 4.9 Chemical structures of important secondary metabolites of ginger.

4.11 NUTMEG *(MYRISTICA FRAGRANS)*

Nutmeg produces two spices, nutmeg seed and mace, the thick fiber like red aril on the kernel (Rombaut et al., 2009). Nutmeg oil possesses strong antibacterial, antifungal, anti-inflammatory, and insecticidal properties mainly for sabinene, β- and α-pinenes, eugenol, isoeugenol, methyl eugenol, safrol, neolignan, myristicin, elemicin, and linalool (Str. 4.10). Myristicin, isolated from the nut, impacts hallucinogenic properties and also reported to be an effective insecticide, while the lignin types of the constituents are anticarcinogenic (Narasimhan & Dhake, 2006). Adewole et al. (2013) examined nutmeg for its secondary metabolites, analyzed antimicrobial activities and characterized the oil using GC–MS. They found various secondary metabolites including flavonoids, tannin, saponin, and alkaloids. Thomas and Krishnakumari (2015) indicated that nutmeg has the potential to act as a source of useful drugs because of presence of various phytochemical components such as carbohydrate, protein, alkaloids, phenols, flavonoids, and tannin.

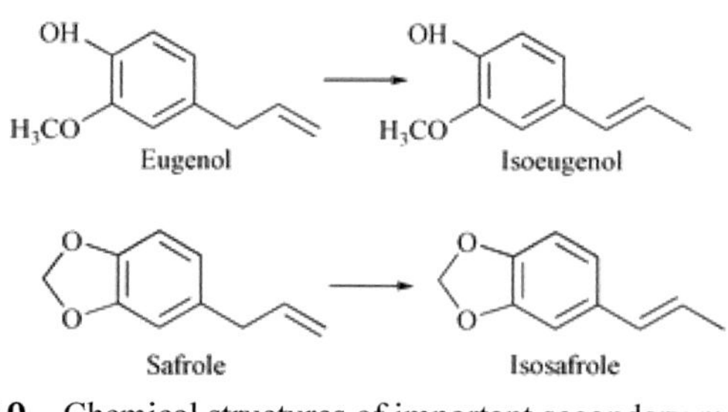

STRUCTURE 4.10 Chemical structures of important secondary metabolites of nutmeg.

4.12 ONION *(ALLIUM CEPA)*

Onions are cultivated both in tropical and temperate regions. It produces several secondary metabolites, such as flavnols (quercitin), flavnoids (kaempferol, luteolin), lignans (lariciresinol), α-tocopherols, quinon (phylloquinone, vitamin K), sterols (campesterol), saponins (tropeosides, ascalonicosides), sapogenins (gitogenin, diosgenin, β-chlorogenin, cepagenin), diallylsulphides, cysteine-sulphoxides, thio-sulphinates (zwiebelanes), vanillic acid, and cinnamic acid (Lanzotti, 2006; Hounsome et al., 2008) (Str. 4.11). Onion and its juice may be used to treat appetite loss, prevention of age-related changes in blood vessels (arteriosclerosis), digestive disturbances, and other traditional uses such as colds, cough, asthmas, and diabetes (Van Wyk & Wink, 2005). Onions undergo enzymatic breakdown of sulfur-containing substances due to damages of tissue to give pungent volatiles that cause weeping (Van Wyk, 2005). The pharmacological activity as well as the pungent smell are due to several sulfur-containing compounds—mainly sulphoxides such as *trans*-5-(1-propenyl)-L-(+)-cysteine sulphoxide) and cepaenes (α-sulphinyl-disulphides) (Van Wyk & Wink, 2005).

Thampi & Shalini (2015) explored the antibacterial activity of fresh leaves of green onions (*A. cepa* var. *cepa*) against specific pathogens and characterization of the active compounds, namely, alcohols, phenols, alkanes, alkynes, alkyl halides, carboxylic acids, and aromatic, aliphatic amines in methanolic extracts and phenols, alkynes, aldehyde, alkyl halides in aqueous extracts. Onions possess a vast array of secondary metabolites, and plant breeders could potentially develop nutraceutical onion cultivars. Modified breeding programs could reveal the hidden biochemical variation for targeted crop improvement and better quality products (Saxena & Cramer, 2013).

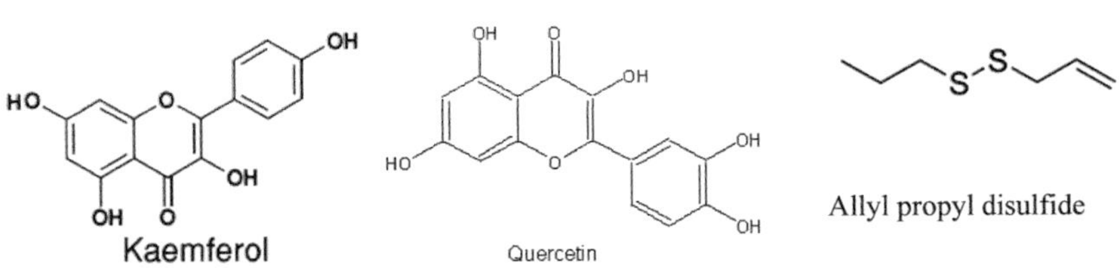

STRUCTURE 4.11 Chemical structures of important secondary metabolites of onion.

4.13 RED PEPPER OR CHILLI *(CAPSICUM ANNUUM)*

Red pepper (*Capsicum annuum* Linn.; Family: Solanaceae) is grown all over the world, primarily in tropical and subtropical countries. The fruits are an excellent source of health-related compounds, such as ascorbic acid, carotenoids, tocopherols, flavonoids, and capsaicinoids (Str. 4.12). It is rich in capsaicin having analgesic and antioxidant activity and used for the treatment of asthma, coughs, sore throats, and toothache (Wahyuni et al., 2013, Wu et al., 2013).

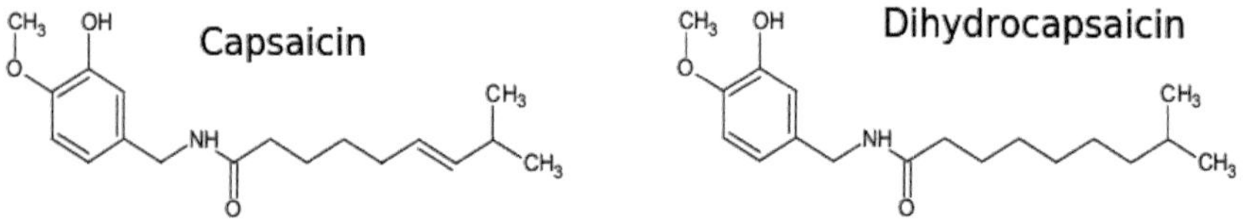

STRUCTURE 4.12 Chemical structures of important secondary metabolites of chilli.

4.14 STAR ANISE *(ILLICIUM VERUM)*

Star anise is an aromatic evergreen tree bearing purple-red flowers and anise-scented star-shaped fruit. It grows almost exclusively in southern China and Vietnam. It contains shikimic acid, a primary precursor in the pharmacological synthesis of anti-influenza drugs namely Oseltamivir (Tamiflu). In star anise, the presence of a prenyl moiety in the phenylpropanoids plays an important role in antitumor-promoting activity (Str. 4.13). Hence, the prenylated phenylpropanoids might act as a potential cancer chemopreventative agent (Padmashree et al., 2007). Aly et al. (2016) characterized star anise and assessed its antioxidant, antifungal, and antimycotoxigenic properties. The major components of star anise EO identified by GC/MS were *trans*-anethole (82.7%), carryophyllene (4.8%), and limonene (2.3%).

The herbal medicines are comparatively safer and cheaper than synthetic drugs. The plant-based traditional knowledge has become a recognized tool in search for new sources of drugs and neutraceuticals. Ethnopharmacological knowledge can bring out many different tools for the development of new drugs to treat various human diseases (Das & Selva Kumar, 2013).

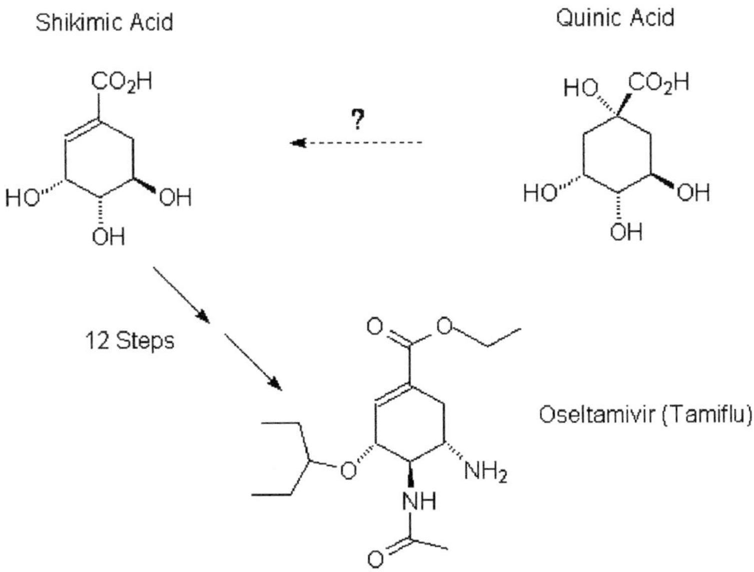

STRUCTURE 4.13 Chemical structures of shikimic acid and oseltamivir present in star anise.

4.15 TAMARIND *(TAMARINDUS INDICA)*

Tamarind (*T. indica*, Fabaceae), a tropical fruit found in Africa and Asia, is highly valued for its pulp. Tamarind fruit pulp has a sweet acidic taste due to a combination of high contents of tartaric acid and reducing sugars. It has numerous traditional uses including that against liver and bile disorders. It has been reported to be among the recipe in the treatment of cold, fevers, stomach disorders, diarrhea, jaundice, and as skin cleanser. The fruit pulp, rich in pectin, monosaccharides and organic acids, is mainly used as drinks. Abubakar et al. (2010) was of opinion that the pulp and especially the leaves of *T. indica* could be a promising antifungal agent and might be used in traditional medicine for the treatment of fungal infections.

4.16 TURMERIC *(CURCUMA LONGA)*

Turmeric is a perennial rhizomatous plant of Zingibraceae, native to South Asia. Curcuminoids and EO in turmeric are used as anti-inflammation, antipoison, antitumor, antifungi, HIV1 virus preventive agents and are useful in Alzheimer, urinal diseases abatement, and diabetes treatment (Javaprakasha et al., 2005). Curcumin (1,7-bis(4-hydroxy-3-methoxy phenyl)-1-6-hepatadine-3-5-dione), demethoxycurcumin, and bisdemethoxycurcumin are members of curcuminoid family; represent one of the yellow pigments isolated from turmeric. Other compounds are bisabolane, guaiane, α- and β-turmerone, curlone, and zingiberene.

Its immunomodulatory properties including antioxidant, anti-inflammatory, and antitumor properties are well documented (Govindrajan, 1980). Turmeric contains EOs, fatty oils, and 2–5% curcuminoids. Turmeric is fairly rich in omega-3 fatty acids. Certain varieties contain up to 9% curcuminoids. It has all proximate principles such as carbohydrates, proteins, and fats and provides all nutrients in small quantities. Curcuminoids are polyphenolic compounds with a β-diketone moiety. The three types of curcuminoids, namely Curcumin I, II, and III differ in their hydroxyl and methyl groups. Current research is focused on evidence-based science to determine the functional benefits of their bioactive compounds.

Tissue culture is an important tool for enhancement of medicinal compound production utilizing various methods including elicitation. It was found that nitrogen stress was successful in upregulating metabolite production to increase antioxidant activity from extracts (Cousins, 2008). Mostajeran et al. (2014) evaluated the effect of salt stress on the growth and chlorophyll synthesis of plant leaves and on the amount of secondary metabolism such as EO components. It may be possible that the EO of turmeric inhibits the growth of *A. niger*, which is toxigenic in nature and produces different mycotoxins like aflatoxin and ochratoxin, and others (Jeswal & Kumar, 2015) (Str. 4.14).

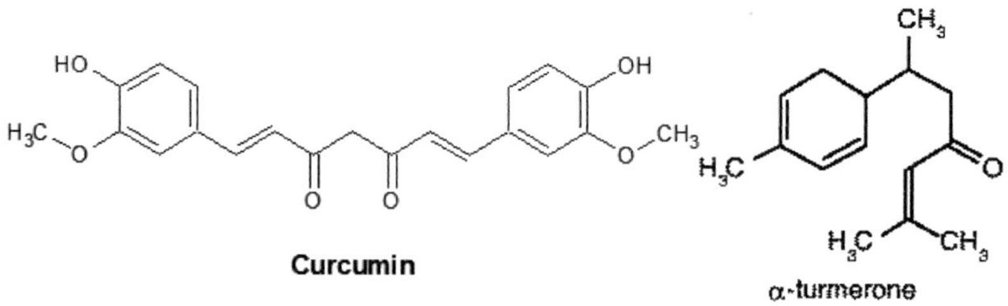

STRUCTURE 4.14 Chemical structures of curcumin and α-turmerone present in turmeric.

Herbs and spices and their extracts are natural antimicrobials and commonly used by the food industry. They are generally containing EOs possessing antimicrobial potentiality. EOs from plants usually has a relatively high vapor pressure and are capable of reaching microbial pathogens through the liquid and gas phases (Du et al., 2011). Herbs and spices can exert direct or indirect effects to extend foodstuff shelf life as antimicrobial agents act against variety of Gram-positive and Gram-negative bacteria. Edible and herbal plants and spices such as oregano, clove, cinnamon, citral, garlic, coriander, rosemary, parsley, lemongrass, sage, and vanillin have been successfully used alone or in combination with other preservation methods (Gutierrez et al., 2008; Lopes-Lutz et al., 2008; Angioni et al., 2004; Proestos et al., 2008). Other spices, such as ginger, black pepper, red pepper, chili powder, cumin, and curry powder showed lower but appreciable antimicrobial properties (Holley & Patel, 2005). The degree of antibacterial action of various spices tested against *Salmonella* and other enterobacteria was reported in the order of clove > kaffir lime peels > cumin > cardamom > coriander > nutmeg > mace > ginger > garlic > holy basil > kaffir lime leaves (Proestos et al., 2008).

CONCLUSION

Spices produce a vast and diverse assortment of organic compounds, the great majority of which do not appear to participate directly in growth and development. These substances, traditionally referred to as secondary metabolites, assume great significance. Presence of these metabolites in spices and herbs has created a wide and productive platform for chemical investigation for decades, driving the frontiers of chemical knowledge forward.

In recent years, the effect of dietary phenolics is of great interest due to their antioxidative and possible anticarcinogenic activities posing a considerable impact on human health. Phenolic acids and flavonoids also function as reducing agents, free radical scavengers, and quenchers of singlet oxygen formation. The area is of utmost interest to the scientific community particularly in the possibility of altering the production of bioactive plant metabolites by means of tissue culture technology. Bioprospecting of selected plants including herbs and spices could also isolate new and novel therapeutic molecules. The mode of action, bioavailability, and interaction of secondary metabolites with drugs need thorough investigations and validation for greater interest and confidence toward formulating future research directions.

KEYWORDS

- **dietary spices**
- **phytochemicals**
- **organic compounds**
- **saponins**
- **drought stress**

REFERENCES

Abubakar M. G.; Yerima, M. B.; Zahriya, A. G.; Ukwuani, A. N. Acute Toxicity and Antifungal Studies of Ethanolic Leaves, Stem and Pulp Extract of *Tamarindus indica*. *Res. J. Pharm., Biol. Chem. Sci.* **2010,** *1*(4), 104–111.

Achinewhu S. C.; Ogbonna C. C.; Hart A. D. Chemical Composition of Indigenous Wild Herbs, Spices, Fruits, Nuts and Leafy Vegetables used as Food. *Plant Foods Hum. Nutr.* **1995,** *48*, 341–348.

Adewole E.; Ajiboye, B. O.; Idris, O. O.; Ojo, O. A.; Onikan, A.; Ogunmodede O. T.; Adewumi, D. F. Phytochemical, Antimicrobial and GC–MS of African Nutmeg (*Monodora myristica*). *Int. J. Pharm. Sci. Invent.* **2013,** *2*(5), 25–32.

Ahmad, N.; Fazal, H.; Abbasi, B. H.; Farooq, S.; Ali, M.; Khan, M. A. Biological Role of *Piper nigrum* L. (Black pepper): A Review. *Asian Pac. J. Trop. Biomed.* **2012,** *2*(3), S1945–S1953.

Al-Gabbiesh, A.; Kleinwächter, M.; Selmar, D. Influencing the Contents of Secondary Metabolites in Spice and Medicinal Plants by Deliberately Applying Drought Stress during their Cultivation. *Jordan J. Biol. Sci.* **2015,** *8*(1), 1–10.

Aly, S. E.; Sabry, B. A.; Shaheen, M. S.; Hathout, A. S. Assessment of Antimycotoxigenic and Antioxidant Activity of Star Anise (*Illicium verum*) In Vitro. *J. Saudi Soc. Agric. Sci.* **2016,** *15*(1), 20–27.

American Cancer Society. *Phytochemicals*. Available at http://www.cancer.org/eprise/main/docroot/ETO/content/ETO_5_3X_Phytochemicals, June 2000.

Angioni, A.; Barra, A.; Cereti, E.; Barile, D.; Coïsson, J. D.; Arlorio, M. Chemical Composition, Plant Genetic Differences, Antimicrobial and Antifungal Activity Investigation of the Essential Oil of *Rosmarinus officinalis* L. *J. Agric. Food Chem.* **2004,** *52*, 3530.

Arekemase, M. O.; Adetitun, D. O.; Oyeyiola,G. P.; Adetitun, D. O.; Oyeyiola, G. P. In-vitro Sensitivity of Selected Enteric Bacteria to Extracts of *Allium sativum* L. *Not Sci Biol.* **2013,** *5*(2), 183–188.

Atoui, A. K.; Mansouri, A.; Boskou, G.; Kefalas, P. Tea and Herbal Infusions: Their Antioxidant Activity and Phenolic Profile. *Food Chem.* **2005,** *89*, 27–36.

Baratta, M. T.; Dorman, H. J. D.; Deans, S. G.; Biondi, D. M.; Ruberto, G. Chemical Composition, Antimicrobial and Antioxidative Activity of Laurel, Sage, Rosemary, Oregano and Coriander Essential Oils. *J. Essential Oil Res.* **1998,** *10*, 618–627.

Behera, B.; Yadav, D.; Sharma, M. C. Effect of an Herbal Formulation (*Indrayanadi yog*) on Blood Glucose Level. *Int. Res. J. Biol. Sci.* **2013,** *2*(4), 67–71.

Chapagain, B.; Wiesman, Z. Variation in Diosgenin Level in Seed Kernels among Different Provenances of *Balanites aegyptiaca* and its Correlation with Oil Content. *Afr. J. Biotechnol.* **2005,** *4*, 1209–13.

Christensen, L. P.; Brandt, K. Bioactive Polyacetylenes in Food Plants of the Apiaceae Family: Occurrence, Bioactivity and Analysis. *J. Pharm. Biochem. Anal.* **2006**, *41*(3), 683–693.

Cousins, M. M. Development of In Vitro Protocols to Enhance Secondary Metabolite Production from Turmeric (*Curcuma longa* L.). M. S. Thesis, Clemson University, 2008, pp 109, Publication number: 1450686.

Croteau, R.; Kutchan, T. M.; Lewis, N. G. Natural Products Secondary Metabolites. In *Biochemistry and Molecular Biology of Plant*; Buchannan, B. B., Gruissem, W., Jones, R. L., Eds.; American Society of Plant Physiologists ; Rockville, MD, 2000; pp 1250–1318.

Das, M. P.; Selva Kumar, S. Preliminary Phytochemical Analysis of *Illicium verum* and *Wedelia chinensis*. *Int. J. PharmTech. Res.* **2013,** *5*(2), 24–29.

DíSouza, P.; Amit, A.; Saxena, V. S.; Bagchi, D.; Bagchi, M.; Stohs, S. J. Antioxidant Properties of Aller-7, a Novel Polyherbal Formulation for Allergic Rhinitis. *Drugs Exp. Clin. Res.* **2004,** *30*(3), 99–109.

Du, W. –X.; Avena-Bustillos, R. J.; Hua, S. S. T.; McHugh, T. H. Antimicrobial Volatile Essential Oils in Edible Films for Food Safety. *Science* **2011**, 1124.

Ehiabhi, O. S.; Edet, U. U.; Walke, T. M.; Schmidt, J.; Setzer, W. N.; Ogunwande, I. A.; Essien, E.; Ekundayo, O. Constituents of Essential Oils of *Apium graveolens* L., *Allium cepa*, and *Voacanga africana* Staph. from Nigeria. *J. Essential Oil Bear. Plants*, **2003,** *9*(2), 126–132.

Elumalai, A.; Dewick, P. M. *Medicinal Natural Products: A Biosynthetic Approach*, 2nd ed. John Wiley and Sons: Chichester, 2002.

Ertas, O. N.; Guler, T.; Cftc, M.; Dalklc, B.; Ylmaz, O. The Effect of a Dietary Supplement Coriander Seeds on the Fatty Acid Composition of Breast Muscle in Japanese Quail. *Rev. Méd. Vét.* **2005,** *156*, 514–518.

Eswariah, M. C. Herbalism—A Review. *Inter. J. Phytother.* **2012,** *2*(2), 96–105.

Farnsworth, N. R.; Loub, W. D. Workshop Proceedings. OTA-BP-F-23. U.S. Congress, Office of Technology Assessment: Washington, DC, 1983, 178–195.

Favarin, D. C.; de Oliveira, J. R.; de Oliveira, C. J. F.; de Paula Rogerio, A. Potential Effects of Medicinal Plants and Secondary Metabolites on Acute Lung Injury. *BioMed. Res. Int.* **2013,** *2013*, 1–12, Article ID 576479.

Ghasemzadeh, A.; Jaafar, H. Z. E. Effect of CO_2 Enrichment on Synthesis of Some Primary and Secondary Metabolites in Ginger (*Zingiber officinale* Roscoe). *Int. J. Mol. Sci.* **2012,** *12*, 1101–1114. doi:10.3390/ijms12021101.

Ghasemzadeh, A.; Jaafar, H. Z. E.; Rahmat, A.; Devarajan, T. Evaluation of Bioactive Compounds, Pharmaceutical Quality and Anticancer Activity of Curry Leaf (*Murraya koenigii* L.). *Evid. Based Complem. Altern. Med.* **2014,** *2014*, 1–8, Article ID 873803.

Ghasemzadeh, A.; Jaafar, H. Z. E.; Rahmat, A.; Wahab, P. E. M.; Halim, M. R. A. Effect of Different Light Intensities on Total Phenolics and Flavonoids Synthesis and Anti-oxidant Activities in Young Ginger Varieties (*Zingiber officinale* Roscoe). *Int. J. Mol. Sci.* **2010,** *11*, 3885–3897.

Govindrajan, V. S. Turmeric. Chemistry Technology and Quality. *Food Sci. Nutr.* **1980,** *13*, 199–301.

Gutierrez, J.; Barry-Ryan, C.; Bourke, P. The Antimicrobial Efficacy of Plant Essential Oil Combinations and Interactions with Food Ingredients. *Int. J Food Microbiol.* **2008,** *124*, 91.

Holley, R. A.; Patel, D. Improvement in Shelf-life and Safety of Perishable Foods by Plant Essential Oils and Smoke Antimicrobials. *Food Microbiol.* **2005,** *22*, 273.

Hounsome, N.; Hounsome, B.; Tomos, D.; Edwards-Jones, G. Plant Metabolites and Nutritional Quality of Vegetables. *J. Food Sci.* **2008,** *73*, R48–R65.

Janovska, D.; Kubikova, K.; Kokoska, L. Screening for Antimicrobial Activity of Some Medicinal Plant Species of Traditional Chinese Medicine. *Czech. J. Food Sci.* **2003,** *21*, 107–111.

Javaprakasha, G. K.; Jagan, L.; Roa, M.; Sakariah, K. K. Chemistry and Biological Activities of *C. longa*. Trends *Food Sci. Technol.* **2005,** *16*, 533–548.

Jeswal, P.; Kumar, D. Assessment of Co-occurrence of Aflatoxin and Ochratoxin A in Medicinal Herbs and Spices from Bihar State (India). *Curr. Res. Microbiol. Biotechnol.* **2015,** *3*(1), 586–592.

Karthikeyan, J., Rani, P. Enzymatic and Nonenzymatic Antioxidants in Selected *Piper* species. *Indian J. Exp. Biol.* **2003,** *41*(2), 135–140.

Karuppiah, P.; Rajaram, S. Antibacterial Effect of *Allium sativum* Cloves and *Zingiber officinale* Rhizomes against Clinical Pathogens. *Asian Pac. J. Tropl. Biomed.* **2012,** *2*, 597–601.

Khan, B. A.; Abraham, A.; Leelamma, S. Biochemical Response in Rats to the Addition of Curry leaf (*Murraya koenigii*) and Mustard Seeds (*Brassica juncea*) to the Diet. *Plant Foods Hum. Nutr.* **1996,** *49*, 295–299.

Kochhar, K. P. Dietary Spices in Health and Diseases (II). *Indian J. Physiol. Pharmacol.* **2008,** *52*(4), 327–354.

Krishnaswamy, K. Traditional Indian Spices and Their Health Significance. *Asia–Pac. J. Clin. Nutr*. **2008,** *17*, 265–268.

Kumar, C. R.; Sarwar, M.; Dimri, B. P. Bulgarian Coriander in India and its Future Prospects in Export Trade. *Indian Perfum.* **1977,** *21*, 146–150.

Lanzotti, V. The Analysis of Onion and Garlic. *J. Chromatogr. A* **2006,** *1112*, 3–22.

Lopes-Lutz, D.; Alviano, D. S.; Alviano, C. S.; Kolodziejczyk, P. P. Screening of Chemical Composition, Antimicrobial and Antioxidant Activities of Artemisia Essential Oils. *Phytochemistry* **2008,** *69*, 1732.

Mallavarapu, G. R.; Ramesh, S.; Syamasunday, K. V.; Chandrasekhara, R. S. Composition of Indian Curry Leaf Oil. *J. Essential Oil Res.* **1999,** *11*, 176–178.

Menon, A. N. The Aromatic Compounds of Pepper. *T. Med. Aromatic Plant Sci.* **2000,** *22*(2/3), 185–190.

Mishra, R. P.; Sharma, K. Antimicrobial Activities of *Syzigium aromaticum* L. (Clove). *Int. Res. J. Biol. Sci.* **2014,** *3*(8), 22–25.

Mohammad Pour, H.; Moghimipour, E.; Rasooli, I.; Fakoor, M. H.; Alipoor Astaneh, S.; Shehni Moosaie, S.; Jalili, Z. Chemical Composition and Antifungal Activity *of Cuminum cyminum* L. Essential Oil from Alborz Mountain Against *Aspergillus* Species. *Jundishapur J. Nat. Pharm. Prod.* **2012,** *7*(2), 50–55. http://dx.doi.org/10.5812/jjpharma.3445.

Mostajeran, A.; Gholaminejad, A.; Asghari, G. Salinity Alters Curcumin, Essential Oil and Chlorophyll of Turmeric (*Curcuma longa* L.). *Res. Pharm. Sci.* **2014,** *9*(1), 49–57.

Narasimhan, B.; Dhake, A. S. Antibacterial Principles from *Myristica fragrans* Seeds. *J. Med. Food* **2006,** *9*(3), 395–399.

Onayade, D. A.; Adebajo, A. C. Composition of the Leaf Volatile Oil of *Murraya koenigii* Growing in Nigeria. *J. Herbs, Spices Med. Plants* **2000,** *7*(4), 59–66.

Ononiwu, I. M.; Ibeneme, C. E.; Ebong, O. O. Effects of Piperine on Gastric Acid Secretion in Albino Rats. *Afr. Med. Sci.* **2002,** *31*(4), 293–295.

Padmashree, A.; Roopa, N.; Semwal, A. D.; Sharma, G. K.; Aganthian, G.; Bawa, A. S. Star-anise (*Illicium verum*) and Black Caraway (*Carum nigrum*) as Natural Antioxidants. *Food Chem.* **2007,** *104*(1), 59–66.

Pratibha, N.; Saxena, V. S.; Amit, A.; DíSouza, P.; Bagchi, M.; Bagchi, D. Anti-Inflammatory Activities of Aller-7, a Novel Polyherbal Formulation for Allergic Rhinitis. *Int. J. Tissue React.* **2004,** *26*(1–2), 43–51.

Proestos, C.; Boziaris, I. S.; Kapsokefalou, M.; Komaitis, M. Natural Antioxidant Constituents from Selected Aromatic Plants and Their Antimicrobial Activity against Selected Pathogenic Microorganisms. *Food Technol. Biotechnol.* **2008,** *46*, 151.

Purseglove, J. W.; Brown, E. G.; Green, G. L.; Robbins, S. R. J. *Spices*, vol. 2. Longman: New York, 1981, pp 447–531.

Rahimi, A. R.; Rokhzadi, A.; Sheno A.; Ezzat, K. Effect of Salicylic Acid and Methyl Jasmonate on Growth and Secondary Metabolites in *Cuminum cyminum* L. *J. Biodiversity Environ. Sci.* **2012,** *3*(12), 140–149.

Rathore, S. S.; Saxena, S. N. and Singh, B. Potential Health Benefits of Major Seed Spices. *Int. J. Seed Spices* **2013,** *3*(2), 1–12.

Rombaut, R.; Clercq, N. D.; Foubert, I.; Dewettinck, K. Triacylglycerol Analysis of Fats and Oils by Evaporative Light Scattering Detection. *J. Am. Oil Chem. Soc.* **2009,** *86*, 19–25.

Rozanida, A. R.; Nurul Izza, N.; Mohd Helme, M. H.; Zanariah, H.; Xanwhite, T. M. A Cosmeceutical Product from Species in the Family Zingiberaceae. Forest Research Institute: Selangor, Malaysia, 2005, pp 31–36.

Saxena, A.; Cramer, C. S. Metabolomics: A Potential Tool for Breeding Nutraceutical Vegetables. *Adv. Crop Sci. Technol.* **2013,** *1*, 106. doi:10.4172/ 2329-8863.1000106.

Saxena, M.; Saxena, J.; Nema, R.; Singh, D.; Gupta, A. Phytochemistry of Medicinal Plants. *J. Pharm. Phytochem.* **2013,** *1*(6), 168–182.

Scognamiglio, M.; D'Abrosca, B.; Esposito, A.; Fiorentino, A. Chemical Composition and Seasonality of Aromatic Mediterranean Plant Species by NMR-Based Metabolomics. *J. Anal. Methods Chem.* **2015,** *2015*, 1–9, Article ID 258570.

Selvan, M. T. Role of Spices in Medicine. *Indian J. Arecanut Spices Med. Plants* **2003,** *5*, 129–133.

Sharopov, F. S.; Zhang, H.; Wink, M.; Setzer, W. N. Aromatic Medicinal Plants from Tajikistan (Central Asia). *Medicines* **2015,** *2*, 28–46. doi:10.3390/medicines2010028.

Sivakumar, Ch. V.; Meera, I. Antioxidant and Biological Activities of Three Morphotypes of *Murraya koenigii* L. from Uttarakhand. *J. Food Process Technol.* **2013,** *4*(7), 1–7.

Sowbhagya, H. B. Chemistry, Technology, and Nutraceutical Functions of Celery (*Apium graveolens* L.): An Overview. *Crit. Rev. Food Sci.* **2014**, *54*, 389–398.

Srinivasan, M. R.; Satyanarayana, M. N. Influence of Capsaicin, Curcumin and Ferulic Acid in Rats Fed High Fat Diets. *J. Biol. Sci.* **1987,** *12*, 1943.

Tapsell, L. C.; Hemphill, I.; Cobiac, L.; Sullivan, D. R.; Clifton, P. M.; Williams, P. G.; Fazio, V. A.; Inge, K. E.; Fenech, M.; Patch, C. S.; Roodenrys, S.; Keogh, J. B. The Health Benefit of Herbs and Spices: The Past, the Present and the Future. *Med. J. Aust. Suppl.* **2006,** *185*(4), S1–S24.

Tham, D. M.; Gardner, C. D.; Haskell, W. L. Potential Health Benefits of Dietary Phytoestrogens: A Review of the Clinical, Epidemiological, and Mechanistic Evidence. *J. Clin. Endocrinol. Metab.* **1998,** *83*(7), 2223–2235.

Thampi, N.; Shalini, J. V. In vitro Antibacterial Studies of Fresh Green Onion Leaves and the Characterization of its Bioactive Compounds Using Fourier Transform Infrared Spectroscopy (FTIR). *J. Chem. Pharm. Res.* **2015,** *7*(3), 1757–1766.

Thomas, R. A.; Krishnakumari, S. Phytochemical Quantification of Primary and Secondary Metabolites of *Myristica fragrans* (H.) Ethanolic Seed Extract. *Int. J. Pharm. Biol. Sci.* **2015,** *6*(1), (B) 1046–1053.

Van Wyk, B. E. *Food Plants of the World—Identification, Culinary Uses and Nutritional Value*. Briza Publications: Pretoria, South Africa, 2005, p 48.

Van Wyk, B. E.; Wink, M. *Medicinal Plants of the World*. Briza Publications: Pretoria, South Africa. 2005, pp 38–349.

Vijayakumar, R. S.; Surya, D.; Nalini, N. Antioxidant Efficacy of Black Pepper (*Piper nigrum* L.) and Piperine in Rats with High Fat Diet Induced Oxidative Stress. *Redox. Rep.* **2004,** *9*(2), 105–110.

Wahyuni, Y.; Ballester, A. R.; Sudarmonowati, E.; Bino, R. J.; Bovy, A. G. Secondary Metabolites of *Capsicum* Species and Their Importance in the Human Diet. *J. Nat. Prod.* **2013,** *76*(4), 783–793.

Wang, L. The Antibacterial Effect of Secondary Compounds in Plants. Paper Presented to 50th Annual Southeastern Michigan Junior Science and Humanities Symposium, March 6–7, 2014. McGregor Memorial Conference Center, Wayne State University, Detroit, MI, 2014, pp 1–28.

Wen, Y. P. Secondary Metabolites from Pepper (*Piper nigrum*) and *Tahitian noni* (*Morinda citrifolia*) and Their Biological Activities. Masters Thesis, Universiti Putra Malaysia, 2009.

Wohlmuth, H.; Smith, M. K.; Brooks, l. D.; Myers, S. P.; Leach, D. N. Essential Oil Composition of Diploid and Tetraploid Dones of Ginger (*Zingiber officinale* Roscoe) Grown in Australia. *J. Agric. Food Chem.* **2006,** *54*(4), 1414–1419.

Wu, H.-M.; Li, H.-T.; Chen, H.-L.; Chen, C.-Y.; Juan, S.-W.; Huang, J.-C.; Kuo, C. N. Secondary Metabolites from the Stems of *Capsicum annuum* var. *longum*. *Chem. Nat. Compd.* **2013**, *49*(4), 765–766.

CHAPTER 5

BACTERIAL CELLULOSE AS SECONDARY METABOLITE: PRODUCTION, PROCESSING, AND APPLICATIONS

IDAYU MUHAMAD IDA[1,2*], PA'E NORHAYATI[1], and AZLY ZAHAN KHAIRUL[1,3]

[1]*Bioprocess Engineering Department, Faculty of Chemical Engineering, Universiti Teknologi Malaysia, 81310 Johor Bahru, Johor, Malaysia*

[2]*IJN-UTM Cardio Engineering Centre, V01 FBME, Universiti Teknologi Malaysia, 81310 Johor Bahru, Johor, Malaysia*

[3]*Section of Bioengineering Technology, Malaysian Institute of Chemical and Bioengineering Technology, Universiti Kuala Lumpur, Lot 1988, Bandar Vendor Taboh Naning, 78000 Alor Gajah, Melaka, Malaysia*

**Corresponding author, E-mail: idaidayu@.utm.my*

CONTENTS

ABSTRACT

Natural polymer, such as cellulose, has attracted much attention due to the effect of environmental pollution of synthetic polymer. Cellulose and its derivatives such as cellulose acetate were known capable of being used as one of biomaterial sources. The development of biomaterial-based film started to gain attention as it has high potential application in many fields such as bioseparation, tissue engineering, and food processing. Previously, plant-derived cellulose was widely used. However, the use of trees for the production of paper and construction materials has continuously depleted forest resources leading to high need of cellulose alternative. In recent times, bacterial cellulose started to be used because of its excellent and promising properties.

5.1 INTRODUCTION

Natural polymer, such as cellulose, has attracted much attention due to the effect of environmental pollution of synthetic polymer (Pei et al., 2013). Cellulose and its derivatives such as cellulose acetate were known capable of being used as one of biomaterial sources. The development of biomaterial-based film started to gain attention as it has high potential application in many field such as bioseparation, tissue engineering, and food processing (Liang et al., 2007). Previously, plant-derived cellulose was widely used. However, the use of trees for the production of paper and construction materials has continuously depleted forest resources leading to high need of cellulose alternative. In recent times, bacterial cellulose started to be used because of its excellent and promising properties (Ashori et al., 2012).

Cellulose from bacteria has more advantages than the cellulose found in plants. These advantages provide plenty of room for the use of bacterial cellulose in various fields. The most significant advantage of bacterial cellulose compared to plant cellulose is its purity. Bacterial cellulose does not have lignin to be removed as in plant cellulose. It is highly hydrophilic with high mechanical strength. As bacterial cellulose is 100% pure and produced in hydrophilic matrix forms, its extensive fibrils and high mechanical strength will be maintained throughout the formation. These properties open the opportunities in many fields such as paper, medical, and audio industries. These were proven by several researchers that successfully produced bacterial cellulose composites membranes such as cellulose acetate membranes reinforced with bacterial cellulose sheet (Gindl & Keckes, 2004). Sokolnicki et al. (2006) suggested novel uses for bacterial cellulose thus may include membrane systems for tissue growth, cell-based therapies, and drug delivery.

Bacterial cellulose can be produced using different methods and substrates, whereby the properties can be modified based on the application. Soluble or insoluble particles in the media, such as living cells and chemicals can readily be incorporated in the growing matrices (Serafica et al., 2002). Additional other materials that are not needed in the growth of bacteria in the fermentation medium are proven to affect the production of cellulose (Ruka et al., 2013). Several researchers have shown the ability to alter the properties of the cellulose by adding certain substrates or by manipulating the operating conditions. For example, adding up magnetite nanoparticles and polyaniline enhanced the thermal stability of bacterial cellulose (Park et al., 2013).

This unique ability of bacterial cellulose allowed a series of potential applications as a new novel bacterial cellulose composite can be produced with modified properties based on its function. These applications range from high mechanical strength of hydrogel with addition of genipin (Nakayama et al., 2004), antimicrobial film with addition of chitosan (Phisalaphong & Jatupaiboon, 2008), incorporation of aloe vera as wound dressing (Saibuatong & Phisalaphong, 2010), and electromagnetic nanocomposite with incorporation of magnetite nanoparticles and polyaniline (Park et al., 2013).

5.2 CELLULOSE

Cellulose is a polysaccharide that is built up from 100 to more than 10,000 linear chains of $\beta(1 \rightarrow 4)$ linked D-glucose units. It is the most common organic compound on earth with formula $(C_6H_{10}O_5)_n$. Cellulose can be found as a structural component of green plants as 33% of plant material is cellulose. Besides, it also can be found in some algae and also in oomycetes. More than that, some species of bacteria were found with ability to produce cellulose as a film. This compound is unscented and has no flavor. It is hydrophilic and cannot be dissolved in water like most organic solvents. Cellulose is also reported as chiral and biodegradable (Nishiyama et al., 2002).

Cellulose is derived from D-glucose units, which condense through β(1→4)-glycosidic bonds. This linkage motif contrasts with that for α(1→4)-glycosidic bond present in starch, glycogen, and other carbohydrates. Cellulose is different with starch as it has a straight-chain polymer with no coiling taking place as in starch. The molecule lay on the configuration of longer and stiff rod-like structure. Hydrogen bonds are formed from binding of multiple hydroxyl groups on glucose and oxygen molecules from different chains. These bonds hold the chains together resulting into microfibrils with high tensile strength. Such strength is indispensable in cell walls, where they are crushed into a carbohydrate matrix, which provides rigidity to the plant cell. Plant-derived cellulose is usually contaminated with hemicellulose, lignin, pectin, and other substances, while bacterial cellulose is quite pure, has much higher water content, and consists of long chains.

5.3 BACTERIAL CELLULOSE

Bacterial cellulose is another source of cellulose other than plant cellulose. It is produced by bacteria from many genera such as *Acetobacter*, *Achromobacter*, *Agrobacterium*, and *Sarcina*. However, from that many list of cellulose producers, the only genus that is well known with its capability to produce cellulose in large quantities is *Acetobacter*. While, within that species, *Acetobacter xylinus* or *Acetobacter xylinum* is the one that is extensively being used in research and studies (Jonas & Farah, 1997).

A. xylinum produced cellulose from a glucan chains. The chains extrude into the fermentation medium from *A. xylinum* pores. These processes repeated until a bundle of microfibrils gathered and form bacterial cellulose. Medium for bacterial cellulose fermentation can be from any kind of carbon sources or sugars. *Acetobacter* needs oxygen in order to produce bacterial cellulose. Therefore, the production will occur mostly at surface of the liquid.

The bacterial cellulose pellicle is pure and extremely hydrophilic. Therefore, it needs no treatment which makes its original high mechanical strength retain. These unique characteristics of bacterial cellulose open many rooms for new applications as the properties of bacterial cellulose can be changed by manipulating the fermentation process.

Bacterial cellulose has many advantages compared to plant cellulose. The most significant advantage is its purity. Bacterial cellulose has no hemicelluloses or lignin as in plant cellulose. Besides, its structure is stronger and finer compared to plant cellulose. This is due to long fiber length in bacterial cellulose. Bacterial cellulose grows following the surface of the vessel. Therefore, it can be grown to almost any desired shape. Besides, it can be produced with variety of carbon sources as the fermentation substrate.

Figure 5.1 shows a comparison between bacterial cellulose and plant cellulose. From the FESEM image, it can be seen that bacterial cellulose has a smooth and finer network of cellulose fibrils while plant cellulose also consist similar fibrils, but larger compared to bacterial cellulose.

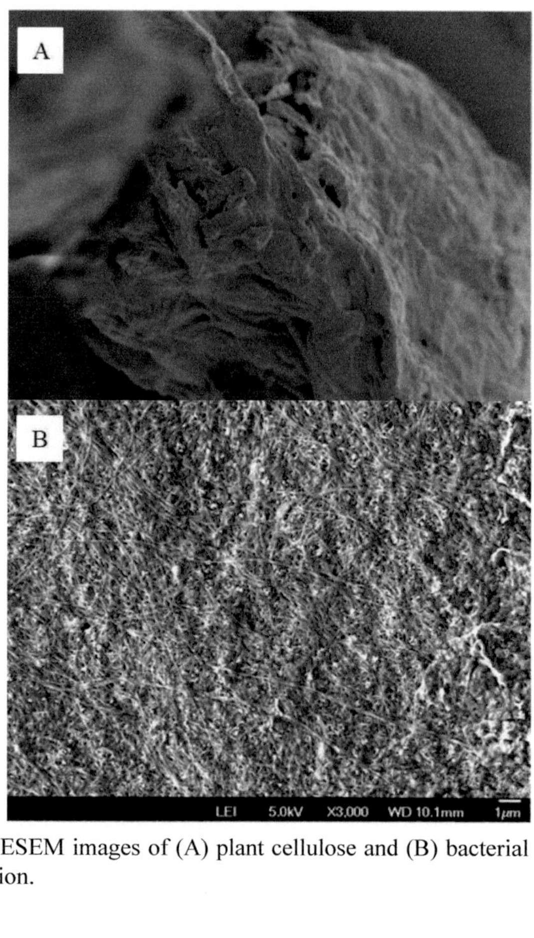

FIGURE 5.1 FESEM images of (A) plant cellulose and (B) bacterial cellulose at 3000 times magnification.

5.3.1 BACTERIAL CELLULOSE CONTENT

Generally, bacterial cellulose contains high fiber content, low calories, and no cholesterol content. Previous research found that film produced by *Acetobacter aceti* subsp. *xylinum* contains water and cellulose as main component in bacterial cellulose-*nata* (Yeoh et al., 1985). Yeoh et al. (1985) also mention that bacterial cellulose consist of 95.4% of moisture, 4.6% of total solid, 2.2% of total sugar and 1.1% of rough fiber. In certain cases, drying is very important in order to produce dried bacterial cellulose. Wei et al. (2011) reported that the dried *nata* was preferred compared with wet form since it is more convenient and portable for practical application. It was proved that different methods of drying will affect the properties of the bacterial cellulose (Pa'e et al., 2014).

5.3.2 FEATURES OF BACTERIAL CELLULOSE

Bacterial cellulose synthesized by *A. xylinum* has many advantages compared to those from plant cellulose. Cellulose from trees consists of lignin and hemicelluloses. Therefore, many costly stages of pulping process were involved in removing those compounds in order to produce pure cellulose especially in paper manufacturing. Bacterial cellulose is pure as it does not have lignin and hemicelluloses as its components. This makes the processing easier with fewer step and lower cost compared to plant cellulose processing with less waste generated.

Bacterial cellulose is very hydrophilic material with high tensile strength and excellent shape. This is because the cellulose fibers remain intact because of the simpler processing which allowed that the cellulose remains stronger and retains its attractive properties. Brown (1991) reported the ability of bacterial cellulose to absorb water from 60 to 700 times of its original weight without any treatment. This is a big advantage as plant cellulose such as wood or cotton must be physically disintegrated in order to increase their hydrophilic properties.

Bacterial cellulose can be produced using many types of substrates and variety types of strains. For industrial application, this happens to be important as cheaper substrates will decrease the production costs. The properties of bacterial cellulose can be changed simply by manipulating the fermentation process. This includes addition of other chemicals during or before fermentation and changing the fermentation condition. For example, the water-holding capacity of bacterial cellulose can be increased by adding carboxymethyl cellulose (CMC) during the fermentation.

In static fermentation, cellulose forms at the air/liquid interface. Therefore, the shape of the interface can be modified to determine the shape of the pellicle. White and Brown (1989) were able to form a seamless cellulose glove using gas permeable mold submerged in an *A. xylinum* culture. This was done in situ with any molds with desired shape as the cells will gather at the interface of the mold and secrete long fibrils to form cellulose layers.

Bacterial cellulose is biopolymer with special features. It can be produced from many substrates, and its properties can be altered and it can be made to just about any shape or size. Those unique properties open the doors for new applications especially in medical, waste water, and paper industries.

5.3.3 BACTERIAL CELLULOSE SYNTHESIS

Cellulose is the main component of plant cell wall. Cellulose that produced by bacteria called biocellulose or bacterial cellulose. Although both bacterial and plant cellulose have same chemical structure their properties are different to one another. For example, the diameter of bacterial cellulose is smaller compared to plant cellulose while its mechanical strength is higher.

Bacterial cellulose is produced by *A. xylinum* which is an acetic acid-producing bacterium. It has superior properties such as ultra-fine network structure, high biodegradability, and unique mechanical strength as compare to plant cellulose (Tsuchida & Yoshinaga, 1997). It is expected to be an innovative biodegradable biopolymer.

During fermentation, only at the upper surface of cellulose does *A. xylinum* exist. The cellulose synthesis will continued until a limited condition such as when not enough carbon sources or when the bacterial cellulose fills the discs if using rotary discs reactor (RDR) (Pa'e, 2009). According to Lee et al. (2002), cellulose synthesized by the bacterium *A. xylinum* and plant cellulose are highly crystalline. However, the arrangements of glucosyl units of the crystallites make them differ to each other.

The cellulose fibrils synthesized from pores at the cell wall gathered to form cellulose ribbon. This will happen at surface of fermentation medium as *A. xylinum* needs oxygen to convert glucose to cellulose.

Most cellulose-producing acetic acid bacteria can convert glucose to gluconic acid and ketogluconic acid. The enzyme responsible for the conversion of glucose to gluconic acid is membrane-bound glucose dehydrogenase (Hwang et al., 1999). This will simply remove glucose from the fermentation medium and avoid the formation of cellulose (Toru et al., 2005). Other than that, the cellulose biosynthesis rate by *A. xylinum* is much faster than the synthesis of plant cellulose.

5.4 BACTERIA *ACETOBATER XYLINUM*

Acetobacter is a genus of acetic acid bacteria which has an ability to convert alcohol (ethanol) to acetic acid in the presence of oxygen. *A. aceti* has four subspecies which are *aceti*, *orleanensis*, *xylinum*, and *liquafaciens*. *A. aceti* is a Gram-negative bacterium, motile by peritrichous flagella, obligate aerobic, does not form any endospores and exist in soil, water, flowers, fruits, and honey bees where sugar fermentation is occurring. In these liquids, acetic acid bacteria grow as a surface film due to their aerobic nature and active motility. Some acetic acid bacteria like *A. xylinum* can synthesize cellulose which is normally only done by plants (Verschuren et al., 1999). *Acetobacter* is very important in the commercial industry such as in production of vinegar and production of bacterial cellulose.

A. xylinum is a Gram-negative and aerobic bacterium that secretes during its normal metabolic activity. It is unique in its prolific synthesis of cellulose; typically a single cell can convert up to 10^8 of glucose molecules per hour into cellulose. This bacterium has an ability to produce multiple poly-β-1,4-glucan chain which is chemically identical with cellulose. Multiple cellulose chain is synthesized at the bacterium surface at sites external to the cell membrane. Their size is 0.6–0.8 μm by 1.0–1.4 μm and usually exists in single, double, or strain form. These bacteria are obligate aerobic bacteria and need enough oxygen to form cellulose membrane and for life (Verschuren et al., 1999). They grow and live on fruits gist, cereal, herbal tea, and spoiling vegetables. Its non-photosynthetic organism can produce glucose, sugar, glycerol, or other organic substrates and convert them into cellulose. *A. xylinum* taxonomy is shown below (Deinema & Zevenhuizen, 1971; Chawla et al., 2008; Yamada & Yukphan, 2008):

Family (group 4)	:	*Acetobacter aceace*
Genus	:	*Acetobacter*
Species	:	*Acetobacter aceti*
Subspecies	:	*Acetobacter xylinum*

A. xylinum is the most common bacteria that has been used as model microorganism for fundamental and advanced studies on the production of cellulose. This is because of its ability to produce relatively high amounts of polymer. It also can grow and produce cellulose from a wide range of substrates. The cellulose produced by *A. xylinum* is in the form of interwoven extracellular ribbons as part of primary metabolite. The microorganism serves as important future source of cellulose in textile, paper, and lumber industries and provided its fermentation can be effectively scaled up.

5.4.1 OPTIMUM CONDITION FOR GROWTH OF ACETOBACTER XYLINUM

A. xylinum is found in a specific environment at which they can grow. It can be influenced by a variety of physical and biochemical factors. Physical factors include pH, temperature, oxygen concentration, moisture, pressure (hydrostatic and osmotic), and radiation. While biochemical factors include availability of carbon, nitrogen, sulfur, phosphorus, traces of elements, and vitamins.

5.4.1.1 pH

Optimum pH is the pH where it can give a best condition for the growth rate of bacteria. The pH scale measures hydrogen ion (H^+) concentration. This hydrogen concentration gives an effect to the enzyme activity, thus influencing the microbe growth. High pHs correspond to low concentration of H^+ while low pHs correspond to high concentration of H^+ and neutral pH is the condition where number of H^+ and OH^- (hydroxyl ions) is equal. *A. xylinum* is an acidophilic bacterium. It is capable to live at pH as low as 3.5 (Coban & Biyik, 2011). Its optimum pH is 5.4 to 6.3 (Sumate et al., 2005).

5.4.1.2 TEMPERATURE

Optimum temperature is the temperature at which the growth rate of the microbe is the best. The optimum temperature for *A. xylinum* growth is between 25 and 30°C. Typically, organisms whose optimum growth temperature is between 20 and 40°C are classified as mesophiles. Therefore, *A. xylinum* is considered as a mesophile bacterium. At temperature 37°C, these bacteria failed to multiply even in an optimal medium. It is because, at a high temperature, the cell component of the microbe such as nucleic acid and protein will be denatured. Therefore, temperature is one of the important factors and it can give a large impact for the microbial growth (Krystynowicz et al., 2002).

5.4.1.3 OXYGEN CONCENTRATION

A. xylinum is an obligate aerobic bacterium that has an oxygen-based metabolism. Thus, it requires oxygen for aerobic respiration to oxidize substrates such as glucose to obtain energy and convert the glucose into cellulose. The bacterium cell can obtain oxygen at the air–liquid interface where the cellulose is produced. The oxygen supply is considered as the limiting factor for their growth and cellulose formation. The oxygen concentration has a limitation, too much dissolved oxygen inside the fermentation medium will increase the level of gluconic acid and too low oxygen concentration could reduce the production of bacterial cellulose due to the insufficient oxygen source for the culture to grow (Krystynowicz et al., 2005).

5.4.1.4 BIOCHEMICAL FACTOR

Every organism must find an environment with all the substances required for energy generation and cellular biosynthesis. The chemicals and elements are utilized for nutrients to grow, repair themselves, and to replicate. The elements are carbon, nitrogen, sulfur, phosphorus, and various trace elements as listed in Table 5.1. *A. xylinum* growth needs some major elements which are found in terms of water, inorganic ions, small molecule, and macromolecule which serve either a structural or functional role in the cells (Chawla et al., 2008).

TABLE 5.1 Major Elements and Their Functions in Bacterial Cellulose Production.

Major elements	Functions
Carbon	• Main constituent of cellular material • It is the most important element • From the source of organic compound, carbon dioxide, sugars, etc.
Nitrogen	• Constituent of amino acid, nucleic acid, nucleotides, and coenzyme • From the source of NH_3, N_2, NO_3, and organic compound
Phosphorus	• Constituent of nucleic acid, phospholipids, and nucleotides. • Exists in inorganic phosphate (PO_4)
Sulfur	• Constituent of cysteine, methionine, glutathione, and several coenzymes • Exists as the growth activation for bacteria • From the source of SO_4, H_2S, S, and organic sulfur compounds
Potassium	• Main cellular inorganic cation and cofactor for certain enzymes • Exists in potassium salt
Magnesium	• Inorganic cellular cation • Cofactor for certain enzymatic reactions • Exists in magnesium salt
Calcium	• Inorganic cellular cation • Cofactor for certain enzymes • A component of endospores • From sources of calcium salt
Iron	• Component of cytochromes and certain non-heme iron-proteins • Cofactor for some enzymatic reactions • From the sources of iron salt

5.5 PRODUCTION OF BACTERIAL CELLULOSE UNDER DIFFERENT CONDITIONS

The production of bacterial cellulose by *A. xylinum* can be done under a variety conditions. Each condition comes with its own advantages and disadvantages. A summary of methods that had been used for production of bacterial cellulose is simplified in Table 5.2.

TABLE 5.2 Production of Bacterial Cellulose Using Different Methods.

Methods	Explanations	Advantages	Disadvantages
Static culture	Using culture medium tray (mold about 5–7 cm) which remains motionless	The simplest and easiest way to produce bacterial cellulose	Only the uppermost layer of film grows (Vandamme et al., 1997)
	Production occurs at air–liquid medium interface, leaving the cellulose at the liquid surface once it is produced.	Produced pellicles cellulose which consists of layer of cellulose fibers	The cells below the surface will die because of lack of oxygen supply
	Doubling time between 8 and 10 h	No high technology equipment used thus reduced production cost	The culture also cannot be disturbed until harvest time once it is inoculated
	All ingredients are mixed together at the early stage	The most well-known method to produce Nata	Not applicable for large-scale production
Shaken culture (fermentation)	Reciprocal shaking at about 90–100 rpm When culture is shaking, cell can grow more rapidly	By increasing the cell concentration, cellulose production will also increase (Ishikawa et al., 1995)	Cellulose fibrils do not form well-organized pellicles
	Doubling time is 2 h Addition of micro particles (glass bead) can increase the level of dissolved oxygen in the medium which can increase cellulose fermentation (Pa'e et al.,2007)	Amount of cellulose produced is associated with increasing the cell growth Applicable for large-scale production	
Agitated culture with swirling motion and aerated system	Czaja et al. (2004) stated that cell in agitated medium is able to produce bacterial cellulose It will form stellate gel bodies which comprise cellulose and cells Doubling time is about 4–6 h	Overcome many limitations in static culture including diffusion, controllability, and scale-up	Product is not in pellicle form but is formed as reticulated cellulose slurry Problem with culture instability which demonstrate by loss of ability to make cellulose
RDR	New alternative using concept of Rotating Biological Contactor (RBC) Alternately soaks the organisms in nutrient medium and exposes them to air (Norhayati et al., 2011)	High production yield Less labor Easy to scale-up	Produces bacterial cellulose with high water content if there is no drier at the reactor

5.5.1 STATIC PRODUCTION

Bacterial cellulose produced by growing *A. xylinum* in a culture medium tray which remains motionless is known as static production. Production of bacterial cellulose occurs as every single cell will constantly create thread-like fibrils at the air–liquid interface of the medium. When a new cellulose layer is formed at the surface, others will be slightly pushed downward resulting in to pellicles with many cellulose layers at the end of fermentation.

During growth, the cellulose propagates from the surface of the culture once the film has formed. So the uppermost layer of the film is the only one growing (Vandamme et al., 1997), and the cells that are left further into the film become inactive or die from lack of oxygen. It is important to know that nutrients diffuse up to the cells, and oxygen must diffuse down. Figure 5.2 shows production of bacterial cellulose in static condition and only top of the cellulose known as active layer is growing because it has access to oxygen.

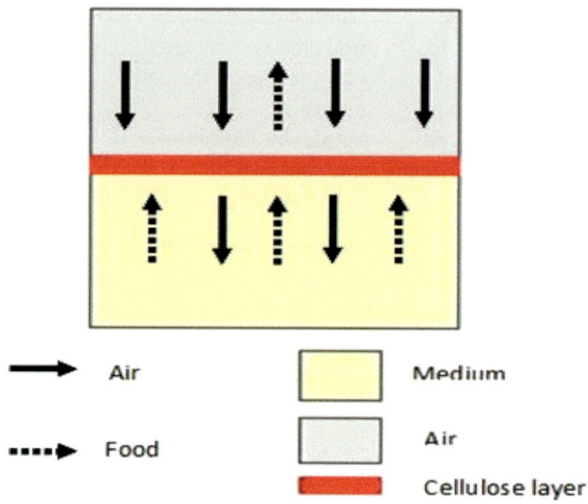

FIGURE 5.2 Cellulose growing in static culture.

Although static fermentation is simple and does not need high technology, there are many disadvantages. Apart from diffusion problem, static fermentation has problem in monitoring the fermentation condition. With this conventional method, once the trays were inoculated, it cannot be disturbed until harvest time. Therefore, controlling and maintaining fermentation condition is nearly impossible in static fermentation. This will not encourage rapid growth of bacteria as the medium condition is not an optimum level.

There is also another big problem when dealing with static fermentation which is scaled up. In order to increase the production, more trays need to be added. It is noneconomic for commercialization. Figure 5.3 shows the bacterial cellulose formed in a tray using static fermentation.

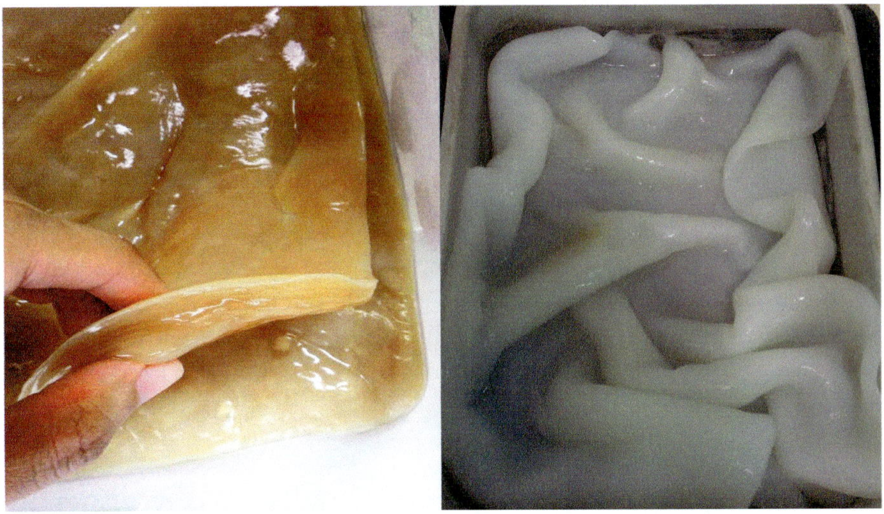

FIGURE 5.3 Bacterial cellulose formed during static fermentation process (before and after treatment).

5.5.2 AGITATED PRODUCTION

When the cells are grown in a condition where the culture medium is agitated with swirling motion, it will form gel bodies which are composed of cellulose and cells. It is stated by Czaja et al. (2004) that bacterial cellulose is able to be produced in an agitated medium. Many of the limitations in static cultures can be overcome through switching to continuous stirrer tank reactor (CSTR) production, including diffusion, controllability, and scale-up. Unfortunately, the product of this new production (agitated medium) is not in pellicle form, but rather is formed as reticulated cellulose slurry as shown in Figure 5.4.

FIGURE 5.4 Cellulose formed in agitated culture fermentation.

The production of cellulose in continuously agitated cultures comes with numerous problems; the most difficult is culture instability. This instability can be confirmed by the failure of the bacteria to produce cellulose and the increase of cellulose non-producer cells (Cel^-) types in the culture.

5.5.3 BACTERIAL CELLULOSE PRODUCTION USING REACTOR

Conventionally, bacterial cellulose has been produced in static and agitated culture method. Static fermentation requires more labor and longer fermentation time which leads to low production yields with high production costs (Norhayati et al., 2011). Meanwhile, shaken or agitated culture method causes mutation to convert cellulose-producing bacteria to cellulose-negative cells. Those mutant cells grow rapidly and become dominant compared to cellulose producers, thus reducing the cellulose production. In airlift reactors and stirred tank, bacterial cellulose production was decreased due to the adhesion of the culture broth to the reactor wall and the upper part of the equipment. A few different methods have been used for production of bacterial cellulose as reported by a few researchers (Prashant et al., 2009).

5.5.3.1 BIOREACTOR WITH SPIN FILTER

Jung et al. (2007) developed and used a fermentation method using a bioreactor equipped with spin filter. Fermentation was done in a bioreactor with six flat-blade turbine impeller and a spin filter which consisted of a cylinder surrounded by stainless steel mesh, attached to the agitator shaft at the bottom. Fermentation had been done using *Gluconacetobacter hansenii*. In periodical perfusion culture, 4.57 g/L cellulose has been produced after 140 h of fermentation, which was 2.9 times higher than that obtained in conventional jar fermentation.

5.5.3.2 REACTOR WITH SILICONE MEMBRANE

In static culture, Yoshino et al. (1996) developed a fermentation system in which the cellulose was formed on a liquid surface and on the oxygen-permeable synthetic membrane in order to increase the production of cellulose using *Acetobacter pasteurianus* AP-1SK. In cylindrical vessels with the bottom ends covered with 100-mm thick silicone sheet, the production rate was doubled. Other than that, a silicone air bag was also used to produce bacterial cellulose. The degree of roughness of silicone membrane surface was strongly affecting the production rate of bacterial cellulose. The production rate on silicone membrane with glossy surface was five times higher than the silicone membrane with embossed surface.

5.5.3.3 ROTARY DISCS REACTOR

RDR is a new method in the production of bacterial cellulose in order to give better aeration for *A. xylinum* so that it can produce higher yield of bacterial cellulose. Figure 5.5 shows bacterial cellulose produced in RDR. RDR consists of a series of discs that are mounted to shaft. The shaft is connected to driven motor so that it can rotate the shaft with the discs. The discs a placed in a horizontal trough that contains a biological medium in which at least a portion of the discs is submerged. The discs then will alternately soak the organisms in nutrient medium and expose them to the air (Norhayati et al., 2011). Table 5.3 shows the important components of RDR which are used in the production of bacterial cellulose.

TABLE 5.3 Components of RDR.

Components	Explanations
Disc	Disc must have appropriate mesh size to allow the growth of bacteria and attachment of bacterial cellulose produced
	The radius must be at optimum length in order to maximize its surface area for the development of cellulose
	The gap between the discs must be as small as possible to maximize the available surface area in the trough
	About 30–40% of the discs surface will be submerged in the culture medium. Figure 5.6 shows the submergence area and active surface of the discs
	Cellulose will form at both sides of the disc
Trough	Need to hold biological medium used in fermentation
	Sufficient length and width is important to ensure that at least half of the discs surface can be submerged into the medium during fermentation
	The body is made from polypropylene as durable and clear material has been used for easy supervision
Driven motor and shaft	The function of the shaft is to hold the discs and it is located at the center of the discs
	Driven motor will be connected with the shaft to give power for the shaft to rotate
	The rotational speed usually depends on the diameter of the discs used

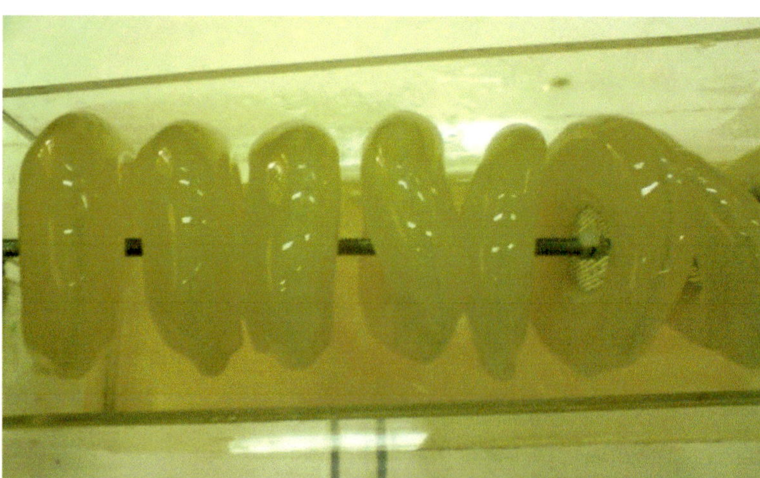

FIGURE 5.5 Wet bacterial cellulose produced after 4 days incubation in RDR.

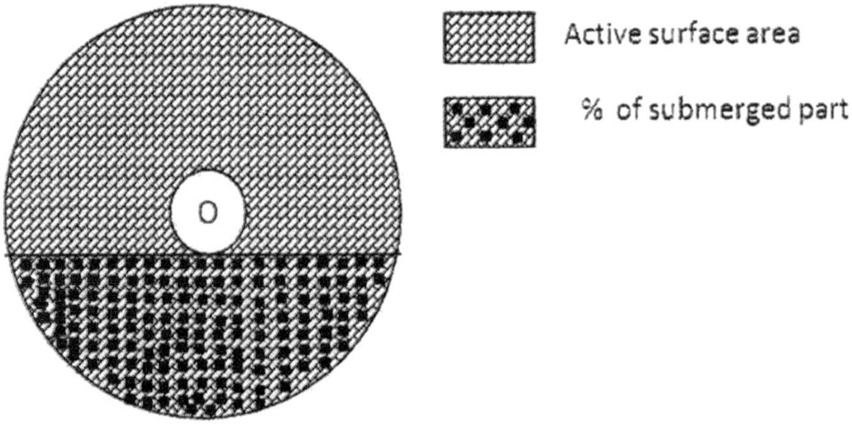

FIGURE 5.6 Active surface area and percentage of submerged area for discs.

The optimum conditions for production of bacterial cellulose in this reactor were investigated by Krystynowicz et al. (2002). During the investigation, he changed the medium volume, rotation speed, and the number of discs. Finally, he stated that the optimum production of bacterial cellulose had been obtained when the ratio of surface area to the medium volume (*S*/*V*) was 0.71–1.00 cm and the rotational speed is 4 rpm. Kim et al. (2007) applied RDR for fermentation of bacterial cellulose using *Gluconace bacter* sp. He stated that, the highest production of cellulose was with eight discs at 5.52 g/L. They also determined that, when the aeration rate was maintained at 1.25 vvm, 5.67 g/L cellulose was produced. Finally, the optimum rotational speed of the discs was 15 rpm.

5.6 FACTORS AFFECTING PRODUCTION OF BACTERIAL CELLULOSE

Critical factors that affect the production of bacterial cellulose by *A. xylinum* can be divided into two major groups (Zahan et al., 2013). First, the component of the fermentation (type/strain of organism used, the medium composition such as carbon source us as substrate and nitrogen source as element used in the medium) and second is the operating conditions (example pH, oxygen, temperature, relative concentration of substrate, and type of culture design used) (Zhang & Greasham, 1999).

Table 5.4 shows some major elements and their physiological function in bacterial cellulose production.

5.6.1 CARBON SOURCE

Carbon source is the main nutrient that provides *A. xylinum* a suitable fermentation condition. *A. xylinum* can produce bacterial cellulose from a variety of inexpensive carbon substrate such as glucose, sucrose, and fructose. However, most of the cellulose synthesis has been done on glucose culture medium only. Keshk and Sameshima (2005) have come out with an overview of various carbon substrates for their efficiency in the production of bacterial cellulose. From their research, only glucose, fructose, inositol, and glycerol give the highest production yield. But glycerol is the best substrate for cellulose production with the efficiency of 28.7% on the added substrate weight and without the sharp drop of pH during incubation (Keshk & Sameshima, 2005). Different carbon sources such as monosaccharide, oligosaccharides, alcohol sugar, and organic acids were also used to develop the bacterial cellulose (Jonas & Farah, 1997). Glucose was recommended as the ideal carbon source for bacterial cellulose production (Mahadevaswamy & Anu, 2011; Son et al., 2001; Masaoka et al., 1993) where the fermentation is most excellently done at low initial glucose as it leads to higher cellulose production (Zahan et al., 2014; Krystynowicz et al., 2002; Son et al., 2001).

5.6.2 NITROGEN SOURCE

Budhiono (1999) reported nitrogen source is also one of the macronutrients that was used by bacteria *A. xylinum* in the fermentation process. Organic sources like yeast and inorganic source like ammonium sulfate provide nitrogen. Normally, ammonium sulfate and ammonium phosphate are used as nitrogen source for bacterial cellulose fermentation. But ammonium phosphate is more effective than ammonium sulfate.

Cellulose production is favored by the use of organic source with combination of peptone and yeast extract which will give highest bacterial cellulose. Jonas and Farah (1997) stated that the medium with added yeast extract and peptone 0.5%, respectively, will produce twice the cellulose yielding. A few amino acids are always cited as necessary such as methionine and glutamate (Gupta et al., 2010). Methionine had an important effect on the cell growth and cellulose production compared to the media without this kind of amino acid. With the presence of methionine, the cell growth and cellulose production achieved almost 90% increase.

5.6.3 pH

The effective pH range for culturing the cellulose microorganism production is between pH of 4.0 and 6.0 and most preferred pH is 5.0. Outside this range, minimal cellulose is produced and growth is also adversely affected, although *A. xylinum* can live at pH as low as 3.5 (Lapuz et al., 1969; Coban & Biyik, 2011). A pH between 4.0 and 4.5 gives better result in bacterial cellulose production and this range of pH is especially to avoid contaminations.

Verschuren et al. (1999) showed that the most favorable pH-values for initial formation of bacterial cellulose are pH 5.0 and 4.0. These pH values allowed optimal bacterial growth and affected the oxygen uptake during fermentation. Most researchers used pH 5.0 (Ishikawa et al., 1995; Joris et al., 1990; Coban & Biyik, 2011). Norhayati et al. (2011) reported that pH 5.0 allows the highest yield for the fermentation in RDR and pH 4.0 for static fermentation.

5.6.4 OXYGEN CONCENTRATION

A. xylinum is an obligate aerobic bacterium. So increasing the oxygen limitation within culture flasks provides an aerated surface for the cells to grow on and form cellulose. Oxygen is important for cellulose production because *A. xylinum* cells fail to form a coherent membrane when grown in nitrogen atmosphere (Jonas & Farah, 1997). The rate of bacterial production increases when they are exposed to a high oxygen tension (100% oxygen atmosphere) while at lower oxygen, about 10–15% levels in the gas phase as compared to 21% at normal atmosphere, enhances bacterial cellulose production with constant cell growth. At concentration higher than this, the production rate drops. It was found that increasing the gaseous oxygen concentration to 30% markedly decreased the bacterial cellulose yield, while the cell growth rate remained constant (Chao et al., 2001).

At high cell concentration, the rate of oxygen consumption may exceed the rate of oxygen supply and will lead to a condition of oxygen limitation. Since *A. xylinum* is an obligate aerobic bacterium and because of the oxygen limitation within the culture flask, an aerated surface for cells grow and bacterial cellulose formation occurs. The growth rate increased, along with production rate, caused a dense film which then restricted the diffusion of both oxygen from above and nutrient from below, leading to a thinner but denser pellicle (Chao et al., 2001). They also stated that if the oxygen is supplied, for example in a CSTR, cellulose is not produced; however, once the air is shut off and mixing is halted, cellulose production will begin again. By supplying oxygen-enriched air to the cultures, the simultaneous increase in flow rate and inlet oxygen concentration will increase the dissolved oxygen concentration of the medium cultures and the bacterial cellulose production will increase (Chao et al., 2001).

5.6.5 TEMPERATURE

Temperature gives an effect to the cellulose synthesis and is much like pH wherein the cells prefer a certain range. The cell growth and pellicle production occurred in the range of 20–30°C. But in Philippines, the maximum bacterial cellulose production is at room temperature which is about 28–31°C. No growth was observed at 10, 15, 35, and 40°C during 48 h after inoculum. The optimum growth temperature for bacterial cellulose is between 25 and 30°C (Jonas & Farah, 1997).

5.6.6 SURFACE AREA

Holmes (2004) reported that at constant volume, the surface area has been proved to be directly proportional to the surface area of air–liquid interface. The experimental result clearly demonstrated that doubling the surface area available also doubled the cellulose production rate. The result also suggests that the surface area up to certain extent has more dominant influences on the cellulose production than absolute number of cells in the medium. Thus, this parameter is very important to be maximized to ensure the cellulose production is at the maximum level. Other than that, by increasing the quantity or size of the discs in the bioreactor, it enlarges the total surface area available for the bacteria, which makes the production easier to scale up.

5.7 CULTURE MEDIUM FOR FERMENTATION OF BACTERIAL CELLULOSE

Bacterial cellulose can form in many media such as coconut water, fruit juice, and domestic waste product or in a nutrient medium. Traditionally, bacterial cellulose has been produced in static trays by using coconut water (also other liquid medium such as sugar cane, pineapple extract, etc.). Table 5.4 shows various media used for fermentation of bacterial cellulose.

TABLE 5.4 Production of Bacterial Cellulose Using Different Media.

Medium	Explanations
Palm oil mill effluent (POME) (Hidayah, 2013)	8.0 g of sucrose, 0.8 g of yeast extract, 0.3 g of KH_2PO_4, and 0.005 g of Mg_2SO_4 in 100% (v/v) of POME concentration Static fermentation process at pH 5.5, temperature 26°C, and 7-day incubations Bacterial cellulose wet weight is 29.25 g
Pineapple waste (Junaidi & Muhammad, 2012)	Optimum conditions are pH 5.50, temperature at 30°C and 80% of pineapple concentration Bacterial cellulose produced is 3.4368 g
Bacterium isolated from rotten fruit (Firdaus et al., 2012)	Used different fruits like apple, pineapple, orange, sweet lime, and pomegranate Condition used is temperature at 30°C
Maple syrup (Zeng et al., 2011)	Fermentation in a rotary shaker with shaking speed 135 rpm, inoculum age 3 days, inoculum volume 6% (v/v) and incubation temperature 25°C Optimal conditions: 30 g carbohydrate/l, $(NH_4)_2SO_4$ 3.3 g/l, KH_2PO_4 1 g/l, yeast extract 20 g/l, citric acid 1.6 g/l, trisodium citrate dehydrate 2.4 g/l, ethanol 0.5% (v/v), acetic acid 0.5 g/l, $MgSO_4$ 0.8 g/l
Sucrose (Norhayati et al., 2011)	Shigeru Yamanaka medium (50 g sucrose, 5 g yeast extract (Bacto), 5 g ammonium sulfate, 3 g potassium dihydrogen phosphate and 0.05 g magnesium sulfate) Fermentation in RDR at pH 5.0 or 4 days at 28°C produced 139.78 g wet cellulose

TABLE 5.4 *(Continued)*

Agricultural waste (Kongruang, 2008)	Used coconut and pineapple juices Conditions used are temperature at 30°C, pH 4.75, and fermentation in 5 L fermenters Gives highest productivity using coconut juice
Glucose, mannitol, and xylose (Barbara et al., 2008)	Using a Herstin–Schramm nutrient medium containing: 2% glucose, 0.5% yeast extract, 0.5% bactopepton, 0.115% citric acid, 0.27% Na_2HPO_4, 0.05% $MgSO_4$, 1% ethanol. Added after sterilization of the base Fermentation at temperature of 30°C

5.8 APPLICATION OF BACTERIAL CELLULOSE

Bacterial cellulose is same as plant cellulose and has higher purity without lignin or hemicelluloses. Therefore, using bacterial cellulose is more economic compared to plant cellulose because it can skip many stages in production of pure plant cellulose (Jonas & Farah, 1997). Table 5.5 shows potential applications for bacterial cellulose.

TABLE 5.5 Products Which Can Be Manufactured from Bacterial Cellulose.

Material	Applications	Patent
BC sheet produced in bioreactor	Implant for treating wound	US20160022867 A1 (Bayon et al., 2016)
BC film	Skin moisturizer, skin exfoliator and for sebum absorption in cosmetic product	US20150216784 A1 (Lin et al., 2015)
BC Soy-based resins	Fiber composites	US20140083327 A1 (Netravali et al., 2014)
BC-poly (2 - hydroxyethyl methacrylate)	Contact lenses and optic component for biosensor	US20130011385 A1 (Li et al., 2013)
BC hydrogel	Cold pack	CN 201806818 U (Li et al., 2011)
BC nanosilver	Mask	CN 101589854 B (Zhong, 2011)
Metalized BC	Fuel cells and electronic devices	US 2011/0014525 A1 (Evans et al., 2011)

TABLE 5.5 *(Continued)*

Poly(vinyl alcohol)-BC	Soft tissue replacement, medical devices, artificial dura mater	CN101053674 A (Ma et al., 2010) US 20050037082 A1 (Wan & Millon, 2005)
BC membrane	Membrane electrode	CN 101320812 B (Xu et al., 2011)
Gelatinous BC	Production of soft and light fibers	US 5962676 (Tammarate, 1999)
Reticulated BC	Replacement for latex binders	US 4919753A (Johnson & Neogi, 1990)
BC sheets	Food and functional materials	US RE38,792 E (Iguchi et al., 1988) US 20040091978 A1 (Ishihara & Yamanaka, 2004)

BC, bacterial cellulose.

This bacterial cellulose is relatively pure and does not contain lignin and hemicelluloses associated with wood. Its unique property is that the chain length is much longer. Therefore, it is able to produce stronger paper. However, using cellulose crops from wood and cotton makes a direct impact on the earth's carbon cycle as it uses up forest resources. Therefore, bacterial cellulose can play a role as another smart alternative for plant cellulose. Bacterial cellulose can also be used as paper. Research shows that its property can increase the toughness of paper.

In food industries, bacterial cellulose is used to increase the viscosities of solutions of guar gum, sodium CMC, and xanthan. Besides, bacterial cellulose does not add undesirable flavors, does not mask desirable flavors and is chemically inert. Moreover, dried cellulose pellicles can serve as acoustic membranes or conductive membranes because of its mechanical strength (Vandamme et al., 1997). The native *Acetobacter* pellicle is able to display Young's modulus value of 30 GPa, approximately 4 times greater than any organic fiber. These properties of bacterial cellulose make it an ideal product for speaker diaphragms. This had been implemented by SONY Corporation which already produced a headphone using bacterial cellulose.

Bacterial cellulose also offers a wide range of applications due to its high purity and special physicochemical characteristics (Vandamme et al., 1997). It has been used in medical field as artificial skin because of its good water absorption capacity. Besides, research shows that bacterial cellulose has the ability to improve growth of human skin cells (Brown, 1996). These examples show that bacterial cellulose has potential to be used in many fields. More researches should be conducted in order to discover new possible application for bacterial cellulose.

5.9 SOLIDS INCLUSIONS DURING BACTERIAL CELLULOSE FORMATION

The most attractive feature of bacterial cellulose is the ability to control and modify the physical characteristics of the cellulose product while being synthesized. With this technology, one can custom-design the properties such as hydrophilicity (Brown, 1996) and also the chemical composition of the cellulose fiber being produced (Shirai et al., 1994). It was found that addition of dyes and CMC can significantly alter the type of cellulose formed during fermentation leading to enhanced product properties (Haigler & Benziman, 1982). Norhayati et al. (2013) found increased conductivity by adding polyaniline during the fermentation. Such an ability to control the cellulose synthesis can enable the manufacturer to alter the properties of the bacterial cellulose to a greater extent than would be possible by post-synthetic processing of other sources of cellulose.

Bungay and Serafica (1999) addressed a new way of producing bacterial cellulose using a rotating disk system. This new process has almost all of the desirable characteristics of a film-producing bioreactor. Not only does the new design increase the surface area for cellulose production by an order of magnitude, the film reactor also allows modification of the film during fermentation by adjusting nutrient delivery and composition. This versatility of changing medium composition offers the possibility to custom-design the cellulose depending on the application. The unique configuration of the rotating disk reactor allows for the manipulation of the growth medium during fermentation. While the cellulose is formed, solid particles can be added to incorporate them into the cellulose matrix. There are numerous potential applications for composite materials, depending on the solid being added.

The concept of immobilizing particles within the cellulose matrix during formation in a rotating disk reactor was first put forth by Serafica et al. (2002). Several different-sized molecules and particles were added to the rotating reactor after film formation. Proteins, dyes, and polyethylene glycol in the size range of 50–500,000 MW diffused into the disk because they were smaller than the pore openings. Bacteria and yeast cells also appeared to enter the matrix through diffusion. Particles, in the range of 20–250 μm, were incorporated by physical method.

5.10 CONCLUSIONS

Bacterial cellulose as secondary metabolite has been produced from fermentation process since many years. The initial interest in bacterial cellulose was mainly for dessert food so-called nata and some applications in electronic devices. It can be observed that the development in its fermentation technology (reactor system, types of medium, bacterial strains, fermentation process) has continuously being researched and reported in scientific literatures. Furthermore, there is also an interesting growth in wider expansion of bacterial cellulose application in many sectors such as food and nutraceuticals, cosmetics, pharmaceuticals, construction and industrial materials, wood and laminates, packaging and plastics, electrical and electronics, textiles, and medicals, which have encouraged more effort toward the scale-up and commercialization process. Hence, these overall combinations of effort can be categorized as second generation of bacterial cellulose in which the production process could be designed conferring the specifically targeted application in lieu of boundless benefits for the betterment of mankind.

KEYWORDS

- **natural polymer**
- **bacterial cellulose**
- **acetic acid bacteria**
- **optimum temperature**
- **carbon**

REFERENCES

Ashori, A.; Sheykhnazari, S.; Tabarsa, T.; Shakeri, A.; Golalipour, M. Bacterial Cellulose/Silica Nanocomposites: Preparation and Characterization. *Carbohydr. Polym.* **2012,** *90*, 413–418.

Barbara, S. S.; Sebastian, P.; Dariusz, D. Characteristics of Bacterial Cellulose Obtained from *Acetobacter xylinum* Culture for Application in Paper Making. *Inst. Pap. Mak. Print.* **2008,** *3*, 50–370.

Bayon, Y.; Ladet, S.; Lefranc, O.; Gravagna, P. Template for Bacterial Cellulose Implant Processed within Bioreactor. U.S. Patent No. 20160022867 A1. 2016.

Brown, Jr. R. M. Advances in Cellulose Biosynthesis. In: *Polymers from Biobased Materials*; Chum, H. L., Ed.; Doyes Data Corp: New Jersey, 1991.

Brown, Jr. R. M. *Microbial Cellulose: A New Resource for Wood, Paper, Textiles, Food and Specialty Products*. University of Texas: Austin, TX, 1996.

Budhiono, A.; Rosidi, B.; Taher, H.; Iguchi, M. Kinetic Aspects of Bacterial Cellulose Formation in *Nata-de-coco* Culture Systems. *J. Carbohydr. Polym.* **1999,** *40*, 137–143.

Bungay, H. R.; Serafica, G. C. Production of Microbial Cellulose Using a Rotating Disk Film Bioreactor. United States Patent 5955326, Sep 21, 1999.

Chao, Y.; Sugano, Y.; Shoda, M. Bacterial Cellulose Production under Oxygen Enriched Air at Different Fructose Concentrations in 50 Liters Internal Loop Airlift Bioreactor. *Appl. Microb. Biotechnol.* **2001,** *55*, 673–679.

Chawla, P. R.; Bajaj, I. B.; Survase, S. A.; Singhal, R. S. Microbial Cellulose: Fermentative Production and Applications. *Food Technol. Biotechnol.* **2008,** *47*(2), 107–124.

Coban, E. P. L.; Biyik, H. Evaluation of Different pH and Temperatures for Bacterial Cellulose Production in HS (Hestrin–Scharmm) Medium and Beet Molasses Medium. *Afr. J. Microbiol. Res.* **2011,** *5*(9), 1037–1045.

Czaja, W.; Krystynowicz, A.; Bielecki, B.; Brown Jr. R. M. Microbial Cellulose—The Natural Power to Heal Wounds. *J. Biomater.* **2006,** *27*(2), 145–151.

Czaja, W.; Romanovicz, D.; Brown, Jr. R. M. *Cellulose II: Structural Investigations of Microbial Cellulose Produced in Stationary and Agitated Culture.* Kluwer Academic Publishers: Netherlands, 2004, pp 403–411.

Deinema, M. H; Zevenhuizen, L. P. T. M. Formation of Cellulose Fibrils by Gram-negative Bacteria and Their Role in Bacterial Flocculation. *Arch. Microbiol.* **1971,** *78*(1), 42–57.

Evans, B. R.; O'Neill, H. M.; Jansen, V. M.; Woodward, J. Metalization of Bacterial Cellulose for Electrical and Electronic Device Manufacture. U.S. Patent No. 20110014525 A1. 2011.

Firdaus, J.; Vinod, K.; Garima, R.; Saxena, R. X. Production of Microbial Cellulose by a Bacterium Isolated from Fruit. *Appl. Biochem. Biotechnol.* **2012,** *167*(5), 1157–1171.

Gindl, W.; Keckes, J. Tensile Properties of Cellulose Acetate Butyrate Composites Reinforced with Bacterial Cellulose. *Compos. Sci. Technol.* **2004,** *64*(15), 2407–2413.

Gupta, G.; Basavaraj, S.; Hungund, S. Production of Bacterial Cellulose from *Enterobacter Amnigenus* GH-1 Isolated from Rotten Apple. *World J. Microbiol. Biotechnol.* **2010,** *26*, 1823–1828.

Haigler, C. H.; Berziman, M. Cellulose and Other Material Polymer System. Plenum Press: New York, 1982.

Hidayah, W. N. A. W. M. Y. Palm Oil Mill Effluent (POME) as Fermentation Medium for Bacterial Cellulose Production Using Static Fermentation Method. Undergraduate Thesis, University Teknologi Malaysia, Malaysia, 2013.

Holmes, D. Bacterial Cellulose. Thesis, Master of Engineering in Chemical and Process Engineer, University of Canterbury: Christchurch, New Zealand, 2004. http://www.rpi.edu/~weberd/HolmesThesisFixed.doc.

Hwang, J. W.; Yang, Y. K.; Hwang, J. K.; Pyun, Y. R.; Kim, Y. S. Effects of pH and Dissolved Oxygen on Cellulose Production by *Acetobacter xylinum* BRCS in Agitated Culture. *J. Biosci. Bioeng.* **1999,** *88*(2), 183–188.

Iguchi M.; Mitsuhashi, S.; Ichimura, K.; Nishi, Y.; Uryu, M.; Yamanaka, S.; Watanabe, K. Bacterial Cellulose-Containing Molding Material Having High Dynamic Strength. U.S. Patent No. 4742164. 1988.

Ishihara, M.; Yamanaka, S. Modified Bacterial Cellulose. U.S. Patent No. 20040091978 A1. 2004.

Ishikawa, A.; Matsuoka, M.; Tsuchida, T.; Yoshinaga, F. Increase in Cellulose Production by Sulfaguanidine-resistant Mutants Derived from *Acetobacter xylinum* subsp. *sucrofermentans. Biosci., Biotechnol. Biochem.* **1995,** *59*, 2259–2262.

Johnson, D. C.; Neogi, A. N. Nonwoven Fabric-Like Product Using a Bacterial Cellulose Binder and Method for Its Preparation. U.S. Patent No. 4919753 A. 1990.

Jonas, R.; Farah L. F. Production and Application of Microbial Cellulose. *J. Polym. Degrad. Stab.* **1997,** *59*, 101–106.

Joris, K.; Biliet, F.; Drieghe, S.; Brack D.; Vandamme, E. Microbial Production of B-1,4-Glucan. *Meded Fac. Landbouwwet-Rijksuniv Gent.* **1990,** *55*, 1563–1566.

Junaidi, Z.; Muhammad, A. N. *Optimization of Bacterial Cellulose Production from Pineapple Waste: Effect of Temperature, pH and Concentration.* Fifth Engineering Conference,10–12th July 2012, Kuching, Sarawak, 2012.

Jung, J. Y.; Khan, T.; Park, J. K.; Chang, H. N. Production of Bacterial Cellulose by *Gluconacetobacter hansenii* Using a Novel Bioreactor Equipped with a Spin Filter. *Korean J. Chem. Eng.* **2007,** *24*, 265–271.

Keshk, S. M. A. S.; Sameshima, K. Evaluation of Different Carbon Sources for Bacteria Cellulose Production. *Art. J. Biotechnol.* **2005,** *4*(6), 478.

Kim, Y. J.; Kim, J. N.; Wee, Y. J.; Park, D. H.; Ryu, H. W. Bacterial Cellulose Production by *Gluconacetobacter* sp. RKY5 in a Rotary Biofilm Contactor. *J. Appl. Biochem. Biotechnol.* **2007,** 136–140.

Kongruang, S. Bacterial Cellulose Production by *Acetobacter xylinum* Strains from Agricultural Waste Products. *Appl. Biochem. Biotechnol.* **2008,** *148*, 245–256.

Krystynowicz, A.; Czaja, W.; Jezierska, A. W.; Miśkiewicz, M. G.; Turkiewicz, M.; Bielecki, S. Factors Affecting the Yield and Properties of Bacterial Cellulose. *J. Ind. Microbiol. Biotechnol.* **2002,** *29*, 189–195.

Krystynowicz, A.; Maria, K.; Aginiezka, W. K.; Stanislaw, B.; Emilia, K.; Aleksander, M.; Andrzej, P. Molecular Basis of Biosynthesis Disappearance in Submerged Culture of *Acetobacter xylinum. J. Ind. Microbiol. Biotechnol.* **2005,** *52*(3), 691–698.

Lapuz, M. M.; Gallardo, E. G.; Palo, M. A. The Nata Organism-Cultural Requirements, Characteristics and Identify. *Philippine J. Sci.* **1969,** *96*, 91–109.

Lee, R. L.; Paul, J. W.; Willem, H. Z; Isak, S. P. Microbial Cellulose Utilization: Fundamentals and Biotechnology. *J. Microbiol. Mol. Biol. Rev.* 2002, 66 (3), 506–577.

Li, X.; Wan, W.; Panchal, C. J. Transparent Bacterial Cellulose Nanocomposite Hydrogels. U.S. Patent No. 20130011385 A1. 2013.

Li, Z.; Zhu, B. J.; Yang, J. X.; Peng, K.; Zhou, B. H.; Xu, R. Q.; Hu, W. L.; Chen, S. Y.; Wang, H. P. Method for Manufacture of Bacterial Cellulose Hydrogel Cold Pack. CN Patent No. 201020239963.4. 2011.

Liang, S.; Zhang, L.; Li, Y.; Xu, J. Fabrication and Properties of Cellulose Hydrated Membrane with Unique Structure. *Macromol. Chem. Phys.* **2007,** *208*(6), 594–602.

Lin, Y.; Wey, Y.; Lee, M. L.; Lin, P. C. Cosmetic Composition Containing Fragments of Bacterial Cellulose Film and Method for Manufacturing Thereof. U.S. Patent No. 20150216784 A1. 2015.

Ma, X.; Wang, R. M.; Guan, F. M.; Wang, T. F. Artificial Dura Mater Made from Bacterial Cellulose and Polyvinyl Alcohol. CN Patent No. 200710015537.5. 2010.

Mahadevaswamy, U. R.; Anu, A. Optimization of Culture Conditions for Bacterial Cellulose Production from *Gluconacetobacter hansenii* UAC09. *Ann. Microbiol.* **2011,** *61*(4), 781–787.

Masaoka, S.; Ohe, T.; Sakota, N. Production of Cellulose from Glucose by *Acetobacter xylinum*. *J. Ferment. Bioeng.* **1993,** *75*(1), 18–22.

Nakayama, A.; Kakugo, A.; Gong, J. P.; Osada, Y.; Takai, M.; Erata, T.; Kawano, S. High Mechanical Strength Double-Network Hydrogel with Bacterial Cellulose. *Adv. Funct. Mater.* **2004,** *14*, 1124–1128.

Netravali, A. N.; Qiu, K. Bacterial Cellulose Based 'Green' Composites. U.S. Patent No. 20140083327 A1. 2014.

Nishiyama, Y.; Langan, P.; Chanzy, H. Crystal Structure and Hydrogen-Bonding System in Cellulose from Synchrotron X-ray and Neutron Fiber Diffraction. *J. Am. Chem. Soc.* **2002,** *124*(31), 907.

Norhayati, P.; Khairul, A. Z.; Ida, I. M. Production of Biopolymer from *Acetobacter xylinum* Using Different Fermentation Method. *Int. J. Eng. Technol. IJET–IJENS.* **2011,** *11*(5), 90–98.

Norhayati, P.; Khairul, A. Z.; Ida, I. M.; Kok, F. S. Modified Fermentation for Production of Bacterial Cellulose/Polyaniline as Conductive Biopolymer Material. *J. Teknol. (Sci. Eng.)* **2013,** *62*(2), 21–23.

Pa'e, N. Rotary Discs Reactor for Enhanced Production Microbial Cellulose. Master Thesis, Universiti Teknologi Malaysia, 2009.

Pa'e, N.; Hamid, N. A.; Khairuddin, N.; Zahan, K. A.; Kok, F. S.; Siddique, B. M.; Muhamad, I. I. Effect of Different Drying Methods on the Morphology, Crystallinity, Swelling Ability and Tensile Properties of *Nata De Coco. Sains Malaysiana.* **2014,** *43*(5), 767–773.

Park, M.; Chenga, J.; Choia, J.; Kimb, J; Hyuna, J. Electromagnetic Nanocomposite of Bacterial Cellulose Using Magnetite Nanoclusters and Polyaniline. *Colloids Surf. B: Biointerfaces*. **2013,** *102*, 238–242.

Pei, Y.; Yang, J.; Liu, P.; Xu, M.; Zhang, X.; Zhang, L. Fabrication, Properties and Bioapplications of Cellulose/Collagen Hydrolysate Composite Films. *Carbohydr. Polym.* **2013,** *92*, 1752–1760.

Phisalaphong, M; Jatupaiboon, N. Biosynthesis and Characterization of Bacteria Cellulose-Chitosan Film. *Carbohydr. Polym.* **2008,** *74*, 482–488.

Prashant, R. C.; Ishwar, B. B.; Shrikant, A. S.; Rekha, S. S. Microbial Cellulose: Fermentative Production and Applications. *Food Technol. Biotechnol.* **2009,** *47*(2), 107–124.

Ruka, D. R.; Simon, G. P.; Deana, K. M. In Situ Modifications to Bacterial Cellulose with the Water Insoluble Polymer Poly-3-Hydroxybutyrate. *Carbohydr. Polym.* **2013,** *92*, 1717–1723.

Saibuatong, O; Phisalaphong, M. Novo *Aloe vera*—Bacterial Cellulose Composite Film from Biosynthesis. *Carbohydr. Polym.* **2010,** *79*, 455–460.

Serafica, G. C.; Mormino, R.; Bungay, H. R. Inclusion of Solid Particles in Bacterial Cellulose. *J. Appl. Microbiol. Biotechnol.* **2002,** *58*, 756–760.

Shirai, A.; Takahashi, M.; Kaneko, H.; Nishimura, S.; Ogawa, M.; Nishi, N.; Tokura, S. Biosynthesis of a Novel Polysaccharide by *Acetobacter xylinum*. *Int. J. Biol. Macromol.* **1994,** *16*(6), 297–300.

Sokolnicki, A. M.; Fisher, R. J.; Harrah, T. P.; Kaplan D. L. Permeability of Bacterial Cellulose Membranes. *J. Membr. Sci.* **2006,** *272*(1–2), 15–27.

Son, H. J.; Heo, M. S.; Kim, Y. G.; Lee, S. J. Optimization of Fermentation Conditions for the Production of Bacterial Cellulose by a Newly Isolated Acetobacter sp. A9 in Shaking Cultures. *Appl. Biochem. Biotechnol.* **2001,** *33*, 1–5.

Sumate, T.; Pramote, T.; Waravut, K.; Pattarasinee, B.; Angkana, P. Effect of Dissolved Oxygen on Cellulose Production by *Acetobacter* sp. *J. Sci. Res. Chulalongkorn Univ.* **2005,** *30*(2), 179–186.

Tammarate, P. Method for the Modification and Utilization of Bacterial Cellulose. U.S. Patent No. 5962676. 1999.

Toru, S.; Kazunori, T.; Masaya, K.; Tetsuya, M.; Takaaki, N.; Shingeru, M.; Kenji, K. Cellulose Production From Glucose Using a Glucose Dehydrogenase Gene (GDH)-Deficient Mutant of *Gluconacetobacter xylinus* and Its Use For Bioconversion of Sweet Potato Pulp. *J. Biosci. Bioeng.* **2005,** *99*(4), 415–422.

Tsuchida, T.; Yoshinaga, F. Production of Bacterial Cellulose by Agitation Culture System. *J. Pure Appl. Chem.* **1997,** *69*(11), 2453–2458.

Vandamme, E. J.; DeBaets, S.; Vanbaelan, A.; Joris, K.; DeWulf, P. Improved Production of bacterial Cellulose and Its Application Potential. *J. Polym. Degrad. Stab.* **1997,** *59*, 93–99.

Verschuren, P. G.; Cardona, T. D.; Robert Nout, M. J.; Gooijer, K. D.; Den Huevel, J. V. Location of Cellulose Production by *Acetobacter xylinum* Established from Oxygen Profiles. *J. Biosci. Bioeng.* **1999,** *89*(5), 414–419.

Wan, W. K.; Millon, L. Poly(vinyl alcohol)-Bacterial Cellulose Nanocomposite. U.S. Patent No. 20050037082 A1. 2005.

Wei, B.; Yanga, G.; Hong, F. Preparation and Evaluation of a Kind of Bacterial Cellulose Dry Films with Antibacterial Properties. *Carbohydr. Polym.* **2011,** *84*, 533–538.

White, D. G.; Brown Jr., R. M. Prospects for the Commercialization of the Biosynthesis of Microbial Cellulose. In: *Cellulose and Wood-Chemistry and Technology*, vol. 573; Schuerech, C., Ed.; John Wiley & Sons: New York, 1989.

Yamada, Y.; Yukphan, P. Genera and Species in Acetic Acid Bacteria. *Int. J. Food Microbiol.* **2008,** *125*, 15–24.

Yeoh Q. L.; Lee G. L.; Fatimah, H. Teknologi Pengeluaran Nata. *J. Teknologi Makanan.* **1985,** *4*(1), 36–39.

Yoshino, T.; Asakura, T.; Toda, K., Cellulose Production by *Acetobacter pasteurianus* on Silicone Membrane. *J. Ferment. Bioeng.* **1996,** *81*, 32–36.

Zahan, K. A.; Pa'e, N.; Kok, F. S.; Muhamad, I. I. Monitoring Initial Glucose Concentration for Optimum pH Control during Fermentation of Microbial Cellulose in Rotary Discs Reactor. *Key Eng. Mater.* **2014,** *594–595*, 319–324.

Zahan, K. A.; Pa'e, N.; Muhamad, I. I. Process Parameter for Fermentation in Rotary Disc Reactor for Optimum Microbial Cellulose Production using Response Surface Methodology. *Bioresources* **2013,** *9*(2), 1858–1872.

Zeng, X.; Darcy, P. S.; Wankei, W. Statistical Optimization of Culture Conditions for Bacterial Cellulose Production by *Acetobacter xylinum* BPR 2001 from Maple Syrup. *Carbohydr. Polym.* **2011,** *85*, 506–513.

Zhang, J.; Greasham R. Chemically Defined Media for Commercial Fermentation. *Appl. Microbiol. Biotechnol.* **1999,** *51*, 407–421.

Zhong, C. Y. Method for Manufacturing Air-Filtering Bacterial Cellulose Face Mask. CN Patent No. 200910149665.8. 2011.

CHAPTER 6

SECONDARY METABOLITES IN HORTICULTURAL CROPS

ANKITA ANU[1], SANGITA SAHNI[2], PANKAJ KUMAR[3], and BISHUN DEO PRASAD[3*]

[1]*Department of Horticulture, Bihar Agricultural University, Sabour, Bhagalpur, Bihar, India*

[2]*Department of Plant Pathology, T.C.A., Dholi, RAU, Pusa, Samastipur, Bihar, India*

[3]*Department of Molecular Biology and Genetic Engineering, Bihar Agricultural College, Sabour, Bhagalpur, Bihar, India*

**Corresponding author, E-mail: dev.bishnu@gmail.com*

CONTENTS

ABSTRACT

Secondary metabolites are products produced by plants that aid in their growth and development but are not required for their survival. Secondary metabolism plays a significant role in keeping all of the systems of plants working properly and makes plants tolerant against abiotic and biotic stress. They are also used in anti-feeding activity, toxicity, or acting as precursors to physical defense systems. Due to above-mentioned properties, secondary metabolites have been in limelight for their beneficial role in crop, especially horticultural crops. Several literatures are available on secondary metabolites produced and their beneficial roles. But detailed and systemic information is still lacking. Therefore, in this chapter, we will discuss the secondary metabolites that are produced by plants and their beneficial roles with more emphasis on horticultural crops.

6.1 INTRODUCTION

Metabolism is defined as the sum of all the biochemical reactions carried out by an organism. Primary metabolic pathways converge into a few end products, while secondary metabolic pathways diverge into many products. Primary metabolites are compounds that have essential roles associated with photosynthesis, respiration, and growth and development. These include phytosterols, acyl lipids, nucleotides, amino acids, and organic acids. Apart from these, many phytochemicals, accumulating in surprisingly high concentrations in some species, are referred to as secondary metabolites. Secondary metabolites are known for their role in the growth and development of plants. However, these secondary metabolites are not required for their survival. Secondary metabolism facilitates the primary metabolism in plants. The first documented report of secondary metabolism dates back to approximately 300 BC by Greek and Roman writers. However, it was not until 1937 that Hans Molisch coined the term "allelopathy" to describe such plant–plant interactions (Rice, 1984; Willis, 2004). Secondary metabolism plays a significant role in keeping all of plants' systems working properly and also helps in defense mechanisms. They fight off herbivores, pests, and pathogens. Secondary metabolites are also used in anti-feeding activity, toxicity, or acting as precursors to physical defense systems. Alkaloids, cyanogenic glycosides, flavonoids, terpenoids, and phenolic compounds are the examples of secondary metabolites (Zhao, 2007). Secondary metabolites are also used as dyes, glues, fibers, oils, waxes, flavoring agents, drugs, and perfumes, and they are viewed as potential sources of new natural drugs, antibiotics, insecticides, and herbicides (Dewick, 2002).

Secondary metabolites can be divided into three major groups: (1) flavonoids and allied phenolic and polyphenolic compounds, (2) terpenoids, and (3) nitrogen-containing alkaloids and sulfur-containing compounds. Secondary metabolites of plants species are synthesized in specific pathway that exists at the sites of production. Moreover, some secondary metabolites can be synthesized in all plant tissues, whereas others are produced in a specific tissue (Yazdani et al., 2011). The site of production and accumulation of secondary metabolites may vary and depends upon the nature of metabolites produced, for example, hydrophilic compounds are mainly stored in the vacuole, while the lipophilic secondary metabolites are commonly sequestered in resin ducts, laticifers, oil cells, trichomes, or in the cuticle (Engelmeier & Hadacek, 2006).

The apparent climate change worldwide leads to extreme temperatures, drought periods, and frequent storms causing a negative influence on the growth and development of all plants including horticultural and vegetative ones. Secondary metabolites present in the epidermis of fruits are not only responsible to give color to the fruit and attract insects and other animals that are important for the pollination and spreading of the seeds, but also protect plants from harmful UV radiation and perform a screening role from diverse forms of stress. The impact of changing environmental factors on the synthesis of phenolics, carotenoids, and other secondary metabolites would be studied. In this chapter, we will discuss different types of secondary metabolites produced in horticultural crops.

6.2 PHENOLIC COMPOUNDS

Phenolic compounds are hydroxyl group (–OH) containing a class of chemical compounds where the –OH bonds directly to an aromatic hydrocarbon group or a hydroxyl functional group of an aromatic ring called phenol. Phenolic compounds are an important part of the defense system of plants against pests and diseases including root parasitic nematodes (Wuyts et al., 2006). Phenolic compounds constitute a large and complex group of chemical found in plants (Walton et al., 2003). Phenolics exhibit several beneficial properties including its antioxidant properties which play an important role in protecting against free radical-mediated disease processes. Flavonoids, phenolic acids, and polyphenols are the three important groups of dietary phenolics. Flavonoids are the largest and the most studied group of plant phenols (Dai & Mumper, 2010). Phenolic acids form a diverse group that includes the widely distributed hydroxybenzoic and hydroxycinnamic acids. Phenolic polymers, commonly known as tannins, are compounds of high molecular weight that are divided into two classes: hydrolyzable and condensed tannins.

6.2.1 FLAVONOIDS

Flavonoids are the well studied and belong to the largest group of plant phenols (Dai & Mumper, 2010). Flavonoids are also known as Vitamin P or citrin. It encompasses more than 10,000 structures and has long been reported as serving multiple functions in plants suffering from a wide array of environmental injuries (Stafford, 1991; Dixon & Paiva, 1995; Cockell & Knowland, 1999; Ferdinando et al., 2012). These metabolites are mostly used in plants to produce yellow and other pigments having a role in the color of the plants. Flavonoids are readily ingested by humans and they seem to display important anti-inflammatory, antiallergic, and anticancer activities. Flavonoids are reported to be the dominant antioxidants and studies are going on to analyze their ability to prevent cancer and cardio-vascular diseases. Flavonoids have indeed the capacity to absorb the most energetic solar wavelengths (i.e., UV-B and UV-A), inhibit the generation of reactive oxygen species (ROS), and then quench ROS once they are formed (Agati et al., 2012).

Flavonoids are present in high concentrations in the epidermis of leaves and the skin of fruits and have important and varied roles as secondary metabolites. In plants, flavonoids are involved in diverse processes such as UV protection, pigmentation, stimulation of nitrogen-fixing nodules, and disease resistance. Flavonoids are also closely related to flavones which are actually a subclass of flavonoids, and are the yellow pigments in plants. In addition to flavones, there are 11 other subclasses of flavonoids including, isoflavones, flavans, flavanones, flavanols, flavanolols, anthocyanidins, catechins (including proanthocyanidins), leukoanthocyanidins, dihydro-chalcones, and aurones. The known dietary sources of flavonoids are all citrus fruits, which contain the specific flavanoids hesperidins, quercitrin, and rutin, berries, tea, dark chocolate, and red wine, and many of the health benefits attributed to these foods come from the flavonoids they contain.

6.3 TERPENES

Terpenes are volatile compounds produced by many plants, as well as some insects. The name "terpene" comes from turpentine, a terpene-based solvent distilled from pinesap. It constitutes the largest class of secondary metabolites which are united by their common biosynthetic origin from acetyl-coA or glycolytic intermediates (Grayson, 1998). Plants that produce terpenes often possess smells and flavors we find pleasing and are known as aromatic herbs. A vast majority of the different structures of terpenes are produced by plants as secondary metabolites that are presumed to be involved in defense against, for example, toxins and feeding deterrents to a large number of plant-feeding insects and mammals (Croteau et al., 2000; Gershenzon & Croteau, 1991; Saviranta et al., 2010). Many of the terpenoids are commercially interesting because of their use as flavors and fragrances in foods and cosmetics, for example, menthol and sclareol or, because they are important for the quality of agricultural products, such as the flavor of fruits and the fragrance of flowers like linalool (Harborne & Tomas-Barberan, 1991). Terpenes are widespread in nature, mainly in plants as constituents of essential oils.

6.3.1 MONOTERPENES

Monoterpenes are present in almost all essential oils of many plants including fruits, vegetables, and herbs. Monoterpenes may be divided into two types, that is, linear (acyclic) or contain rings, for example, geranyl pyrophosphate, eucalyptol, limonene, citral, camphor, and pinene. In gymnosperms (conifers) like pine and fir, monoterpenes accumulate in resin ducts found in the needles, twigs, and trunks mainly as α-pinene, β-pinene, limonene, and myrecene, all being toxic to numerous insects including bark beetles, serious pest of conifer species, and so on (Turlings et al., 1995). Many derivatives are important agents of insect toxicity, for example, pyrethroids (monoterpene esters) occurring in the leaves and flowers of chrysanthemum species show strong insecticidal responses (neurotoxins) toward insects like beetles, wasps, moths, bees, and others, and are popular ingredient in commercial insecticides because of low persistence in the environment and low mammalian toxicity (Grayson, 1998). Monoterpenes are shown to prevent the carcinogenesis process at both the initiation and promotion/progression stages (Gould, 1997).

Monoterpenes such as limonene and perillyl alcohol have been shown to prevent mammary, liver, lung, and other cancers (Gould, 1997).

6.3.2 *SESQUITERPENES*

Sesquiterpenoids are defined as the group of 15-carbon compounds derived by the assembly of 3 isoprenoid units, and they are found mainly in higher plants and also in invertebrates. Sesquiterpenes, with monoterpenes, are an important constituent of essential oils in plants. Sesquiterpenes have three isoprene units, for example, artemisinin, bisabolol, and fernesol, oil of flowers, or as cyclic compounds, such as eudesmol, found in Eucalyptus oil. Abscisic acid (ABA) belong to sesquiterpene, play primarily regulatory roles in the initiation and maintenance of seed and bud dormancy and plants' response to water stress by modifying the membrane properties and act as transcriptional activators (McCarty et al., 1991; Giraudat et al., 1992). A number of sesquiterpenes that have been till now reported for their role in plant defense, such as costunolides, are antiherbivore agents of family Compositae and have strong feeding repellence to many herbivorous insects and mammals (Dzantor & Woolston, 2001).

6.3.3 *DITERPENES*

It is composed of four isoprene units. They are derived from geranyl pyrophosphate. There are some examples of diterpenes such as cembrene, kahweol, taxadiene, and cafestol. Retinol, retinal, and phytol are the biologically important compounds while using diterpenes as a base. Abietic acid is a diterpene found in pines and leguminous trees. It is present in or along with resins in resin canals of the tree trunk. When these canals are pierced by feeding insects, the outflow of resin may physically block feeding and serve as a chemical deterrent to continued predation. Abietic acid is a diterpene found in pines and leguminous trees. It is present along with resins in resin canals of the tree trunk. When these canals are pierced by feeding insects, the outflow of resin may physically block feeding and serve as a chemical deterrent to continued predation (Bailey & Mansfield, 1982).

6.3.4 TRITERPENE

The triterpenes are one of the most numerous and diverse groups of plant natural products. They are complex molecules having a wide range of applications in the food, health, and industrial biotechnology sectors. Simple triterpenes are components of surface waxes and specialized membranes and may potentially act as signaling molecules, whereas complex glycosylated triterpenes (saponins) provide protection against pathogens and pests. Limnoid triterpene is a group of bitter substances found in citrus fruits which acts as antiherbivore compounds. Azadirechtin, a complex limnoid from *Azadirachta indica*, commonly known as "neem" in India, acts as a feeding deterrent to some insects and exerts various toxic effects (Mordue & Blackwell, 1993).

6.3.5 POLYTERPENES

Several high molecular weight polyterpenes occur in plants. Larger terpenes include the tetraterpenes and the polyterpenes. The principal tetraterpenes are carotenoids family of pigments. The other one is rubber, a polymer containing 1500–15,000 isopentenyl units, in which nearly all the C–C double bonds have a *cis*(Z) configuration while in gutta rubber has its double bond in *trans*(E) configuration. Rubber found in long vessels called laticifers, provide protection as a mechanism for wound healing and as a defense against herbivores (Eisner & Meinwald, 1995; Klein, 1987).

6.3.6 HEMITERPENOIDS

It consists of a single isoprene unit, the only hemiterpene is the isoprene itself, but oxygen-containing derivatives of isoprene such as isovaleric acid and prenol are classified as hemiterpenoid.

6.3.7 TETRATERPENOIDS

It contains eight isoprene units which may be acyclic like lycopene, monocyclic like gamma-carotene, and bicyclic like alpha- and beta-carotenes. Carotenoids are a group of naturally occurring fat-soluble pigments consisting of more than 700 compounds, which are mainly found in plants, algae, and several lower organisms. Carotenoids are tetraterpenes formed by eight isoprenoid units and can be divided into two groups: carotenes and xanthophylls. Carotenes contain only carbon and hydrogen, while xanthophylls (e.g., lutein and zeaxanthin) are oxygenated carotenoids containing an alcohol, carbonyl, or other functional groups. Carotene plays an essential role as a source of vitamin A. The most active role is protection against serious disorders such as cancer, heart diseases, and degenerative eye diseases.

6.4 SULFUR-CONTAINING SECONDARY METABOLITES

Sulfur-containing secondary metabolites include glutathione (GSH), glucosinolate (GSL), phytoalexins, thionins, defensins, and allinin which have been linked to the defense responses of plants against pathogens (Hell, 1997; Crawford et al., 2000; Halkier & Gershenzon, 2006).

6.5 NITROGEN-CONTAINING SECONDARY METABOLITES

Nitrogen-containing secondary metabolites include alkaloids, cyanogenic glucosides, and nonprotein amino acids (NPAAs). Most of them are biosynthesized from common amino acids. Their role in the antiherbivore defense and toxicity to humans are well known.

6.5.1 ALKALOIDS

Alkaloids are a group of naturally occurring chemical compounds (natural products) that contain mostly basic nitrogen atoms. Nitrogen-containing secondary metabolites are mainly found in different species of vascular plants (Hegnauer, 1988). Usually most of alkaloids are toxic in nature and involved in defense against microbial infection and herbivoral attack.

6.5.2 CYANOGENIC GLUCOSIDES

Cyanogenic glucosides constitute a group of nitrogen-containing protective compounds other than alkaloids. They usually occur in members of families, namely, Graminae, Rosaceae, and Leguminosae (Seigler, 1981). They are not themselves toxic but are readily broken down to give off volatile poisonous substances like hydrogen cyanide and hydrogen sulfide. When the plant is crushed, their presence deters feeding by insects and other herbivores (Ballhorn et al., 2009). More than 2000 plant species contain cyanogenic glucosides. Past research suggests that cyanogenic glucosides stored in the seed of the plant are metabolized during germination to release nitrogen for seedling to grow. Therefore, it can be concluded that cyanogenic glycosides play various roles in plant metabolism.

6.5.3 GLUTATHIONE

GSH is one of the major forms of organic sulfur in the soluble fraction of plants and has an important role as a mobile pool of reduced sulfur in the regulation of plant growth and development, and as a cellular anti-oxidant in stress. Specialized cells such as trichomes exhibit high activity of enzymes for synthesis of GSH and other phytochelatins necessary for detoxification of heavy metals.

6.5.4 NPAAS

Many plants also contain unusual amino acids called nonprotein amino acids which act as protective defensive substance. Some block the synthesis of or uptake of protein amino acid while others can be mistakenly incorporated into proteins. After ingestion, canavanine is recognized by herbivorous enzyme that normally binds arginine to the arginine transfer RNA molecule and so become incorporated into proteins in place of arginine; this results in a nonfunctional protein because either its tertiary structure or catalytic site is disrupted (Rosenthal & Berenbaum, 1991). Plants that synthesize NPAAs are not susceptible to the toxicity of these compounds but gain defense against herbivorous animals, insects, and pathogenic microbes; also, a number of plants such as *Arabidopsis* use arginine as a storage and transport from nitrogen and proline as a compatible solute in the defense against abiotic stresses causing water deprivation.

6.6 SECONDARY METABOLITES IN HORTICULTURAL CROPS

Plants produce several types of secondary metabolites which play an important role in shaping interactions and communities due to their bioactivity (Latkowska et al., 2008). For example, the phenolic compounds distributed in plants parts are important precursors of humic substances in soils (John & Sarada, 2012). Phenolic acids like benzoic, hydrobenzoic, vanillic, and caffeic were known to have antimicrobial and antifungal activities due to their enzyme-inhibition activities. Secondary metabolites like hydroxycinnamic acids and their derivatives possess strong antioxidant activity due to their inhibitory activities of lipid oxidation and scavenging ROS (Hounsome et al., 2008). The release of secondary metabolites, produced inside the plant, to their surrounding environment mediate allelopathic interactions among plants. The realising nature of some secondary metabolites is important for the pharmaceutical industry (Chen et al., 2009).

Fruits are rich sources of secondary metabolites. The resveratrol present in grape contains the preventive and therapeutic abilities of wine. Citrus fruit species are especially known for the production of volatile terpenoids, mainly mono- and sesquiterpenes. Flavonoids are other phytochemicals produced by citrus fruits. The flavonoids have strong inherent ability to modify the body's reaction to allergens, viruses, and carcinogens. They show antiallergic, anti-inflammatory, antimicrobial, and anticancer activity. The bitter taste of some fruits like grape, lemons, oranges, and others, are due to presence of flavonoids like quercetin, myricitin, rutin, tangeritin, naringin, and hesperidin. Bael, a minor fruit crop, contains different biochemicals, namely, alkaloids, coumarin, and steroid.

The flavors and aromas of fruits are determined by complex mixtures of often hundreds of volatile compounds. More than 300 compounds contribute in different levels to the characteristic flavor associated with ripe strawberry fruit (Honkanen & Hirvi, 1990) and more than 400 volatiles contribute to the characteristic aroma of tomatoes and their products (Petro-Turza, 1987). These volatile compounds include metabolites of different chemical groups that include acids, aldehydes, ketones, alcohols, esters, sulfur compounds, furans, phenols, terpenes, epoxides, and lactones and are derived from different biosynthetic pathways (Schwab et al., 2008).

Plum fruits known for its natural antioxidant properties contain profuse amounts of natural phenolic phytochemicals, such as flavonoids, phenolic acids, anthocyanins, and other phenolics (Weinert et al., 1990; Kristl et al., 2011). Wang et al. (1996) reported that the antioxidant capacity of plum is several-fold higher compared to apple.

The horticultural plants are also a good source of numerous metabolites which are being used in the pharmaceutical industry (Krzyzanowska et al., 2010). Due to various side effects of synthetic chemicals, the demand for natural products has increased tremendously in recent years (Fucile et al., 2008; Vogt, 2010). Green chili contains natural antioxidants such as vitamin C, carotenoids, and phenols. Moreover, the metalloenzyme, superoxide dismutase, which is universally present in all plants and imparts defense against oxidative stress, converts superoxide radical anion into hydrogen peroxide (Matsufuji et al., 2007). Chili pepper (*Capsicum annuum*) is reported to contain moderate to high levels of neutral phenolics or flavonoids and phytochemicals that are important antioxidant components of a plant-based diet. Mango, king of fruits in India, is rich in a variety of phytochemicals and nutrients that qualify it as a model "superfruit." The fruit is high in prebiotic dietary fiber, vitamin C, polyphenols, and carotenoids. The edible mango peel has considerable value as a source of dietary fiber and antioxidant pigments (Rocha et al., 2007; Ajila & Prasada, 2008). The peel and leaves of mango contain significant amount of polyphenols, including xanthones, mangiferin, and gallic acid (Barreto et al., 2008). Strawberry fruits are well known for their richness in flavors. The unique flavor of strawberry is due to combination of nearly 400 volatile compounds. The major compound that generates strawberry flavor is 4-hydroxy-2,5-dimethyl-3(2*H*)-furanone (furaneol), a compound that possesses a strong strawberry aroma.

KEYWORDS

- **metabolism**
- **allelopathy**
- **flavonoids**
- **phenolic compounds**
- **terpenes**

REFERENCES

Agati, G.; Azzarello, E.; Pollastri, S.; Tattini M. Flavonoids as Antioxidants in Plants: Location and Functional Significance. *Plant Sci.* **2012,** *196*, 67–76.

Ajila, C. M.; Prasada Rao, U. J. Protection against Hydrogen Peroxide Induced Oxidative Damage in Rat Erythrocytes by *Mangifera indica* L. Peel Extract. *Food Chem. Toxicol.* **2008,** *46*(1), 303–309.

Bailey S. A.; Mansfield J. W. *Phytoalexins*. Wiley: New York, 1982.

Ballhorn, D. J.; Kauts, S.; Heil, M.; Hegeman, A. D. Cyanogenesis of Wild Lima Bean (*Phaseolus lunatus* L.) is an Efficient Direct Defense in Nature. *Plant Signal. Behav.* **2009,** *4*(8), 735–745.

Barreto J. C.; Trevisan M. T.; Hull W. E. Characterization and Quantitation of Polyphenolic Compounds in Bark, Kernel, Leaves, and Peel of Mango (*Mangifera indica* L.). *J. Agric. Food Chem.* **2008,** *56*(14), 5599–5610, 261–268.

Chen F.; Liu C.; Tschaplinski T. J.; Zhao N. Genomics of Secondary Metabolism in Populus: Interactions with Biotic and Abiotic Environments. *Crit. Rev. Plant Sci.* **2009,** *28*, 375–392.

Cockell, C. S.; Knowland, J. Ultraviolet Radiation Screening Compounds. *Biol. Rev.* **1999,** *74*, 311–345.

Crawford N. M.; Kahn M. L.; Leustek T.; Long S. R. Nitrogen and Sulfur. In *Biochemistry and Molecular Biology of Plants*; Buchanan, B. B., Gruissem, W., Jones, R. L., Eds.; American Society of Plant Biologists: Rockville, MD, 2000; pp 824–849.

Croteau, R.; Kutchan, M.; Lewis, N. G. Natural products (secondary metabolites). In *Biochemistry and Molecular Biology of Plants*; Buchanan, B. B., Gruissem, W., Ones, R. L. J., Eds.; American Society of Plant Physiologists: Rockville, MD, 2000; pp 1250–1318.

Dai, J.; Mumper, R. Plant Phenolics: Extraction, Analysis and their Antioxidant and Anticancer Properties. *Molecules* **2010,** *15*, 7313–7352.

Dewick, PM. *Medicinal Natural Products: A Biosynthetic Approach*, 2nd ed. John Wiley and Sons Ltd.: Chichester, UK, 2002, ISBN 0-471-49641-3, 507 p.

Dixon, R. A.; Paiva, N. L. Stress-induced Phenylpropanoid Metabolism. *Plant Cell.* **1995,** *7,* 1085–1097.

Dzantor, E. K.; Woolsten, J. E. Enhancing Dissipation of Arocalor 1248 (PCB) Using Substrate Amendment in Rhizospere Soil. *J. Environ. Sci. Health* **2001,** *36*, 1861–1871.

Eisner, T.; Meinwald J. *Chemical Ecology: The Chemistry of Biotic Interaction*. National Academy Press: Washington, DC, 1995.

Engelmeier, D.; Hadacek, F. Antifungal Natural Products: Assays and Applications. In *Naturally Occurring Bioactive Compounds*; Rai and Carpinella (Eds.); Elsevier Sci. Ltd. P.: Amsterdam, 2006; pp 423–467.

Ferdinando, Di.; Brunetti, M.; Fini, C. A.; Tattini, M. Flavonoids as Antioxidants in Plants Under Abiotic Stresses. In *Abiotic Stress Responses in Plants: Metabolism, Productivity and Sustainability*; Ahmad, P., Prasad, M. N. V., Eds.; Springer: New York, 2012; pp 159–179.

Fucile, G.; Falconer, S.; Christendat, D. Evolutionary Diversification of Plant Shikimate Kinase Gene Duplicates. *PLoS Genet.* **2008,** *4,* 1–6.

Gershenzon, J.; Croteau, R. Terpenoids. In *Herbivores their Interaction with Secondary Plant Metabolites, Vol. I: The Chemical Participants*, 2nd ed.; Rosenthal, G. A., Berenbaum, M. R., Eds.; Academic Press: San Diego, CA, 1991; pp 165–219.

Giraudat, J.; Hauge, B. M.; Valon, C.; Smalle, J.; Parcy, F.; Goodman, H. M. Isolation of the Arabidopsis AB13 Gene by Positional Cloning. *Plant Cell* **1992,** *4*, 1251–1261.

Gould, N. N. Cancer Chemoprevention and Therapy by Monoterpenes. *Environ. Health Perspect.* **1997**.

Grayson, D. H. Monoterpenoids. *Nat. Product Rep.* **1998,** *5*, 497–521.

Halkier, B. A.; Gershenzon, J. Biology and Biochemistry of Glucosinolates. *Ann. Rev. Plant Biol.* **2006,** *57*, 303–333.

Harborne, J. B.; Tomas-Barberan, F. A. *Ecological Chemistry and Biochemistry of Plant Terpenoids*. Clarendon: Oxford, 1991.

Hegnauer, R. Biochemistry, Distribution and Taxonomic Relevance of Higher Plant Alkaloids. *Phytochemistry* **1988,** *27*, 2423–2427.

Hell, R, Molecular Physiology of Plant Sulfur Metabolism. *Planta* **1997,** *202*, 138–148.

Honkanen, E.; Hirvi, T. The Flavour of Berries. In *Food Flavours*; Morton, I. D., Macleod, A. J., Eds.; Elsevier Scientific Publications: Amsterdam, 1990; pp 125–193.

Hounsome, N.; Hounsome, B.; Tomos, D.; Edwards-Jones, G. Plant Metabolites and Nutritional Quality of Vegetables. *J. Food Sci.* **2008,** *73*, 48–65.

John, J.; Sarada, S. S. Role of Phenolics in Allelopathic Interactions. *Allelopathy J.* **2012,** *29*, 215–229.

Klein, R. M. *The Green World: An Introduction to Plants and People*. Harper and Row: New York, 1987.

Kristl, J.; Slekovec, M.; Tojnko, S.; Unuk, T. Extractable Antioxidants and Non-extractable Phenolics in the Total Antioxidant Activity of Selected Plum Cultivars (*Prunus domestica* L.): Evolution During On-tree Ripening. *Food Chem.* **2011,** *125*, 29–34.

Krzyzanowska, J.; Czubacka, A.; Oleszek, W. Dietary Phytochemicals and Human Health. In *Bio-Farms for Nutraceuticals: Functional Food and Safety Control by Biosensors*, Vol. 698, Chapter 7; Giardi, M. T., Rea, G., Berra, B., Eds.; Springer US: New York, 2010, pp 74–99.

Latkowska, E.; Lechowski, Z.; Białczyk, J. Responses in Tomato Roots to Stress Caused by Exposure to (+)-Usnic Acid. *Allelopathy J.* **2008,** *21*, 239–252.

Matsufuji, H.; Ishikawa, K.; Nunomura, O.; Chino, M.; Takeda, M. Anti-oxidant Content of Different Colored Sweet Peppers, White, Green, Yellow, Orange and Red (*Capsucum annuum* L.). *Int. J. Food Sci. Technol.* **2007,** *42*, 1482–1488.

McCarty, D.; Hattori, T.; Carson, C. B.; Vasil, V.; Lazar, M.; Vasil, I. K. The Vivporous-1 Development Gene of Maize Encodes a Novel Transcription Activator. *Cell* **1991,** *66*, 895–905.

Mordue A. J.; Blackwell A. Azadirachtin: An Update. *J. Insect Physiol.* **1993,** *39*, 903–924.

Petro-Turza, M. Flavor of Tomato and Tomato Products. *Food Rev. Int.* **1987,** *2*, 309–351.

Rice, E. L. *Allelopathy (First Edition, November 1974)*, 2nd ed. Academic Press, University of Minnesota, 1984, , p 422.

Rocha Ribeiro, S. M.; Queiroz, J. H.; Lopes Ribeiro de Queiroz, M. E.; Campos, F. M.; Pinheiro Sant'ana, H. M. Antioxidant in Mango (*Mangifera indica* L.) *Pulp. Plant Foods Human Nutr.* **2007,** *62*, 13–17.

Rosenthal, G. A.; Berenbaum, M. R., Eds. *Herbivores: Their Interaction with Secondary Plant Metabolites, Vol. I: The Chemical Participants*. Academic Press: San Diego, CA, 1991, pp 165–219.

Saviranta, N. M.; Jukunen-Tihho, R.; Oksanen, E; Karjalainen, R. O. Leaf Phenolic Compounds in Red Clover (*Trifolium pratense* L.) Induced by Exposure to Moderately Elevated Ozone. *Environ. Pollut.* **2010,** *158* (2), 440–446.

Schwab, W.; Davidovich-Rikanati, R.; Lewinsohn, E. Biosynthesis of Plant-derived Flavor Compounds. *Plant J.* **2008,** *54* (4), 712–732.

Seigler, D. S. Secondary Metabolites and Plant Systematic. In *The Biochemistry of Plants, Vol. 7. Secondary Plant Products*; Conn, E. E., Ed.; Plenum: New York, London, 1981; pp 139–176.

Stafford, H. A. Flavonoid Evolution: An Enzymic Approach. *Plant Phys.* **1991,** *96*, 680–685.

Turlings, T. C. J.; Loughrin, J. H.; Mccall, P. J.; Roese, U. S. R.; Lewis, W. J.; Tumlinson, J. H. How Caterpillar-Damaged Plants Protect Themselves by Attracting Parasitic Wasps. *Proc. Natl. Acad. Sci. U.S.A.* **1995,** *92*, 4169–4174.

Vogt, T. Phenylpropanoid Biosynthesis. *Mol. Plant* **2010,** *3*, 2–20.

Walton, N. J.; Mayer, M. J.; Narbad, A. Molecules of Interest: Vanillin. *Phytochemistry* **2003,** *63*, 505–515.

Wang H.; Cao G.; Prior R. L. Total Antioxidant Capacity of Fruits. *J. Agric. Food Chem.* **1996,** *44*, 701–705.

Weinert, I.; Solms, J.; Escher, F. Diffusion of Anthocyanins During Processing and Storage of Canned Plums. *Food Sci. Technol.—Leb.* **1990,** *23*, 396–399.

Willis, R. J. *Justus Ludewig von Uslar and the First Book on Allelopathy*. Springer Publications: Dordrecht, the Netherlands, 2004, p 54.

Wuyts N.; Waele D.; Swennen R. Extraction and Partial Characterization of Polyphenol Oxidase from Banana (*Musa acuminata*) Rots. *Plant Physiol. Biochem.* **2006,** 44, 308–314.

Yazdani D.; Tan Y. H.; Zainal Abidin M. A.; Jaganath I. B. A Review on Bioactive Compounds Isolated from Plants against Plant Pathogenic Fungi. *J. Med. Plants Res.* **2011,** *30*, 6584–6589.

Zhao J. Nutraceuticals Nutritional Therapy, Phytobnutrents and Phytotherapy for Improvements of Human Health: A Perspective on Plant Biotechnology Applications. *Rec. Patients Biotechnol.* **2007,** *1*, 75–97.

CHAPTER 7

PULSE SECONDARY METABOLITES: A PERSPECTIVE ON HUMAN AND ANIMAL HEALTH

RAFAT SULTANA[1*], RAVI S. SINGH[1], P. RATNAKUMAR[2], NIDHI VERMA[3], S. K. CHATURVEDI[4], A. K. CHAUDHARY[5], C. V. SAMEER[6], and MOHAMMED WASIM SIDDIQUI[7]

[1]*Department of Plant Breeding and Genetics, Bihar Agricultural University, Sabour, Bhagalpur 813210, Bihar, India*

[2]*National Institute of Abiotic Stress Management (NIASM), Baramati, Pune 413115, Maharashtra, India*

[3]*National Bureau of Plant Genetic Resources, New Delhi, India*

[4]*Indian Institute of Pulse Research, Kanpur 208024, India*

[5]*ICAR-RCER Regional Centre, Darbhanga 846005, India*

[6]*International Crops Research Institute for the Semi-Arid Tropics (ICRISAT), Patancheru, Tilangana, India*

[7]*Department of Food Science and Postharvest Technology, Bihar Agricultural University, Sabour, Bhagalpur 813210, Bihar, India*

**Corresponding author, E-mail: rafat.hayat@gmail.com*

CONTENTS

ABSTRACT

The edible legumes are major dietary protein in vegetarian diets. However, in most of the legumes, the biologically active secondary metabolites (SMs) such as flavonoids, tannins, alkaloids, saponins, trypsin (protease) inhibitors, phytates, hemagluttinins (lectins), oxalates, cyanogenic glycosides, cardiac glycosides, coumarins, gossypol, and others hinder the efficient utilization of health-promoting food nutrients. These metabolites show numerous activities in human and animals, from anticancer, antidiabetes, reducing risk of cardiovascular disease, antioxidants to several others. In plants, their role of disease and pest resistance and signaling network are well established. Besides good metabolites, there are some bad metabolites with respect to human health and commonly referred to as toxic or antinutritional factors. Keeping in view the metabolites profile in legumes and their planned and unintended biological effects, the plant breeding strategies should be adopted. The observation on the literatures related to SMs of pulses revealed that lot of efforts have been made in this aspect; however, no single report sums up the wide range of information.

7.1 INTRODUCTION

Food legumes are an important source of dietary proteins in developing countries where dietary preference of people is vegetarian. Most of the legume grains synthesize various secondary metabolites (SMs) such as tannins, saponins and lectins, phytates, gossypol pigments, oxalates, glucosinolates, mycotoxins, mimosine, cyanogens, nitrates, alkaloids, photo sensitizing agents, phyto-estrogens, and others; these metabolites show certain biological activities. Many of these metabolites referred to as antinutritional factor (ANF). ANF hinders the nutrient utilization and/or food intake of plants or plant products used as food or feed; hence, they play a vital role in determining the use of plants for humans and animals (Soetan, 2008). In addition to that, it contains phytochemicals, including phytosterols, natural antioxidants, and bioactive carbohydrates (Amarowicz & Pegg, 2008). Epidemiological and intervention studies indicated that consumption of legume is inversely associated with the risk of coronary heart disease (Bazzano et al., 2001), type II diabetes mellitus (Bazzano et al., 2008), and obesity (Anderson & Major, 2002).

Consumption of legumes also results in lower LDL cholesterol and higher HDL cholesterol (Kurien, 1981).

Legumes commonly used as food crop include soybean (*Glycine max* L.), Chickpea (*Cicer arietinum* L.), blackgram (*Phaseolus mungo* Roxb.), cowpea (*Vigna unguiculata* L.), drybean (*Phaseolus vulgaris* L.), winged bean (*Psophocarpus tetragonolobus* L.), horse gram (*Dolichos biflorus* L.), pigeon pea (*Cajanus cajan* L.), moth bean (*Vigna aconitifolia* Jacq.), faba bean (*Vicia faba* L.), grain amaranth (*Amaranthus* spp.), lentil (*Lens culinaris* Medik), Jack bean (*Canavalia ensiformis* L.), sword bean (*Canavalia gliadata* DC.), and grass pea (*Lathyrus sativus* L.).

Food and Agricultural Organization (FAO) of the United Nations defined pulses as annual leguminous crops having 1–12 grains or seeds of variable size, shape and, color within a pod. The term pulses, defined by the FAO, are held reserved for crops harvested exclusively for the dry grains. Therefore, it excludes vegetable crops (French green beans and green peas) as well as those crops which are mainly grown for its oil extraction (soybeans and peanuts). In general, due to small amounts of sulfur-containing amino acids, and low protein digestibility coupled with presence of antinutritional factors, legumes have been reported to have low nutritive value.

It has been reported that plant contains thousands of compounds, but all are not leading to positive effects on the organisms consuming them; its effects are either beneficial or deleterious, depending upon the situations it can have. These compounds with the exception of nutrients are referred as allelochemicals (Rosenthal & Janzen, 1979). A class of these compounds, which are in general not lethal, is termed as ANFs. Although these ANFs vary in their individual effect, a huge quantity of them can be inactivated/destroyed simply by processes of heat treatment during cooking. The presence of ANF in higher dietary concentration in untreated foodstuffs results in anorexia, reduced growth, and poor food conversion efficiency. Hence, it is required to evaluate the newly developed or exotic foods for their usefulness, as it may contain natural toxicants (Osagie, 1998). The introduction of new plant varieties into human and animal's diet may expose them to new toxic factors with unsuspected biological effects. In addition to that inappropriate processing of plant food like beans and pulses may also expose humans and animals to high concentrations of these toxic factors. An example is the production of soy milk and its use as an alternative to cow's milk in infant formula. It is reported that improperly processed soy milk, without iodine supplementation, causes goiter in infants (Hydowitz, 1960). Raw soybeans also cause goiter in rats and chicks.

SMs present in legume crops may be beneficial because of its role in defense, signaling network, biotic and abiotic stresses, and qualitative enrichment of seeds, and also harmful due to presence of antinutritional activity. Therefore, the plant breeders should always be cautious about the possibility of production of undesirable components, like antinutritional factors (Osagie, 1998), at the time of developing higher yielding or disease-resistant crop varieties. It has been reported that certain food substances that are relatively safer when consumed individually, however, can have serious and even fatal effects when taken together (Osagie, 1998), for example, presence of tannins in a protein-marginal diet.

7.2 CLASSIFICATION OF SMS

SMs are categorized into five major categories: isoprenoids (terpenoids and steroids), polyketides, phenylpropanoids, flavonoids, and alkaloids (Oksman-Caldentey & Inzé, 2004). Of these, isoprenoids are the most prevalent SMs followed by alkaloids and flavonoids in plants. Many of these metabolites are found in legumes in varying amount; detailed description is given in the next section.

7.3 DISTRIBUTION OF SMS IN LEGUMES

7.3.1 PHENOL COMPOUNDS

Tannins are water soluble phenolic compound (molecular weight ranging from 500 to >20,000) that may form a less digestible complex with dietary proteins and may bind and inhibit the endogenous protein such as digestive enzymes. They constitute plants' protection mechanisms against external factors and follow shikimic acid pathway, also known as the phenylpropanoid pathway. Tannins are more common in dicotyledonous plant than monocot and are widespread in fruits, vegetables, pulses, and in some cereals. They cause plants to have a pungent taste and also they reduce bioavailability of some minerals (especially zinc). The major categories of tannins that have an impact on animal nutrition are hydrolyzable tannins and condensed tannins that are resistant to hydrolytic degradation. Tannins reduce the intake by decreasing palatability, effecting pH mechanism, and negatively effecting digestion of protein.

7.3.2 GOSSYPOL

Gossypol is another phenolic compound produced by pigment glands of cotton plant (*Gossypium* spp.), commonly found in its stems, leaves, seeds, and flower buds. Dietary gossypol is a by-product of cotton that is used for animal feeding because it is not only rich in oil and proteins (lysine), but also for iron which it can chelate (Aletor, 1993). Gossypols are reported to cause animal and human toxicity and high incidence of irreversible impairment of male (testicular damage) and female reproductive system which limit its use as animal feed. At the physiological level, gossypol reduces oxygen availability in the blood, and leads to impaired growth, apathy, weakness, hypertrophy, and dilution of heart muscles and changes in electrocardiogram. Skutches and Smith (1974) reported that proteins were reduced by approximately 20% in pigs fed 0.06% free gossypol in their diet.

7.3.3 RICIN

Ricin is a naturally occurring lectin (a carbohydrate-binding protein) in castor beans (*Ricinus communis*) which have been reported to cause poisoning in all classes of livestock. The lethal dose (LD_{50}) of ricin is around 22 μg/kg of body weight (1.78 mg for an average adult). However, the mature leaves of castor bean have been found suitable for feeding to the sheep (Behl et al., 1986), necessary precautions against bean contamination are required. Rao et al. (1988) reported that the castor bean meal can be detoxified by autoclaving it at 20 psi for 60 min.

7.3.4 OLIGOSACCHARIDES AND ISOFLAVONOIDS

In general, legume seeds are rich (up to 20%) in oligosaccharides (stachyose, raffinose, and verbascose) (Bisby et al., 1994; Aksar, 1986) and they produce flatulence and other disturbances in human and animals. Flatulent substances belong to the indigestible fiber group, which reduces the risk of intestinal cancer, increases excretion frequency, and weight as well as HDL cholesterol level. However, symptoms of flatulence are unpleasant, which (oligosaccharide content) can be reduced by increasing duration of cooking. Moreover, a decrease in oligosaccharide content also occurs when soaking water is poured, seeds are washed a second time, or seeds are germinated. Isoflavonoids have been detected in soybean, lupins, and several other legumes (Bisby et al., 1994) and are involved in plant defence mechanism against fungi, bacteria, viruses, and nematodes (phytoalexins, phytoanticipins). It also acts as signal molecule in legume—Rhizobium interactions and exhibit estrogenic activities (Dakora & Phillips, 1996).

7.3.5 PHYTOHEMAGGLUTININS

Lectins, also referred to as phytohemagglutinins, are glycoprotein compounds (molecular weight of 10,000–124,000) which have been shown to agglutinate red blood cells *in vitro* (Gatel, 1994) by affecting erythrocytes. Lectin activity has been determined in more than 800 varieties of the legume family, which includes 600 genera. It has also been reported that 2–10 % of the total protein in legume seeds are lectins. Soybean, chickpea, faba bean, pea, lentil, common bean, jack bean, and peanut are good sources of lectins. Peas generally have higher lectin activities than faba beans, but both show considerably lower amounts in comparison to raw defatted soybeans. One of their most important characteristics is that they prevent absorption of digestive end products in the small intestine. Apart from that lectins also possess some other interesting chemical and biological properties, some of which are as follows: they interact with specific blood groups, perform a range of functions in mitotic division, knock down cancerous cells, and have toxic effects in some animals. If some of the bean varieties are consumed raw, they may cause shock cramps (Saldamlı, 1998). Besides these characteristics, lectins can easily disintegrate (El-Adawy, 2002; Mubarak, 2005).

7.3.6 AMYLASE INHIBITORS

Amylase inhibitors are present in many types of beans; commercially available amylase inhibitors are extracted from white kidney beans. It prevents the action of enzymes that break the glycosidic bonds of starches and other complex carbohydrates, for the release of simple sugars and thus reduces its absorption by the body. Hence, amylase inhibitors, like lipase inhibitors, have been used as a diet aide and obesity treatment.

7.3.7 CYANOGENIC GLUCOSIDE

The cyanide contents of some legumes have been investigated for long years. Cyanogens are glycosides of 2-hydroxynitriles and are widely distributed among plants, known to be present in more than 2500 plant species, for example, the Rosaceae, Leguminosae (*Phaseolus lunatus* and *Vicia sativa*), Gramineae, and Araceae (Conn, 1980; Bisby et al., 1994; Conn, 1981; Rosenthal, 1991). Amygdalin is the cyanogenic glycoside responsible for the toxicity of the seeds of many species of Rosaceae, such as bitter almonds, peaches, and apricots. However, in sweet almonds, amygdalin contents are low, which have been developed through conventional breeding method.

Cyanogen compounds of tall plants are of two types: cyanogens glycosides and cyanogen lipids. Both groups contain cyanohydrins and free carbonyl. Since cyanogenic glycosides, which consist of HCN (hydrocyanohydric acid), can come out as a result of hydrolysis, they are potentially toxic. HCN is highly toxic for animals or microorganisms because it inhibits enzymes of the mitochondrial respiratory chain (i.e., cytochrome oxidases). The cyanide ions also inhibit several enzyme systems and reduce growth through interference with certain essential amino acids. They also cause acute toxicity, neuropathy, and death (Osuntokun, 1972; Fernando, 1981).

The lethal dose of HCN for cattle and sheep is 2.0–4.0 mg/kg body weight. The lethal dose for cyanogens would be 10–20 times greater because the HCN comprises 5–10% of their molecular weight (Conn, 1979). Cyanogens have also been suspected to have teratogenic effects (Keeler, 1984). Postharvest wilting of cyanogenic leaves may reduce the risk of cyanide toxicity. In addition to the toxic effects, cyanogens can serve as mobile nitrogen storage compounds in seeds which is important during germination.

Animals suffering from cyanide poison must be immediately treated by injecting a suitable dose of sodium nitrate and sodium thiosulphate. In case of emergency, when plants are wounded by herbivores or other organisms, the cellular compartmentation breaks down and cyanogenic glycosides come into contact with an active glucosidase, which hydrolyses them to yield 2-hydroxynitrile. This is further cleaved into the corresponding aldehyde or ketone and HCN by hydroxynitrile lyase.

7.3.8 PYRIMIDINE GLYCOSIDES

Vicine and convicine are β-glycosides that constitute about 0.5% of the *V. faba* seeds (wlw) and can serve as the source of the noxious agents such as pyrimidines divicine and isouramil. Favism is an acute hemolytic crisis induced by ingestion of meals prepared from *V. faba* by individuals deficient in the nicotinamide adenine dinucleotide phosphate hydrogen (NADPH) producing enzyme glucose-6-P-dehydrogenase (G6PD) in the red blood cells (Mager et al., 1980; Marquardt, 1989). Several genetic variants of deficiency of G6PD enzyme, known to confer an adaptive advantage against malaria, are identified to occur worldwide (Vulliamy et al., 1992). It is particularly prevalent in some Mediterranean and southwest Asian populations. Vicine and convicine have been implicated in favism because their hydrolysis products are unstable and form radicals, deplete reduced glutathione (GSH) in G6PD-deficient red blood cells. A lack of sufficient NADPH due to G6PD-deficiency impedes GSH replenishment and, ultimately, results in a hemolytic crisis. Antioxidants have been successfully employed to reduce the effects of vicine in animal diets (Mager et al., 1980; Marquardt, 1989; and references therein). Efforts are underway to develop cultivars of *V. faba* with zero levels in their seeds and some promising material has already been identified (Griffiths & Ramsay, 1992).

7.3.9 LUPIN ALKALOIDS

The main SMs of lupins are quinolizidine alkaloids accompanied by piperidine alkaloids, namely, ammodendrine or indole alkaloids, for example, gramine (Bisby et al., 1994). After synthesis of lupin in chloroplasts, alkaloids are exported through the phloem all over the plant and accumulate in the epidermis of leaves, stems, and in reproductive organs (2–6% dry weight) (Wink, 1987, 1992, 1993b).

Lupin alkaloids were deterrents and lethal for a number of insects, especially aphids (Berlandier, 1996), moth, butterfly larvae, beetles, grasshoppers, flies, bees, ants, invertebrates, and vertebrates. Furthermore, these alkaloids are inhibitory for competing plants, viruses, bacteria, and fungi (Wink, 1985, 1988, 1992, 1993a). Since the alkaloids are toxic to humans and animals, breeders have selected lupin varieties with very low alkaloid contents called "sweet" lupins (Wink, 1985, 1988, 1992, 1993a,b). The genus *Lupinus* contains several hundred species and their evolution has recently been described by Käss and Wink (1997) using nucleotide sequences of marker genes. Since some species have a lager and protein-rich seeds (up to 50% protein), these lupins are of considerable agricultural interest and presented in cultivated species like *L. albus*, *L. luteus*, *L. angustifolius*, and *L. mutabilis*.

7.3.10 SAPONINS

Saponins are glycosides with a distinctive foaming characteristic, bitter in taste, causing hemolysis in RBC. They are found in many plants, but their name is derived from the Latin word *sapo* from the soapwort plant (*Saponaria*). They are either a choline steroid or triterpenoid attached via C3 and an ether bond to a sugar side chain. Saponins are glycosides with a polycyclic aglycone moiety of either C27 steroid attached to a carbohydrate. They are widely distributed in soybean, chickpea, faba bean, pea, lentil, and peanuts, in different form of saponins.

Erythrocytes compounds lyse in saponin solution and are toxic (Khalil and El-Adawy, 1994) and their antinutritional effects have been studied using *alfalfa* saponins. Saponins cause hypocholesterolemia by binding cholesterol, making it unavailable for absorption and also cause hemolysis of red blood cells (Johnson et al., 1986). Saponins significantly inhibited acrosine activity of human sperms and the spermicidal effect was attributed to strong damage of the spermal plasma membrane and cause infertility (Su & Guo, 1986; Olayemi, 2007).

Soybean, lupins, and several other legumes have triterpene saponins that are amphiphilous compounds (Price et al., 1987; Bisby et al., 1994). Saponins interact with biomembranes of animals, fungi, and even bacteria. The hydrophobic part of the molecule complex cholesterol inside the membrane and their hydrophilic sugar side chain binds to external membrane proteins. This fluidity of biomembranes is disturbed, leading to holes and pores. As a consequence, cells become leaky and die. Saponins can be considered as a resistance factor in legumes against microbial infection and herbivores. Legume saponins have moderate toxicity and are present in the diet at higher concentrations.

7.3.11 MIMOSINE

Mimosine [β-*N*-(3-hydroxy-4-pyridone)-*a*-aminopropionic acid] is a nonprotein amino acid (NPAA) structurally similar to tyrosine found in the genera *Leucaena* and *Mimosa*. The mimosine contents vary in different parts of plant, as in seeds it ranges from 4 to 5% on a dry-weight basis, the root contains 1–1.5%, and in shoot 1–12%, while old stems contain the minimum amount as compared to growing tips (Jones & Lowry, 1984). The mechanism of action of mimosine is not clear, but it may act as an amino acid antagonist or may complex with pyridoxal phosphate, leading to disruption of catalytic action of B6-containing enzymes such as *trans*-aminases, or may complex with metals such as zinc (Hegarty et al., 1964). The poor growth, loss of hair and wool, swollen and raw coronets, depressed serum thyroxine level, and goiter are symptoms of toxicity in ruminants (Jones & Hegarty, 1984).

7.3.12 PHYTATES

Phytates are the calcium–magnesium–potassium salt of inositol hexaphosphoric acid commonly known as phytic acid. Phytate is the main storage form of phosphorus in soybean and the content ranges from 1.0 to 1.47% on a dry matter (DM) basis. Phytate is located in the protein bodies, mainly within their globoid inclusions. Phosphorus in the phytate form could not be absorbed by monogastric animals, due to lack of phytase, the digestive enzyme required to release phosphorus from the phytate molecule. Phytic acid could form protein phytate or protein–phytate–protein complexes; this has more resistance to digestion by proteolytic enzymes. Phytic acid has also a strong binding affinity to important minerals such as calcium, magnesium, iron, and zinc. Once after a mineral binds to phytic acid, it becomes insoluble, precipitates, and is not absorbed in the intestines.

In food industry, the presence of phytic acid in high concentration is undesirable. In the feed industry, the unabsorbed phytate passes through the gastrointestinal tract of monogastric animals, elevating the amount of phosphorus in the manure. Excess phosphorus excretion can lead to environmental problems such as eutrophication. With the pressure on the swine industry to reduce the environmental impact of pork production, it is important to use feed ingredients that can minimize this influence. The ability of the molds for oriental fermented soybean food to produce phytase has been investigated. Phytase is an enzyme, hydrolyzing phytic acid to inositol and phosphoric acid and thereby removing the metal-chelating property of phytic acid. Inositols can reach concentrations higher (10% of DM) with 4, 5, or 6 phosphate groups in the seeds of many of grain legumes (Bisby et al., 1994), as stores for phosphate and mineral nutrients that are important for valuable plant nutrition during germination. Since phytates complex with iron, zinc, magnesium, and calcium ions in the digestive tract, they cause mineral ion deficiency in animals and humans (Nelson et al., 1968). These compounds serve double purpose, that is, defense, phosphate, and mineral store. (Bardocz et al., 1996).

7.3.13 OXALATES

Oxalates, like phytates, bind minerals like calcium and magnesium and interfere with their metabolism. Oxalates cause the muscular weakness and paralysis, gastrointestinal tract irritation, blockage of the renal tubules by calcium oxalate crystals, development of urinary calculi, hypocalcemia (Oke, 1969; Blood & Radostits, 1989), and nephrotic lesions in the kidney. Oxalate, phytate, and tannins are antinutrients toxic in an unprocessed food (Ojiako & Igwe, 2008). The bioavailability of the essential nutrients in plant foods could be reduced by the presence in these plants of some antinutritional factors, such as oxalates and cyanogenic glycosides (Akindahunsi & Salawu, 2005). Too much of soluble oxalate in the body prevents the absorption of soluble calcium ions as the oxalate binds to the calcium ions to form insoluble calcium oxalate complexes as a result form kidney stones (Adeniyi et al., 2009).

7.3.14 TOXIC AMINO ACIDS

Certain amino acids in legume plants, which are not of protein in nature, reduce the nutritious value and cause toxic effects. Dihydroxyphenylalanine is the most common example of such kind of toxic amino acid found in legumes (*Lathyrus* and broad beans). Although these amino acids do not display a direct toxic effect, the plant first takes on a black color due to these substances and then withers. Moreover, the nutritional value of plants that contain such amino acids (broad beans, Lathyrus) decreases substantially. Toxic amino acids are believed to combine causes of metabolic favism. Despite all these, this substance cannot harm, due to the need to be in large quantities in the plant to pose a risk. In addition to this group of toxic substances, some legumes may contain sparing amounts of anti-vitamin substances and estrogen factors.

Substances of this kind may be activated with heat and cause serious harm. Extensive studies are reported in the relevant literature on the elimination of these substances in order to reduce their harmful effects on plants (Desphande & Cheryan, 1984). When it is taken into account that pulses are the sources of the highest quality vegetable proteins, the importance of studies on the mechanisms of toxic amino acids that have an unfavorable effect on the quality of this protein and the degree of their potential harm become obvious.

7.3.15 GOITROGENS

Soybean, a kind of oil seed crop, contains glycosides called goitrogens. These glycosides contain sulfur, and cause the thyroid gland to grow by inhibiting the iodine intake of the thyroid gland. These toxic effects can be reduced with the addition of iodine to the diet.

7.4 PULSES AND SMS

7.4.1 CHICKPEA

The chickpea seed contains a range of chemical substances that are known to cause digestive disturbances when ingested by humans or animals. The nutritive value and protein digestibility of legumes are generally poor unless the seeds are cool, mainly because of protein inhibitors and heat labile compounds. Of the various AFN, protease inhibitors, amylase inhibitors, oligosaccharides (mainly trisaccharides), and polyphenols of chickpea have been the subject of several studies.

TABLE 7.1 Number of Plant Secondary Metabolites (De Luca & St Pierre, 2000; Wink, 2003).

Type of SMs	Number*
Isoprenoids	30,000
Alkaloids	12,000
Flavonoids	4000
Phenylpropanoids	2500
Polyacetylenes, fatty acids, waxes	1000
Polyketides	750
Nonprotein amino acids	600
Carbohydrates	200
Amines	100
Cyanogenic glycosides	100
Glucosinolates	100

*approximate number of known structures.

7.4.2 FIELD PEA

Trypsin inhibitor and lectins are main antinutritional factors in pea belonging to albumins. Apart from these ANFs, tannins and lectins are also present in pea cultivars having colored flower and are inactivated by heating or soaking in formaldehyde. It may be used for both forage and grain production. As per the activity of trypsin inhibitor expressed by trypsin inhibitor unit (TIU) per DM, field pea cultivars are grouped into four major classes: very low (2–4 TIU mg-1 DM), low (4–7 TIU mg-1 DM), medium (7–10 TIU mg-1 DM), and high activity (10–13 TIU mg-1 DM). However, achievements were made in breeding field pea cultivars with very low trypsin inhibitor activity on the demand of animal husbandry, which made farmers independent from processing industry and provided them with an excellent source of quality plant protein (Mikić et al., 2003). However, it is confirmed that the environment plays an important role in the expression of trypsin inhibitor activity, hence, it is obvious that the genotype remains the most important factor, thus underlining the role of breeding.

7.4.3 FABA BEAN AND OTHER FOOD LEGUMES

In faba bean, tannins are main antinutritional factors. The cultivars without tannins, commonly known as zero-tannin or tannin-free faba bean, have found a wide application for both human and animal consumption and essentially ensured the place of faba bean in feed production. The negative correlation between the zero tannins content and winter hardiness in faba bean proved a possible cause and has resulted in the development of winter tannin-free cultivars (Link et al., 2008). It has also been reported that in pea there is a strong positive correlation between white color of the flowers and reduction of tannin content, which is oligogenic and controlled by recessive genes (Duc, 1997).

The main antinutritional factors in lupins are alkaloids, together with phytates, protease inhibitors, and lectins. The development of sweet types of white and other lupins with low alkaloid content is one of the major achievements in lupin domestication. Modern sweet lupin cultivars usually have an alkaloid content of less than 200 mg/kg (Cowling et al., 1998), with a strict regulation on largest lupin producer countries such as Australia and Poland. However, the cultivars with high alkaloid content, called bitter lupins, have their role as forage or green manure crops.

7.4.4 LATHYRUS

It has been reported that epidemiological association exists between the consumption of grass pea (*L. sativus*) and lathyrism. A neurotoxin, β-*N*-oxalyl-L-a, β-diaminopropionoc acid (ODAP) plays the main antinutritional role in grass pea and causes a neurotoxic disease called lathyrism. The ODAP is present in all parts of the plants and its quantity depends on both genotype and environment. Breeding cultivars having low ODAP content is the best solution. The alternate solution to reduce the toxicity is using various methods of detoxifications like cold and hot water, steaming, roasting, and fermentation.

7.5 EFFECTS OF SMS ON HUMAN AND ANIMAL HEALTH

Many SMs play important role in human and animal health. For example, soybean is a rich source of isoflavones, reported to show putative anti-estrogenic effects and also reduces breast cancer and prostate cancer risk (Messina, 1999). The weak estrogenic effects of isoflavones and the similarity in chemical structure between soybean isoflavones and the synthetic isoflavone ipriflavone, which was shown to increase bone mineral density in postmenopausal women, suggest that soy or isoflavones may reduce the risk of osteoporosis (Messina, 1999). Another legume, *Phaseolus vulgaris* L., commonly known as Mexican common beans, is an excellent source of nutraceutical enriched with fiber, phytic acid, protease inhibitors, flavonoids, lignans, and tannins. The seed coat color depends on the presence of polyphenolic compounds such as flavonol, glycosides, tannins, and anthocyanins, having antioxidant, antimutagenic, and anticarcinogenic activities (Espinosa-Alonso et al., 2006).

Isoflavones also possess biological functions other than estrogen-related activity, for example, antioxidant activity. Pulses contain antioxidants such as tocopherols, flavonoids, and isoflavonoids. The ascorbic acid, also having antioxidant activity, is found in germinated pulses and immature green pulses that increase two to three times after germination (Sharma, 1981). The important isoflavones in common legumes such as soybean, Bengal gram, green gram, red gram, and black gram are daidzein, genistein, daidzin, genistin, biochinin A, protensin, formononetin (Narasinga Rao, 2002). There are earlier reports on the blood cholesterol-lowering potential of chickpea (Bengal gram) and soybean. The isoflavone content is believed to impart hypocholesterolemic properties in these legumes (Sharma, 1979). The content of isoflavone in pulses is reported to increase post germination; however, the potency of isoflavones decreases after cooking (Sharma, 1981).

There are SMs (ANFs) acutely toxic, for example, lectins, cyanogenic glycosides, NPAAs or alkaloids, and unpalatable saponins, tannins, bitter alkaloids or antinutritive, that could reduce the growth and fitness of humans as well as animals. These SMs impair several biological activities, such as nutrient complication (phytates), metabolic inhibition (NPAAs, cyanogenic glycosides, isoflavines, alkaloids), and reduction of digestion (protenase inhibitors, lectins, or oligosaccharides). These SMs are prevalent in different pulses in varying quantities, most common being lectins and protease inhibitors, which are high molecular weight toxins found in seeds. Protease inhibitors are known for blocking the function of digestive enzymes (proteases) in animals, thereby leading to malnutrition and other related disturbances. As reported, feeding of raw soybean with trypsin inhibitor enlarges the pancreas (Gilani et al., 2012), while lectin binds to intestinal track and inhibits protein biosynthesis. In general, plants use these compounds for nitrogen storage and defense chemicals against animals. Saponins, a triterpene or steroid derivatives (attached to one or more sugar moieties), have moderate toxicity, and health problems occur when consumed in high amounts. It affects lipid membranes and cause hemolysis *in vitro* by a mechanism, involving cholesterol binding (Lin & Wang, 2010). Once hydrolyzed in the intestine track, some saponins may cause systemic toxic effects (e.g., weight loss, gastroenteritis) in ruminants, because of microbial fermentation as well as synthesis in the rumen gets inhibited (Lu & Jorgensen, 1987).

The oligosaccharides such as stachyose and raffinose content in the legume seeds are up to 20% (Bisby et al., 1994). In animals, these contents cause flatulence and other complications. Besides, they also serve as carbon sources during germination in soybeans. Vicine is a β-glycoside present in the seeds of *V. sativa* and *V. faba*. The consumption of seeds of *V. faba* causes an acute hemolytic disease called favism that lacks activity of the NADPH in RBCs (Mager et al., 1980, Marquardt, 1989). Favism in sensitive individuals with consumption of broad beans is found widely in people living in the Mediterranean countries and known to be of genetic source (Liener, 1983).

The structure changes in hemoglobin, a primary carrier of oxygen in blood. Dizziness, vomiting, feeling of tiredness, and dark orange urine are the symptoms of this disease. The disease disappears soon, but fatal incidences may be encountered in chronic condition. Favism also causes high fever and jaundice. NPAAs are especially abundant in certain legumes (Vicieae, Phaseoleae, Mimosoideae, etc.). Consumption of plants rich in NPAAs leads to formation of defective proteins; also effects may include fetal malformations, neurotoxins, hallucinogenic effects, hair loss, diarrhea, paralysis, liver cirrhosis, hypoglycemia, and arrhythmia.

7.6 INTERFERENCE OF ANFS IN BIOAVAILABILITY OF NUTRIENTS

Dietary ANFs have been reported to adversely affect the digestibility of amino acids, minerals, fats, and protein quality of food (Gilani et al., 2012). Poor digestibility of protein-containing less-refined legumes and cereals as a major source of protein, found in the diet of developing countries, is due to the presence of less digestible protein fractions, high levels of insoluble fiber, and high concentration of ANFs (Gilani et al., 2012). Phytic acid may influence digestive enzymes by binding minerals in the gut before they are absorbed. Phytate reduces mineral bioavailability, as this is very reactive with other positively charged ions, such as calcium, magnesium, zinc, and iron. It forms insoluble complexes with these ions that are less available for digestion and absorption in the small intestine (Cheryan, 1980; Sandberg, 2002; Champ, 2002; Gebrelibanos et al., 2013).

Phytic acid ability to bind Zn reduces the Zn:Cu ratio in plasma, and this lower ratios tend to predispose man to cardiovascular disease (Klevay, 1977). Moreover, phytic acid reduces glucose and insulin concentrations in the plasma in humans, which may lead to a reduced stimulus for the lipid synthesis in the liver (Wolever, 1990). Like phytates, too much of soluble oxalate prevents the absorption of soluble calcium ions; it binds to the calcium ions to form insoluble complexes of calcium oxalate. This is the one reason oxalate-rich foods are harmful for and avoided by the people with the tendency to form kidney stones (Adeniyi et al., 2009). Phytates also reduce the digestibility of starches, proteins, and fats, and bioavailability of iron to the body. For example, 5–10 mg of phytic acid intake can reduce absorption of iron by 50%. Even phytic acid can also bind with the minerals such as zinc and manganese, besides iron in the intestines (Biehl et al., 1995). Once bound, they are then excreted in waste. Phytic acid may also have a negative effect on amino acid digestibility, and the content of phytates in cereals and legumes can reduce protein and amino acid digestibility. The high levels of trypsin inhibitors in diet from soybeans, kidney beans, or other grain legumes have been reported to cause considerable reductions in protein and amino acid digestibility (up to 50%) and protein quality (up to 100%) in rats and/or pigs (Gilani et al., 2012). Protease inhibitors retard growth by interfering with dietary protein digestion, and as a mechanism to adapt to this situation, the secretory activity of the pancreas can be stimulated and its size enlarged, leading to an overall physiological effect of impaired growth and enlarged pancreas (Gebrelibanos et al., 2013). Lectins strongly resist proteolytic degradation in the gut and bind to surface receptors on the cell linings of the small intestines and interfere with the absorption of nutrients; the result is a failure in growth and eventual death (Gebrelibanos et al., 2013).

Tannins are other phenolic compounds that show antinutritional and toxic effects in the form of depression of food intake, formation of the less digestible enzymes, increased excretion of endogenous protein, digestive tract malfunctions, and toxicity (Jansman & Longstaff, 1993). Tannins also form complexes with divalent metals and reduce mineral absorption (Gebrelibanos et al., 2013). Alpha-amylase inhibitors form a complex with amylase which can inactivate amylase, thereby reducing starch digestion and can cause pancreatic hypertrophy (Champ, 2002).

7.7 ROLE OF SMS IN PULSES

SMs are involved in various functions of plants, directly or indirectly such as hormone, components of cell membrane, enhance the mechanical barrier functions of cell membranes, communication, defense, and enhance thermo-tolerance or quench oxidative stress. In pulses, flavanoids are widespread SMs of polyphenolic group of compounds that also includes tannins and take part in plants' defense against fungi and insects (Butler, 1989, Beecher, 2003). Several plant chemical components show insecticidal properties either as whole leaves, powders, or aqueous or alcoholic extracts (Aletor, 1999). Some of the compounds with insecticidal activities comprise tannins, flavanoids, alkaloids, terpenoids and others, which do not let many polyphagous insects to feed on plants that accumulate such compounds. Thus, these plants are protected against a wide range of insect pests (Nahrstedt, 1988). Soybean is one notable example, known to possess several insecticidal compounds and shows resistance to many insects (Jones & Sullivan, 1979, Sharma & Norris, 1991). The hemagglutenins have been reported as the active principle having insecticidal activity in the winged bean (*Psophocarpus tetragonolobus*) against the larvae of the seed beetle (*Callosobruchus maculatus*) (Gatehouse et al., 1991). Several other leguminous seeds contain range of SMs like ANFs that also protect them against insect attacks (Gate House, 1989). For example, a lectin reported from *Phaseolus vulgaris*, an insecticidal to *C. maculatus*, and trypsin inhibitors from cowpea (*Vigna unguiculata*), an antimetabolic to *C. maculatus* (Gatehouse et al., 1979; Gatehouse et al., 1984). Sharma and Norris (1991) isolated two flavonoids and glyceolins-like ANFs that act against the larvae of the cabbage looper, *Trichoplusia ni,* from soyabens. In the chemical industry, flavanoids are used in the manufacture of insecticides using the isoflavanoid rotenone (Harborne, 1967). Pulses contain several SMs, such as protease inhibitors (trypsin and chymotrypsin), lectins, polyphenols, flatulence factors, lathyrogens, saponins, antihistamines, and other allergens. The protease inhibitors, lectins, and other antinutrients cause toxicity. Protease inhibitors are molecules that inhibit the function of proteases. The protease inhibitors are known to link with one disease known as alpha 1-antitrypsin deficiency.

Lectins are glycols, carbohydrate-binding proteins, ubiquitous in nature. They are widely distributed among plant families and consumed in appreciable amounts by both humans and animals on daily basis. These lectins have nutritionally important features like their ability to survive digestion by the gastrointestinal tract of consumers by binding to membrane glycosyl groups of the cells, lining the digestive tract. They play important role in the immune response by recognizing carbohydrates adherent on pathogens or that are inaccessible on host cells. While the high concentration of lectins in plant seeds decreases with growth, it is suggestive of its role in plant germination, and thereby seed's survival of itself. Other important function of glycoproteins in host–parasite interaction is their binding on the surface of parasitic cells. Some of the plant lectins have been found to recognize non-carbohydrate ligands of primarily hydrophobic in nature, for example, adenine, auxins, and cytokinins as well as water-soluble porphyrins. Such interactions may be physiologically important, as some of these molecules play a role of phytohormones (Komath et al., 2006). These phytohormones also help in increasing disease and insect resistance. This results in interaction of a series of harmful local and systemic reactions, placing these molecules as antinutritive and toxic ones. They are also detrimental to numerous insect pests of crop plants, although the insecticidal mechanisms of action are largely unknown (Vasconcelos & Oliveira, 2004).

The biological attributes of lectins can also be realized as in the development of the biochemical warfare agent that binds cell surface galactosyl residues and enables the insect and fungal protein to enter cells leading to inhibition of protein synthesis and cell death. Among grain legumes, soybean seeds contain the high activity of lectins.

Trypsin inhibitors a which reduce the availability of trypsin, are essential enzymes responsible for the nutrition of many animals and human. This is now possible to eliminate such ANFs using genetic engineering, but the other problems that may result must be looked into, like the beneficial effects of these molecules in plants against disease and insecticidal resistance. Flavanoids are the most polyphenolic compounds ANFs in pulses as secreted by the root of the host plant and help rhizobia to establish symbiotic relationship, in crops like peas, beans, clover, and soybean. The rhizobia living in the soil sense these flavanoids and as a result trigger the secretion of nod factors that are recognized by the host plant. This can also lead to root hair deformation and several cellular responses such as ion fluxes and the formation of root nodule. These flavonoids have also been shown to have antifungal activity *in vitro*. *In vitro* studies of flavanoids have displayed antimicrobial activities (Bjerg, 1984; Cushnie & Lamb, 2005).

Polyphenols are ubiquitous in pulses that affect several physiological processes, including plant defense against pathogens and insects (Piyada et al., 2007). The fungicidal activity of isoflavones from soybeans and chickpeas on three food containing fungi, *Aspergillus ochracens*, *Penicillium digitatum*, and *Fusarium culmorum* (Krammer et al., 1984) are reported. Thus, the plants often defend themselves against pathogens through the production and or accumulation of ANFs. These ANFs inhibit the germination of fungal spores and microorganisms and protect the plants. Plant biotechnology techniques like tissue culture, genetic manipulation, and other modern plant breeding methods may play a crucial role in optimizing the beneficial effects of these ANFs in pulses.

In addition to being perfect sources of vegetable protein, pulses contain nutrients with high fiber content and lower blood cholesterol levels, thereby benefiting the human health. It is generally recommended on cold days to increase legume consumption which balances the need of increased energy demand. Soaking before cooking is the most traditional, economic, and appropriate method.

Pulses contain many ANFs such as trypsin inhibitors, phytic acid, tannins, and oligosaccharides that limit their utilization. Trypsin inhibitors inactivate the digestive enzyme, trypsin (Dave Omaha et al., 2011); phytic acid lowers the bioavailability of minerals (Reddy et al., 1988); tannins decolorize the seed of lentils, thereby reducing their nutritional quality. Flavonoids of plants may be beneficial, in a sustainable way for their ability to fix atmospheric nitrogen in symbiosis with rhizobia (Velazquez et al., 2010).

CONCLUSIONS

Pulses are the cheapest source of protein and other health-promoting content providing nutrition for human beings. The edible legumes are an important source of dietary protein in vegetarian diets. However in most of the legumes, the biologically active SMs, such as saponins, tannins, flavonoids, alkaloids, protease inhibitors, oxalates, phytates, lectins (hemaglut-tinins), cyanogenic glycosides, cardiac glycosides, coumarins, gossypol, and others, hinder the efficient utilization of health-promoting food nutrients. Nevertheless, addition to these health-promoting components in the form of SMs show a range of activity in human and animals, from anti-cancer, antidiabetes, reducing risk of cardiovascular disease, antioxidants to many others. In plants, their role of disease and insect pest resistance and signaling network are well established. Besides good metabolites, there are some bad with respect to animal and human health, which are referred to as toxic or ANFs. Pulses also contain such biologically active ANFs that bring out physiological effects beyond those linked with vital human sustenance. The literature published over the last 50 years reveals that plenty of research has been carried out in the direction of identification, elucidation of properties, and mode of action of SMs including nutritional factors as well as ANFs of pulses with regard to their role *in planta* and on animal and human. Moreover, few efforts have also been made to downgrade and to mitigate the effects of ANFs, with limited success. It is essential to note that ANFs are the natural defense system in pulses to protect the plants from several biotic and abiotic stresses. Therefore, keeping in view the metabolites profile in legumes in general, and pulses in particular, and their intended and unintended biological effects, the plant breeding strategies should be adopted accordingly, and further research is needed.

KEYWORDS

- **food legumes**
- **secondary metabolites**
- **antinutritional factor**
- **soybeans**
- **tannins**

REFERENCES

Adeniyi, S. A.; Orjiekwe, C. L.; Ehiagbonare, J. E. Determination of Alkaloids and Oxalates in Some Selected Food Samples in Nigeria. *Afr. J. Biotechnol.* **2009,** *8*(1), 110–112.

Akindahunsi, A. A.; Salawu, S. O. Phytochemical Screening and Nutrient and Anti-Nutrient Composition of Selected Tropical Green Leafy Vegetables. *Afr. J. Biotechnol.* **2005,** *4*, 497–501.

Aksar, A. Faba Beans (*Vicia faba* L.) and Their Role in the Human Diet. *Food Nutr. Bull.* **1986,** *8*, 15–24.

Aletor, V. A. Allelochemicals in Plant Foods and Feeding Stuffs. Part I. Nutritional, 2. Biochemical and Physiopathological Aspects in Animal Production. *Veterinary Hum.* 3 *Toxicol. J.* **1993,** *35*(1), 57–67.

Aletor, V. A. Antinutritioanl Factors as Nature's Paradox in Food and Nutrition Securities. Inaugural Lecture Series Delivered at the Federal University of Technology, Akure on Thur, August 12, 1999.

Amarowicz, R.; Pegg, R. B. Legumes as a Source of Natural Antioxidants. *Eur. J. Lipid Sci. Technol.* **2008,** *110*, 865–878.

Anderson, J. W.; Major, A. W. Pulses and Lipaemia, Short and Long-term Effect Potential in the Prevention of Cardiovascular Disease. *Br. J. Nutr.* **2002,** *3*(88), S263–S271.

Bardocz, S.; Gelencser, E.; Pusztai, A. *Effects of Antinutrients on the Nutritional Value of Legume Diets*, Vol. 1. ESSE-EC-EAEC: Brussels, 1996.

Bazzano, L. A.; He, J.; Ogden, L. G.; Loria, C.; Vupputuri, S.; Myers, L. Legume Consumption and Risk of Coronary Heart Disease in US Men and Women, Epidemiologic Follow-up Study. *Arch. Intern. Med.* **2001,** *161*, 2573–2578.

Bazzano, L. A.; Tees, M. T.; Nguyen, C. H. Effect of Non-soy Legume Consumption on Cholesterol Levels: A Meta-Analysis of Randomized Controlled Trials, Abstract 3272, *Circulation* **2008,** *118*, 1122.

Beecher, G. R. Overview of Dietary Flavanoids Nomenclature Occurrence and Intake. *J. Nutr.* **2003,** *133*(10), 3248s–3254s.

Behl, C. R.; Pande, M. B.; Pande, D. P.; Radadia, N. S. Nutritive Value of Wilted Castor (*Ricinus communis* L.) Leaves for Crossbred Sheep. *Indian J. Animal Sci.* **1986,** *56*, 473–474.

Berlandier, F. A. Alkaloid Level in Narrow Leafed Lupin, *Lupinous angustifolius* Influences Green Peach Aphid Reproductive Performance. *Entomol. Exp. Appl.* **1996,** *79*, 19–24.

Biehl, R. R.; Baker, D. H.; De Luca, H. F. 1a-Hydroxylated Cholecalciferol Compounds Act Additively with Microbial Phytase to Improve Phosphorus, Zinc and Manganese Utilization in Chicks Fed Soy-Based Diets. *J. Nutr.* **1995,** *125*, 2407–2416.

Bisby, F. A.; Buckinham, J.; Harborne, J. B., Eds. *Phytochemical Dictionary of the Leguminosae*. Vol. 1, Vol. 2. Chapman & Hall: London, 1994.

Bjerg, B.; Heide, M.; Norgaad Knudsen, J. C.; Sorensen, H. Inhibitory Effects of Convicine, Vicine and Dopa from *Vicia faba* on the In Vitro Growth Rates of Fungal Pathogens. *Zeitschr. Pflanzenkr. Pflanzensc.* **1984,** *91*, 483–487.

Blood, D. C.; Radostits, O. M. *Veterinary Medicine*, 7th ed. Balliere Tindall: London, 1989, pp 589–630.

Butler, L. G. Effects of Condensed Tannin on Animal Nutrition. In *Chemistry and Significance of Condensed Tannins*; Hemingway, R. W., Karchesy, J. J., Eds.; Plenum Press, New York, 1989, 391–402.

Champ, M. M. J. Non-Nutrient Bioactive Substances of Pulses. *Br. J. Nutr.* **2002,** *88*(Suppl. 3), S307–S319.

Cheryan, M. Phytic Acid Interactions in Food Systems. CRC *Crit. Rev. Food Sci.* **1980,** *13*, 297–335.

Conn, E. E. Cyanide and Cyanogenic Glycosides. In *Herbivores: Their Interaction with Secondary Plant Metabolites*; Rosenthal, G. A., Janzen, D. H., Eds.; A. P.: New York, 1979, pp 387–412.

Conn, E. E. Cyanogenic Glycosides. In *Encyclopedia of Plant Physiology* Bell, E. A., Charlwood, B. V., Eds. New Series 8. Springer: Berlin Heidelberg New York. C, 1980, pp 461–491.

Conn, E. E. Secondary Plant Products. *The Biochemistry of Plants*, vol. 7, 1981, Academic Press: New York.

Cowling, W. A.; Buirchell, B. J.; Tapia, M. E. Lupin. *Lupinus L.* Institute of Plant Genetics and Crop Plant Research: Gatersleben, Germany–International Plant Genetic Resources Institute: Rome, Italy, 1998, 105.

Cushnie, T. P. T.; Lamb, A. J. Antimicrobial Activity of Flavanoids. *Int. J. Antimicrob. Agents* **2005,** *26*(5), 343–356.

Dakora, F. D.; Phillips, D. A. Diverse Functions of Isoflavonoids in Legumes Transcend Anti-Microbial Definitions of Phytoalexins. *Physiol. Molecular Plant Pathol.* **1996,** *49*, 1–20.

Dave Omaha, B.; Fr Adeniyi, S. A.; Orjiekwe, C. L.; Ehiagbonare, J. E. Determination of Alkaloids and Oxalates in Some Selected Food Samples in Nigeria. *Afr. J. Biotechnol.* **2011,** *8*(1), 110–112.

De Luca, V.; St Pierre, B. The Cell and Developmental Biology of Alkaloid Dehulling on Nutritional Composition of Several Varieties of Lentils (*Lens culinaris*), **2000,** *5*(4), 168–173.

Desphande, S.; Cheryan, M. Effect of Phytic acid, Divalent Cations and Their Digestibility and Retention of Some Vitamins in Two Varieties of Chickpea. Food Directions in Plant Lectin Research. *Org. Biomol. Chem.* **1984,** *4*(6), 973–988.

Duc, G. *Faba bean (Vicia faba L.). Field Crops Res.* **1997,** *53*, 99–109.

El-Adawy, T. A. Nutritional Composition and Anti-Nutritional Factors of Chickpeas (*Cicer arietinum L.*) Undergoing Different Cooking Methods and Germination. *Plant Foods Hum. Nutr.* (formerly *Qualitas Plantarum*) **2002,** *57*(1), 83–87.

Espinosa-Alonso, L. G.; Lygin, A., Widholm, J. M.; Valverde, M. E.; Paredes-Lopez, O. Polyphenols in Wild and Weedy Mexican Common Beans (*Phaseolus vulgaris* L.). *J. Agric. Food Chem.* **2006,** *54*, 4436–4444.

Fernando, R. Plant Poisoning in Sri Lanka. In *Progress in Venom and Toxin Research.* Proc. of the 1st Asia-Pacific Congress in Animal, Plant and Microbial Toxins, 1981, pp 624–627.

Gatehouse, A. M. R.; Deurey, F. M.; Dove, J.; Fenton, K. A.; Puszai, A. Effect of Seed Lectins from *Phaseolus vulgaris* on the Development of *Callosobruchus maculatus* Mechanism of Toxicity. *J. Sci. Food Agric.* **1984,** *25*14, 373–380.

Gatehouse, A. M. R.; Gatehouse, J. A.; Bodie, P.; Kilmnoster, A. M.; Boulter, D. Biochemical Basis of Insect Resistance in Vigna Unguiculata. *J. Sci. Food Agric.* **1979,** *30*, 948–958.

Gatehouse, A. M. R.; Home, D. S.; Elemming, J. E.; Hilder, U. A.; Gatehouse, J. A. Biochemical Basis of Insect Resistance in Winged Bean (*Phosphocarpus tetragonolobus*) Seeds. *J. Sci. Food Agric.* **1991,** *55*, 63–74.

Gatel, F. Protein Quality of Legume Seeds for Non-ruminant Animals: A Literature Review. *Anim. Feed Sci. Technol.* **1994,** *45*, 317–348.

Gebrelibanos, M.; Tesfaye, D.; Raghavendra, Y.; Sintayeyu, B. Nutritional and Health Implications of Legumes. *Int. J. Pharm. Sci. Res.* **2013,** *24*(4), 1269–1279.

Gilani, S. G.; Xiao, C. W.; Cockell, K. A. Impact of Antinutritional Factors in Food Proteins on the Digestability of Protein and the Bioavailability of Amino Acids and on Protein Quality. *Br. J. Nutr.* **2012,** *108*, S315–S332.

Griffiths, D. W.; Ramsay, G. The Concentration of Vicine and Convicine in *Vicia faba* and Some Related Species and Their Distribution with in Mature Seeds. *J. Sci. Food Agric.* **1992,** *59*, 463–468.

Harborne, J. B. Comparative Biochemistry of the Flavanoids. Academic Press: London, 1967, pp 80–303.

Hegarty, M. P.; Schinckel, P. G.; Court. R. D. Reaction of Sheep to the Consumption of *Leucaena glauca* Benth. and to its Toxic Principle Mimosine. *Austr. J. Agric. Res.* **1964,** *15*, 153–167.

Hydowitz, J. D. Occurrence of Goitre in an Infant on a Soybean Diet. *New Engl. J. Med.* **1960,** *262*, 351–353.

Jansman, A. J. M.; Longstaff, M. Nutritional Effects of Tannins and Vicine/convicine in Legume Seeds. In *Recent Advances of Research in Antinutritional Factors in Legume Seeds.* Proceedings of the 2nd International Workshop on 'Antinutritional Factors (ANFs) in Legume Seeds', Wageningen, The Netherlands; Van der Poel, A. F. B., Huisman, J., Saini, H. S., Eds.; Wageningen Press: Wageningen, The Netherlands, EAAP Publication, No. 70, 1993; pp 301–316.

Johnson, I. T.; Gee, J. M.; Price, K.; Curl, C.; Fenwick, G. R. Influence of Saponin on Gut Permeability and Active Nutrient Transport *in vitro*. *J. Nutr.* **1986,** *116*, 2270–2277.

Jones, R. J.; Hegarty, M. P. The Effect of Different Proportion of *Leucaena leucocephala* in the Diet of Cattle on Growth, Feed Intake, Thyroid Function and Urinary Excretion of 3-hydroxy-4-(1*H*)-pyridone. *Austr. J. Agric. Res.* **1984,** *35*, 317.

Jones, R. J.; Lowry, J. B. Australian Goats Detoxify the Goitrogen 3-hydroxy-(1*H*) Pyridone (DHP) after Rumen Infusions from an Indonesian Goat. *Experientia* **1984,** *40*, 1435–1436.

Jones, W. A.; Sullivan, M. J. Soybean Resistance to the Southern Green Stink Bug *Nezera viridula*. *J. Econ. Entomol.* **1979,** *72*, 628–632.

Käss, E.; Wink, M. Phylogenetic Relationships in the Papilionoideae (Family Leguminosae) Based on Nucleotide Sequences of cpDNA (rbcL) and ncDNA (ITS 1 and 2) *Mol. Phylogenet. Evol.* **1997,** *8*, 65–88.

Keeler, R. W. Teratogenes in Plants. *J. Animal Sci.* **1984,** *58*, 1029–1039.

Khalil, A. H.; El-Adawy, T. A. Isolation, Identification and Toxicity of Saponin from Different Legumes. *Food Chem.* **1994,** *50*, 197–201.

Klevay, L. M. Hypocholesterolemia Due to Sodium Phytate. *Nutr. Rep. Int.* **1977,** *15*, 587–593.

Komath, S. S.; kavitha, M.; Swamy, M. J. Beyond Carbohydrate Binding: New Directions in Plant Lectin Research. *Org. Biomol. Chem.* **2006,** *4*(6), 973–988.

Krammer, R. P.; Hindorf, H.; Jha, H. C.; Kallage, J.; Zilliken, F. Antifungal Activity of Soyabean and Cowpea Isoflavones and their Reduced Derivatives. *Phytochemistry* **1984,** *23*(10), 2203.

Kurien, P. P. Advances in Milling Technology of Pigeonpea. *Proceedings of the International Workshop on Pigeon peas*, 15–19 December, ICRISAT Centre: Patancheru, Hyderabad, India, 1981, 321–328.

Liener, I. E. Toxic Constituents in Legumes. In *Chemistry and Biochemistry of Legumes*; Arora, S. K., Ed.; Edward Arnold: London, 1983; pp 217–257.

Lin, F.; Wang, R. Hemolytic Mechanism of Dioscin Proposed by Molecular Dynamics Simulations. *J. Mol. Model.* **2010,** *16*(1), 107–118.

Link, W.; Balko, C.; Stoddard, F. L. Winter Hardiness in Faba Bean: Physiology and Breeding. *Field Crop Res.* **2008**. DOI:10.1016/j.fcr.2008.08.004.

Lu, C. D.; Jorgensen, N. A. Alfalfa Saponins Affect Site and Extent of Nutrient Digestion in Ruminants. *J. Nutr.* **1987,** *117*, 919–927.

Mager, J.; Chevion, M.; Glaser, G. In *Toxic Constituents of Plant Foodstuffs*; Liener, I. E., Ed.; Academic Press: New York, 1980; pp 265–294.

Marquardt, R. R. In *Toxicants of Plant Origin. Vol. II. Glycosides,* CRC Press: Boca Raton, Florida, 1989, pp 161–200, Chapter 6.

Messina, M. J. Legumes and Soybeans: Overview of their Nutritional Profiles and Health Effects. *Am. J. Clin. Nutr*. **1999,** *70* (Suppl), 439S–450S.

Mikić, A.; Mihailović, V.; Katić S.; Karagić, Đ.; Milić, D. Protein Pea Grain—A Quality Fodder. *Biotechnol. Anim. Husbandry* **2003,** *19*(5–6), 465–471.

Mubarak, A. E. Nutritional Composition and Antinutritional Factors of Mung Bean Seeds (*Phaseolus aureus*) as Affected by Some Home Traditional Processes. *Food Chem.* **2005,** *89*, 489–495.

Nahrstedt, A. The Significance of Secondary Metabolites for Interaction between Plants and Insects. *Plant Med.* **1988,** *35*, 333–338.

Narasinga Rao, B. S. Pulses and Legumes as Functional Foods. *Bull. Nutr. Found. India* **2002**, *23*, 1.

Nelson, T. S.; Shieh, T. R.; Wodzinski, R. J.; Ware, J. H. The Availability of Phytate Phosphorus in Soybean Meal before and after Treatment with Mold Phytase. *Poultry Sci.* **1968,** *47*, 1842–1848.

Ojiako, O. A.; Igwe, C. U. The Nutritive, Anti-nutritive and Hepatotoxic Properties of *Trichosanthes anguina* (Snake Tomato) Fruits from Nigeria. *Pak. J.* **2008,** *7*(1), 685–689.

Oke, O. L. Oxalic Acid in Plants and Nutrition. *World Rev. Nutr. Diet.* **1969,** *10*, 262–302.

Oksman-Caldentey, K. M.; Inzé, D. Plant Cell Factories in the Post-Genomic Era: New Ways to Produce Designer Secondary Metabolites. *Trends Plant Sci.* **2004,** *9*, 433–440.

Olayemi, F. O. Evaluation of the Reproductive and Toxic Effects of *Cnestis ferruginea* (de Candolle) Root Extract in Male Rats. Ph. D Thesis, Department of Physiology, University of Ibadan, Nigeria, 2007, pp 46–51.

Osagie, A. U. Antinutritional Factors. In *Nutritional Quality of Plant Foods*. University of Benin: Ambik, 1998.

Osuntokun, B. O. Cassava Diet and Cyanide Metabolism in Wistar Rats. *Br. J. Nutr.* **1972,** *24*, 797–805.

Thipyapong, P.; Stout, M. J.; Attajarusit, J. Functional Analysis of Polyphenols Oxidase by Antisense/sense Technology. *Molecules* **2007,** *12*(8), 1569–1595.

Price, K. R.; Johnson, I. T.; Fenwick, G. R. Chemistry and Biological Significance of Saponins in Food and Feeding Stiffs. *CRC Crit. Rev. Food Sci. Nutr.* **1987,** *26*, 27–135.

Rao, N. P.; Reddy, B. S.; Reddy, M. R. Utilization of Autoclaved Castor Bean Meal in Concentrate Ration for Sheep. *Indian J. Anim. Nutr.* **1988,** *5*, 121–128.

Reddy, N. R.; Sathe, S. K.; Pierson, M. D. Removal of Phytate from Great Northern Beans (*Phaseous vulgaris* L.) and its Combined Density Fraction. *J. Food Sci.* **1988,** *53*, 107–110.

Rosenthal, G. E.; Janzen, D. H., Eds). *Herbivores: Their Interaction with Secondary Plant Metabolites*. Academic Press: New York, 1979, pp 531.

Saldamlı, Đ. Gıda kimyası. Doğal toksik maddeler ve Kontaminantlar. Acar J. ve Uygun Ü. Hacettepe Üniversitesi Mühendislik Mimarlık Fakültesi Gıda Mühendisliği Bölümü-Ankara. 1998, 399–433.

Sandberg, A. S. Bioavailability of Minerals in Legumes. *British J. Nutr.* **2002,** *88*, Suppl. 3, S281–S285.

Sharma, H. C.; Norris, D. M. Chemical Basis of Resistance in Soya Bean to Cabbage Looper, Trichoplusiani. *J. Sci. Food Agric.* **1991,** *55*, 353–364.

Sharma, R. D. Isoflavones and Hypocholesterolemia in Rats Lipids. **1979,** *14*, 535–540.

Sharma, R. D. Isoflavone Content of Bengal Gram at Various Stages of Germination. *J. Plant Foods*, **1981,** *3*, 259–264.

Skutches, C. L.; Smith, F. H. Effect of Phenobarbital on the Level of Gossypol in the Liver and the Effect of Gossypol and Phenobarbital on Liver Microsomal *O*-Demethylation and Lipid Peroxidation Activities in the Rat. *J. Nutr.* **1974,** *104*(12), 1567–1575.

Soetan, K. O. Pharmacological and other Beneficial Effects of Antinutritional Factors in Plants. *Afr. J. Biotechnol.* **2008,** *7*(25), 4713–4721.

Su H.; Guo R. Inhibition of Acrosine Activity of Human Spermatozoa by Saponins of *Bulbostermma paniculatum* Xtian Yike Daxue Xuebae 7, 225. *Chem. Abstr.* **1986,** *10*(1008), 49459.

Vasconcelos, I. M.; Oliveira, J. T. Antinutritional Properties of Plant Lectins *Toxicon* **2004,** *44*(4), 385–403.

Velazquez Encarna, Luis R. Silva, Alvaro Peix. Current Nutrition & Food Science. *Legumes: Healthy Ecol. Sour. Flav.* **2010,** *6*(2), 109–144.

Vulliamy, T.; Mason, P.; Luzzatto, L. The Molecular Basis of Glucose-6-Phosphate Dehydrogenase Deficiency. *Trends Genetics* **1992,** *8*, 138–143.

Wink, M. Chemical Defense of Lupins—Biological Function of Quinolizidine Alkaloids. *Plant Syst. Evol.* **1985,** *150*, 65–81.

Wink, M. Quinolizidine Alkaloids: Biochemistry, Metabolism, and Function in Plants and Cell Suspension Cultures. *Planta Med.* **1987,** *53*, 509–514.

Wink, M. Plant Breeding: Importance of Plant Secondary Metabolites for Protection against Pathogens and Herbivores. *Theor. Appl. Genet.*, **1988,** *75*, 225–233.

Wink, M. In *Insect Plant Interactions*; Bernays, E. A., Ed.; CRC Press: Boca Raton, FL, 1992; pp 131–166.

Wink, M. In *The Alkaloids*; Cordell, G., Ed.; Academic Press: New York, 1993a; vol. 43, pp 1–117.

Wink, M. In *Phytochemistry and Agriculture.* Proceedings of the Phytochemical Society of Europe; van Beek, T. A., Breteler, H., Eds.; Oxford University Press: Oxford, 1993b, *34*, pp 171–213.

Wink, M. Evolution of Secondary Metabolites from an Ecological and Molecular Phylogenetic Perspective. *Phytochemistry* **2003,** *64*, 3–19.

Wolever, T. M. S. The Glycemic Index. *World Rev. Nutr. Dietetics* **1990,** *62*, 120–125.

CHAPTER 8

SPIRULINA: FUNCTIONAL COMPOUNDS AND HEALTH BENEFITS

ASHRAF MAHDY SHAROBA

Food Science Department, Faculty of Agriculture, Benha University, Khalifa, Egypt

Corresponding author, E-mail: ashraf_sharoba@yahoo.com

CONTENTS

ABSTRACT

Spirulina is a "superfood." It is the most nutritious, concentrated whole food known to humankind. It has a rich, vibrant history and occupies an intriguing biological and ecological niche in the plant kingdom. Spirulina is truly an amazing food, full of nutritional wonders. Various vitamins like A, C, and E and phenolic substances present in plants have antioxidant properties. These antioxidants can become prooxidants under certain conditions in certain concentrations in the body and also if other antioxidants are lacking. Natural foods rich in these antioxidants are a better alternative than vitamins in the form of tablets. Very high amounts of beta-carotene, tocopherol, and combined form of these antioxidants make spirulina a very good source of natural antioxidant along with high protein.

8.1 INTRODUCTION

Spirulina is a genus of blue-green algae belonging to the family of Oscillatoriaceae. The two species which are most commonly utilized are *Spirulina platensis* and *Spirulina maxima*; in India, *Spirulina jusi* Jormzs is also regarded as a source plant. Spirulina (*Arthrospira platensis*) is a ubiquitous spiral-shaped, blue-green microalgae, most commonly found in seawater and brackish water. Among the various species, *S. platensis* and *S. maxima* are the only two used as food. The blue-green color of the organism is due to the presence of various types of photosynthetic pigments like chlorophyll, carotenoids, phycocyanin (PC), and phycoerythrin. PC is responsible for the blue color of the organism.

Spirulina, a blue-green alga, is now becoming a healthy food worldwide. It is a multicellular, filamentous cyanobacterium belonging to algae of the class Cyanophyta. The United Nations world food conference declared spirulina as "the best food for tomorrow," and it is gaining popularity in recent years as a food supplement (Kapoor & Mehta, 1993). The United Nations World Health Organization (WHO) has confirmed that spirulina represents an interesting food for multiple reasons; rich in iron and protein, it can be safely administered to children without any risk. The spirulina ability as a potent antiviral (Hayashi et al., 1996), anticancer (Schwartz et al., 1988; Babu, 1995), hypocholesterolemic (Nakaya et al., 1988), and health improvement agent is gaining attention as a nutraceutical and a source of potential pharmaceutical.

Spirulina has found wide applications in the areas of agriculture, food, pharmaceuticals, perfumeries, medicine, and science. Nowadays, this organism is used as a food supplement and is marketed in the form of pills, capsules, and powder or incorporated into various types of food like cakes, biscuits, noodles, and health drinks, and others. Various countries are developing strategic programs for the production and use of spirulina. In the world, the spirulina industry is developing rapidly with several factories producing hundreds of tons of spirulina in the form of dried powder that is used as food, fodder, and medicine.

Though many reports are anecdotal, dietary supplementation with spirulina may benefit diseases such as gastrointestinal ulcers, arthritis, allergies, diabetes, obesity, and hypertension. By searching such ways, we can truly reduce our national medical care expenditures.

The blue-green alga spirulina has been used by mankind as food and drug since ages. However, the last few decades have witnessed the unprecedented momentum in research on nutritional and medicinal potency of this unicellular alga. It has emerged as an undisputed medical food with the discovery and validation of a litany of health benefits ranging from antioxidant, anti-inflammation, hypolipemic, antithrombotic, antidiabetic, anticancer, immunestimulatory, antimicrobial, cardioprotective, hepatoprotective, antianaemic, neuroprotective, tissue engineering to aquaculture, and livestock feed. Many hitherto unknown pharmacological properties are coming forth and myriad research projects are revolving around this miraculous cyanobacterium. Safety regulations recommend its inclusion in nutritional regimen for proofing body against ailments and augmenting vitality. Advances in effective cultivation, drying, extraction, and purification techniques have been summarized. This review outlines the recent progresses and therapeutic possibilities of this spirulina.

Spirulina is a "superfood." It is the most nutritious, concentrated whole food known to humankind. It has a rich, vibrant history and occupies an intriguing biological and ecological niche in the plant kingdom. Spirulina is truly an amazing food, full of nutritional wonders.

Various vitamins like A, C, and E and phenolic substances present in plants have antioxidant properties. These antioxidants can become prooxidants under certain conditions in certain concentrations in the body, and also if other antioxidants are lacking. Natural foods rich in these antioxidants are a better alternative than vitamins in the form of tablets. Very high amounts of beta-carotene, tocopherol, and combined form of these antioxidants make spirulina a very good source of natural antioxidant along with high protein.

Imagine a food that can help regulate blood sugar, blood pressure, and cholesterol; a food that can alleviate pain from inflammation and deliver antioxidant activity to ward off life-threatening diseases like cancer, Alzheimer's, heart disease, and stroke; a food that helps and protects the liver and kidneys and removes radiation from the body; a food that improves the immune system, alleviates allergies, and has been proven to fight many different viruses; a food that helps your eyes and brain; a food that can actually help you lose weight, increase friendly flora in the intestines, and improve digestion. Scientific research shows that spirulina may help in all of these areas and more.

Spirulina's concentrated nutrition makes it an ideal food supplement for people of all ages and lifestyles. Spirulina is about 60% complete, highly digestible proteins, containing every essential amino acids. It contains more beta-carotene than any other whole food; it is the best whole food source of gamma linolenic acid (GLA); it is rich in B vitamins, minerals, trace elements, chlorophyll, and enzymes; and it is abundant in other valuable nutrients about which scientists are learning more each year, such as carotenoids, sulfolipids, glycolipids, PC, superoxide dismutase, RNA, and DNA.

Spirulina–cyanobacteria have been used by different populations as protein source and other nutritional requirements. *Spirulina*–cyanobacteria have been used as food for centuries by different populations. Concentrated nutrition of spirulina makes it an ideal food supplement for people of all ages and lifestyles.

The WHO prognosticates that in the 21st century, spirulina will become one of the most important curative and prophylactic components of nutrition and recommends it also in children's nutrition.

At present, the high cost of packaged product inhibits people from using spirulina as a supplement or a therapeutic agent in many parts of the world. Lack of awareness, palatability of the biomass, and limited availability of commercial food products containing spirulina are other factors to consider. More aggressive promotion through government programs is needed to make spirulina—"The Food for the 21st Century."

8.2 CHEMICAL COMPOSITION AND NUTRITION VALUES OF SPIRULINA

The chemical and nutritional composition of spirulina may vary according to the growing conditions. For example, the iodine content will vary as a function when the spirulina is grown in sea water versus fresh water. The chemical and nutritional composition of dried powdered spirulina grown in fresh water is summarized in Tables 8.1–8.3 from Sharoba (2014). It should be noted that, the cell wall of spirulina is composed of protein, carbohydrates, and fat. Therefore, the bioavailability of nutrients from spirulina might be more than from other food sources, especially plant food sources.

Spirulina is the richest nutrient and complete food source found in the world. It contains over 100 nutrients, more than any other plants, grains, or herbs. Today, spirulina is widely used as a food supplement to maintain health, boost energy, and reduce weight. Spirulina contains 62.84% protein, higher than any other natural food. Spirulina contains all the essential amino acids in fairly high amounts, which makes it a complete protein; other protein sources have negative properties as well, such as animal fat and cholesterol. Spirulina contains essential minerals like calcium, magnesium, potassium, phosphorus, iron, and zinc as well as complete vitamin B groups and many important antioxidants (which protect cells). The antioxidant PC can only be found in spirulina. It is the richest natural source of vitamin E and beta-carotene. The protein and B-vitamin complex in spirulina makes a major nutritional improvement in an infant's diet. It is the only food source other than breast milk containing substantial amounts of essential fatty acids, essential amino acids, and GLA that helps to regulate the entire hormone system.

TABLE 8.1 Chemical Composition and Physical Properties of Spirulina (g/100 g Sample, on Dry Weight Basis).

Chemical composition	Values (%)	Physical properties	Values
Moisture content	4.74 ± 0.84	pH	6.84 ± 0.14
Total solids	95.36	Bulk density	0.82 kg/l
Protein content	62.84 ± 1.38	Particle size	100% 60 mesh
Lipid	6.93 ± 0.57	Appearance	Fine, uniform powder
Ash content	7.47 ± 0.39	Color	Blue green to green
Crude fiber	8.12 ± 0.28	Odor and taste	Mild like sea weed
Starch	3.56 ± 0.27	Consistency	Powder

TABLE 8.2 Amino Acids and Fatty Acids Content of Spirulina (mg/100 g).

Values	Fatty acids	Values	Amino acids
0.46	Myristic (C14:0)	%	Essential amino acids
40.65	Palmitic (C16:0)	6.49	Isoleucine
6.38	Palmitoleic (C16:1 omega-6)	7.89	Leucine
1.92	Stearic (C18:0)	4.73	Lysine
1.64	Oleic (C18:1 omega-6)	2.34	Methionine
17.95	Linoleic (C18:2 omega-6)	4.42	Phenylalanine
24.49	Gamma-linolenic (C18:3 omega-6)	4.58	Threonine
traces	Alpha-linolenic (C18:3 omega-3)	1.93	Tryptophan
5.33	Erucic acid (C22:1)	6.08	Valine
1.18	Lignoceric acid (C24:0)	38.46	Total
44.21	Total saturate fatty acid	%	Non-essential amino acids
55.79	Total unsaturated fatty acid	7.52	Alanine
		7.51	Arginine
		11.17	Aspartic
		1.11	Cysteine
		13.69	Glutamic
		5.24	Glycine
		2.78	Histidine
		4.35	Proline
		4.56	Serine
		3.61	Tyrosin
		61.54	Total
		100%	Total amino acids
		62.84 ± 1.38	% Protein

TABLE 8.3 Vitamins, Phytopigments, and Minerals in Spirulina.

Components	Values	Components	Values
Vitamins	(Values/100 g)	Minerals	(mg/100 g)
Vitamin B_1 (thiamine)	5.8 mg	Calcium (Ca)	922.278
Vitamin B_2 (riboflavin)	4.65 mg	Potassium (K)	2085.28
Vitamin B_3 (niacin)	15.35 mg	Magnesium (Mg)	1.1902
Vitamin B_6 (pyridoxine)	0.94 mg	Sodium (Na)	1540.46
Vitamin B_{12} (analog)	175 μg	Phosphorus (P)	2191.71
Folic acid	9.92 mg	Copper (Cu)	1.2154
Inositol	60.45 mg	Iron (Fe)	273.197
Vitamin E	9.86 mg	Manganese (Mn)	5.6608
Vitamin K	1095 μg	Zinc (Zn)	3.6229
Pantothenate	108 μg	Chromium (Cr)	0.325
Biotin	8 μg	Selenium (Se)	0.0394
Phytopigments	(%)	Boron (B)	2.875
Total carotenoids	0.573	Molybdenum (Mo)	0.372
Beta carotenoids	0.2527		
Xanthophylls	0.2818		
Zeaxanthin	0.1331		
Chlorophyll	1.5609		
Phycocyanin	14.647		

8.3 PHYSICAL PROPERTIES OF SPIRULINA

Spirulina offers a convenient solution to the pH problems of most diets as it is very alkaline. It is an alkaline food (pH 6.84) that counters the acidic foods and helps raise the pH level toward the alkaline side of the scale. This, in turn, promotes increased bone mass (since your body does not have to sacrifice calcium to balance its pH), and vastly improves metabolic functions. Consuming more alkaline foods has been strongly linked with improved immune system function, mental function, kidney function, and higher levels of energy, among other important benefits. Acidic body condition may cause many modern diseases like hypertension, cancer, diabetes, heart disease, gout, and rheumatism.

8.3.1 ADJUST THE BODY'S pH VALUE

The ideal healthy human body's pH level should remain on low alkaline about pH 7.35–7.45. Modern day people indulge in too much acidic food like soft drinks, meat, cheese, eggs, and ham. These cause our body to become acidic (pH < 7). Many medical research reports have proven that acidic bodies will have more chance of getting diseases or cancer. Regular use of spirulina can help keep your body alkaline that will help you reduce this risk and is the ideal food supplement for the weight reducer.

Data in Table 8.1 also shows the bulk density of spirulina (0.82 kg/l). The bulk density of the product is affected by particle size distribution, type of agglomeration, particle porosity, and, to a certain extent, the moisture content. Particle size distribution is affected by the initial size of the trichomes as they are fed to the dryer and the pore diameter of the atomizer. The final quality of the product with respect to bulk density is, therefore, dependent on culturing, harvesting, and drying conditions. To a certain extent, all these factors are harnessed in order to obtain a product that meets the requirements of formulated babies' food formulas. The color of spirulina in the powder form appears to be a blue-green to green color.

Spirulina is called a super food because its nutrient profile is more potent than any other food, such as plants, grains, or herbs. These nutrients and phytonutrients make spirulina a whole food alternative to isolated vitamin supplements: protein and amino acids, vitamins and minerals, essential fatty acids and phytonutrients, compared to other foods. Spirulina can renourish our bodies and renew our health. Spirulina has been used in preparation of baby foods because of its therapeutic properties and the presence of antioxidant compounds. Babies can eat spirulina in complete safety and assimilate its nutrients without difficulty. Even malnourished babies with diminished capacity for nutrient absorption could assimilate spirulina and recover from malnutrition.

Advantage of spirulina is that it has 60% protein by weight that is higher than any other food source. It is the most easily digestible form of protein food especially important for malnourished people (Bucaille, 1990). There are very many strategies that can be adopted for diet supplementation, which in turn contributes to immune ability of the children.

Spirulina offers remarkable health benefits to undernourished children. It is rich in beta-carotene that can overcome eye problems caused by vitamin A deficiency. It provides the daily dietary requirement of beta-carotene which can help prevent blindness and eye diseases.

Research shows it to stimulate the immune system, build both red and white blood cells (WBCs) and assist detoxification (Lisheng, 1991). Spirulina also is a best balanced highly efficient dietary supplement, which satisfies the demands of all systems of the organism and, what is most important, improves the condition of its immune system and is a source of easily available iron. The literature data show that spirulina enhances the hemopoietic system and increases resistance to hypoxia (Milasius et al., 2009).

The protein and B-vitamin complex make a major nutritional improvement in an infant's diet. It is the only food source other than breast milk containing substantial amounts of essential fatty acid, essential amino acids, and GLA that helps to regulate the entire hormone system (Ramesh et al., 2013).

Spirulina is rich in high-quality protein, vitamins, minerals, and many biologically active substances. Its cell wall consists of polysaccharide which has a digestibility of 86%, and could be easily absorbed by human body. The Wuhan Botanical Institute has collaborated with the Changde Central School of Physical Training to study the effect of oral intake of spirulina pill on the physical status of athletes (Li & Qi, 1997). After taking 10 g spirulina pills per day for 4 weeks, female athletes showed an increase in their hemochrome level. The lung capacity of juvenile weight-lifting and jujutsu athletes was improved (Desai & Sivakami, 2004).

Trace metals are essential nutrients required in very small amounts in the daily diet. They play key roles in various activities of the cell. One of the important trace metals is selenium. Selenium bioeffects are mainly involved in immune function, reproduction, cardiovascular disease, cancer, viral infection control, and metal toxicity. In fact, selenium deficiency is endemic in parts of China where the soil lacks selenium and higher rates of cancer have been reported in the United States in places where the cereals are deficient in selenium. Another essential trace element is iodine, whose deficiency affects thyroid function, cardiovascular function, intelligence quotient, and other brain disorders. *S. platensis* has the potential to be used as a matrix for the production of selenium- and iodine-containing compounds in order to treat diseases deficient in these elements. The advantage of this type of matrix is that being a living organism, spirulina can accumulate these elements along with vitamin E (tocopherol) and beta-carotene in a form easily available for human consumption in required amounts. Besides, it can also avoid hyperthyroidism caused by the over-use of iodinated salts.

Spirulina and their use as therapeutic agent were studied by Desai and Sivakami (2004); apart from being a health food, spirulina has invaluable medical applications. Dietary supplementation of this organism showed protective effect toward food allergy. Spirulina has two types of water-soluble polysaccharides called "calcium spirulian" and "Immunila." These showed inhibitory effects against some viruses like HIV and they activate the immune system during cancer chemotherapy. Spirulina extracts prevent the formation of tumors and exhibit hypocholesterolemic and antidiabetic properties. Supplementation of spirulina in patients with oral cancer prevented further damage in these patients. Dietary spirulina can act as an effective chemo-preventive agent against many carcinogens and mutagens.

Spirulina protein is superior to all standard plants and the nutritive value of a protein has higher quality of amino acids, digestibility coefficient as well as the biological value. Khan et al. (2005) state that spirulina has three natural compounds which are potent in inhibiting HIV activity. These three natural compounds are calcium–spirulan, cyanovirin–N, and sulfolipids (Khan et al., 2005).

Spirulina reduces ischemic brain damage in rats, and that these rats had improved post-stroke locomotor activity. This study compared spirulina to blueberries, spinach, and a control group. Both blueberries and spinach reduced brain damage caused by ischemia by roughly 30%; spirulina reduced it by 70%. And what is even more amazing is that blueberries and spinach were given to the rats at a level of 2% of their total diet, while spirulina was given at only 0.33% of their total diet (Wang et al., 2005).

It has been recognized by researchers that spirulina exhibits various bioactivities. Many studies depict the bioactive compounds that are possessed by spirulina as having potential, such as antibacterial, anti-inflammatory, antioxidant, immune-stimulant, anticancer, and anti-HIV bioactivities. The main focus of spirulina used in this thesis is on the therapeutic agents in HIV treatment. Several authors in particular have made contributions toward needs in this area. Spirulina has been shown to be an excellent source of proteins, vitamins, and minerals, with a low content of nucleic acid.

8.3.2 LIVER TOXICITY

Several other animal studies show that spirulina is protective against damage to the liver due to heavy metals like lead and mercury (Ponce-Canchihuaman et al., 2010). Diabetes mellitus, a metabolic disorder, is becoming a major health problem. Although there are a number of drugs available on the market, long-time use may cause a number of side effects. Hence, a large number of studies are in progress to find natural sources, which are effective in reducing the intensity of diabetes. The present study was undertaken to evaluate the antidiabetic effect of spirulina on streptozotocin-induced (45 mg/kg body weight) diabetes in male albino Wistar rats. Blood glucose levels were elevated in diabetic rats. The levels of blood glucose, plasma insulin, and serum C-peptide, and activities of the glucose-metabolizing enzymes hexokinase and glucose-6-phosphatase were estimated using standard protocols. Oral administration of spirulina was carried out for 45 days.

8.3.3 NON-ALCOHOLIC FATTY LIVER DISEASE

Spirulina was shown to reduce liver triglycerides in mice with experimental diabetes. In an animal study, fatty liver was produced in mice by administration of simavastin (statin), ethanol, and hypercholesterolemic diet. Administration of spirulina prevented the increase in liver total lipids and liver triglycerides. Spirulina showed therapeutic effect in patients with non-alcoholic fatty liver disease (NAFLD) as evidenced by ultrasonography (Ferreira-Hermosillo et al., 2010).

Spirulina is a potent mixture of antioxidants and most of spirulina's health benefits are associated with its antioxidant pigments. These are carotenoids (mixture of carotenes and xanthophylls), chlorophyll, and the unique blue pigment PC. Among these pigments, the xanthophyll zeaxanthin is rare in nature, and 3 g spirulina can provide adequate quantities of this very valuable xanthophyll. PC is even rarer in our diet and 3 g spirulina can provide adequate quantities of this highly beneficial pigment. PC can provide excellent health benefits as discussed in this article, and supplementing the diet with spirulina is the easiest way to obtain it. Spirulina supplementation may help in managing age-related health conditions such as age-related macular degeneration (ARMD), type 2 diabetes, NAFLD, neurodegenerative disorders, and cerebrovascular disease (Thomas, 2010).

Spirulina is now widely used as nutraceutical food supplement worldwide. Recently, great attention and extensive studies have been devoted to evaluate its therapeutic benefits on an array of diseased conditions including hypercholesterolemia, hyperglycerolemia, cardiovascular diseases, inflammatory diseases, cancer, and viral infections. The cardiovascular benefits of spirulina are primarily resulted from its hypolipidemic, antioxidant, and anti-inflammatory activities. Data from preclinical studies with various animal models consistently demonstrate the hypolipidemic activity of spirulina. Although differences in study design, sample size, and patient conditions result in minor inconsistency in response to spirulina supplementation, the findings from human clinical trials are largely consistent with the hypolipidemic effects of spirulina observed in the preclinical studies. However, most of the human clinical trials are suffered with limited sample size and some with poor experimental design. The antioxidant and/or anti-inflammatory activities of spirulina were demonstrated in a large number of preclinical studies. However, a limited number of clinical trials have been carried out so far to confirm such activities in human. Currently, our understanding on the underlying mechanisms for spirulina's activities, especially the hypolipidemic effect, is limited. Spirulina is generally considered safe for human consumption supported by its long history of use as food source and its favorable safety profile in animal studies (Kumari et al., 2011, Table 8.4).

Spirulina has demonstrated a bewildering array of food and therapeutic properties. The key implications backed by scientific findings are as functional food and additives, antioxidant, anti-inflammatory, hypolipemic and antihypertensive, antidiabetic, anticancer, immunestimulant, antimicrobial, hepatoprotective, neuroprotection, antianemic, and antileukopenic and tissue engineering. Spirulina biomass has also proved suitable as a nutritive aquaculture feed (Patel & Goyal, 2013).

In recent years, the consumer concern in relation to health and safety issues about the use of synthetic colorants in foods has increased. The Food and Drug Administration (FDA) in the USA, the European Food Safety Authority (EFSA) in Europe, and many other national authorities around the world have restricted the use of synthetic colorants in foods, confectioneries, and beverages because of their confirmed or suspected association with increased cancer development or induction of allergic reactions. For the same reasons, other colorants are under study and are only provisionally allowed. The tendency in food manufacturers is,

TABLE 8.4 Hypolipidemic Effects of Spirulina in Human Clinical Studies.

Subject	Sample size	Dose of spirulina	Duration	Effects of spirulina	Reference
Healthy volunteers (male)	30	4.2 g, daily	4 or 8 weeks	Total serum cholesterol and LDL were reduced significantly. Triglyceride levels decreased slightly whereas HDL-cholesterol showed no significant changes. The reduction of serum cholesterol was even greater in those men with the highest cholesterol levels	Nakaya et al. (1988)
Patients with ischemic heart disease	30	2 or 4 g, daily	3 months	Total plasma cholesterol, LDL, VLDL, and triglycerides were significantly reduced by 22.4% (2 g group)/33.5% (4 g group), 31%/45%, 22%/23%, and 22%/23%, respectively. HDL was significantly increased by 11.5%/12.8%. In addition, a significant reduction in body weight was achieved in both treatment groups	Ramamoorthy et al. (1996)
Patients with type 2 diabetes mellitus	15	2 g, daily	2 months	A significant reduction was detected in triglycerides, total cholesterol, and free fatty acid levels. LDL and VLDL were also decreased. In addition, blood sugar and glycated serum protein levels were significantly decreased	Mani et al. (2000)
Patients with type 2 diabetes mellitus	25	2 g, daily	2 months	Total serum cholesterol and LDL fraction were reduced whereas HDL was slightly increased. As a result, a significant decrease in atherogenic indices and the ratios of total cholesterol/HDL and LDL/HDL was observed. Furthermore, triglycerides and fasting and postprandial blood glucose levels were significantly reduced. Finally, the level of apolipoprotein B showed a significant fall with a concurrent significant increase in the level of apolipoprotein A1	Parikh et al. (2001)
Patients with nephrotic syndrome	23	1 g, daily	2 months	Total serum cholesterol, LDL cholesterol, and triglycerides were all significantly decreased by 46 mg/dL, 33 mg/dL, and 45 mg/dL, respectively. The ratios of LDL/HDL and total cholesterol/HDL were also decreased significantly	Samuels et al. (2002)

TABLE 8.4 *(Continued)*

Subject	Sample size	Dose of spirulina	Duration	Effects of spirulina	Reference
Healthy elderly volunteers	36	4.5 g, daily	6 weeks	Total plasma cholesterol and triacylglycerols were significantly reduced by 10% and 28%, respectively. HDL was significantly increased by 15% whereas LDL cholesterol was significantly decreased. In addition, both systolic and diastolic blood pressures were significantly reduced in both men and women	Torres-Duran et al. (2007)
Healthy elderly volunteers	12	7.5 g, daily	24 weeks	The plasma concentrations of triglycerides, total cholesterol, and LDL cholesterol were decreased after 4 weeks of the supplementation. No differences in hypolipidemic effects of spirulina were observed between mild hypercholesterolemia (cholesterol at or above 200 mg/dL and normocholesterolemic subjects	Park and Kim (2003)
Elderly women with hypercholesterolemia	51	7.5 mg, daily	8 weeks	Serum levels of total cholesterol, LDL cholesterol, and oxidized LDL were significantly reduced. In addition, apolipoprotein B, IL-6, and IL-6 production by peripheral blood lymphocyte were also decreased	Kim et al. (2005)
Patients with type 2 diabetes mellitus	37	8 g, daily	12 weeks	Total serum cholesterol, LDL fraction, and triglycerides were reduced with the subjects with higher initial total cholesterol, LDL-cholesterol, and triglycerides showing higher reduction. In addition, blood pressure and IL-6 levels were also decreased. Finally, a significant reduction in plasma malondialdehyde level was observed	Lee et al. (2008)
Patients with type 2 diabetes mellitus	60	1 or 2 g, daily	2 months	A significant decrease was observed in serum total cholesterol, triglycerides, LDL, and VLDL cholesterol in spirulina treatment groups. In addition, both fasting and postprandial blood glucose levels were also significantly decreased accompanied with decreased mean carbohydrate and protein intake	Kamalpreet et al. (2008)

TABLE 8.4 *(Continued)*

Subject	Sample size	Dose of spirulina	Duration	Effects of spirulina	Reference
Healthy elderly volunteers	78	8 g, daily	16 weeks	Total plasma cholesterol and LDL fraction were significantly reduced in female subjects whereas a significant lowering effect on plasma total cholesterol by repeated test for treatment was observed in male subjects. However, no significant effect was detected in LDL fraction in male subjects. The levels of HDL fraction and triglycerides did not change after the intervention in both men and women	Park et al. (2008)

therefore, going progressively toward the use of natural additives. Among the different colors, the confectionary and drinks industry has a high demand in blue colorants; however, they are uncommon in nature thus leading to the use of synthetic ones. For this reason, the food industry is now expressing a growing interest in the search, use, and stabilization of natural blue colorants. The cyanobacterium *Arthrospira (Spirulina) platensis*, known mainly as a source of nutraceuticals, has recently gained considerable attention also as a source of blue pigment thanks to its accessory photosynthetic blue-protein PC. The great commercial interest on PC is mainly due to the high protein yield and the relatively easy extraction procedures (Martelli et al., 2014).

The food applications, therapeutically important compounds, and health benefits are illustrated in Figures 8.1 to 8.2.

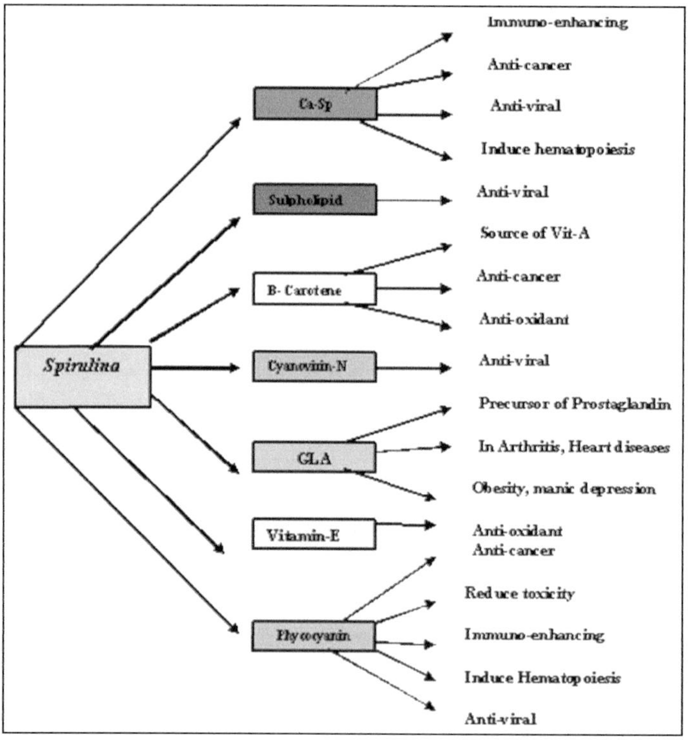

FIGURE 8.1 Therapeutically important compounds of spirulina and its effect (From: Khan, Z.; Bhadouria, P.; Bisen, P. S. Nutritional and Therapeutic Potential of Spirulina. Current Pharmaceutical Biotechnology, 2005, 6, 373–379. Reprinted by permission of Eureka Science Ltd.)

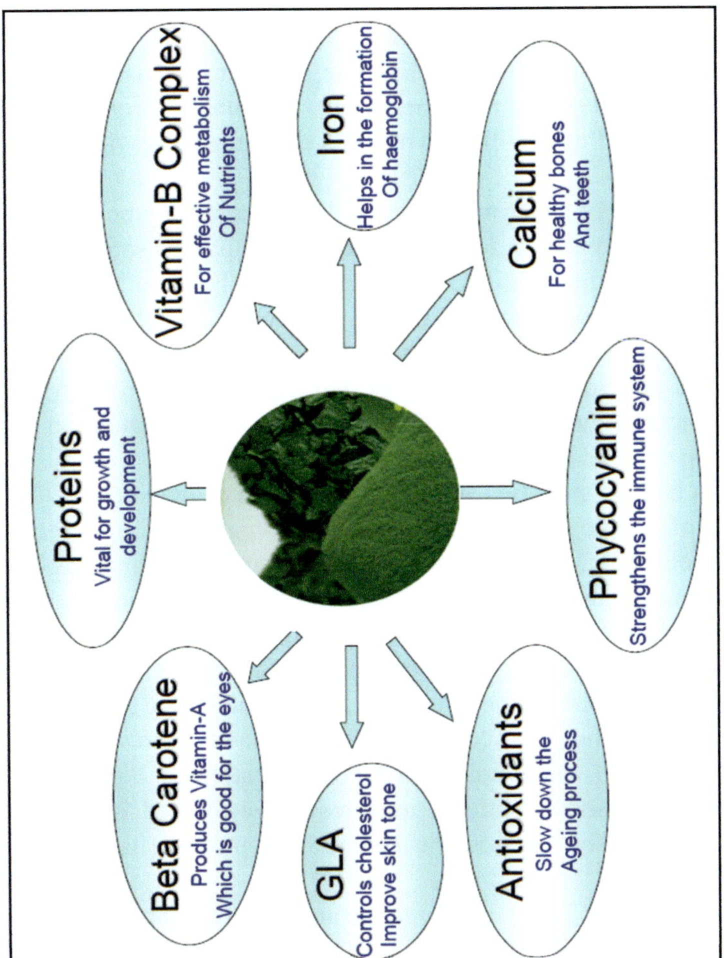

FIGURE 8.2 Therapeutically important compounds of spirulina and its effect (From: Dolly W. D. Biotechnological Potentials and Role of Cyanobacteria in Agriculture and Industry. Division of Microbiology, Indian Agricultural Research Institute: New Delhi, India, 2014. Reprinted with permission from the author.)

8.4 SPIRULINA AND DIABETES

Type 2 diabetes is a multifactorial disease including a cluster of pathologies—insulin resistance, obesity, dyslipidemia, and hypertension.

Diabetes and geriatrics represent two sections of population whose numbers are rapidly increasing in the world. Egypt has the dubious distinction of having the largest number of cases of diabetes. This dramatic rise is largely explained by the change in lifestyle, the ever-increasing problem of weight and obesity and, to some extent, by genetic predisposition. According to WHO, Egypt will have over 3.5 million people with diabetes by 2025.

Excessive intake of sugar and in particular fructose may be an important cause of type 2 diabetes. Fructose causes fatty liver, vascular inflammation, endothelial dysfunction, and increases triglyceride levels. In a preclinical study, rats were fed a diet high in fructose or high in glucose. Significant increase (five fold) was observed in hepatic triglyceride levels in fructose-fed rats compared to glucose-fed rats. This fructose-induced triglyceride increase was reduced significantly by supplementation with spirulina (Gonzalez de Rivera et al., 1993).

There can be no innovation in nutrition unless a change occurs to the composition of people's diets which have the potential to improve their health and well-being. In this health conscious age, spirulina has been found to be the nature's most nutritious wholesome organic food with proven beneficial therapeutic properties. Spirulina is fast emerging as a whole answer to the varied demands due to its impressive nutrient composition which can be used for therapeutic uses. Spirulina has a unique blend of nutrients that no single source can provide (Venkataraman, 1998).

In another human study, spirulina supplementation at 2 g/day for 2 months given to diabetic patients improved glycosylated hemoglobin (HbA1C) and favorably altered the serum–lipid profile (Parikh et al., 2001).

Layam and Reddy (2006) evaluated the antidiabetic property of spirulina; experimentation was studied in an animal model before proceeding to clinical trials. The levels of blood glucose and plasma insulin were estimated and studied in streptozotocin diabetic rats. The findings were compared between normal diabetic, and spirulina-supplemented diabetic rats. The findings indicated that the administration of spirulina tended to bring the parameters significantly toward the normal. The effect of spirulina at a dose of 15 mg/kg body weight yielded a higher level of significance than the doses of 5 and 10 mg/kg body weight; therefore, the former was used in further biochemical and clinical studies.

8.4.1 DYSLIPIDEMIA

The hypocholesterolemic action of spirulina may be due to its PC content (Nagaoka et al., 2005). In a rat study, it was shown that PC and a glycolipid from spirulina inhibited pancreatic lipase activity. This resulted in higher excretion of triglycerides in the feces of the experimental animals and postprandial increase in serum triglycerides was lowered (Li-Kun et al. 2006). In a human study, the effect of orally supplemented spirulina (4.5 g/day, for 6 weeks) was evaluated in Mexican population. Spirulina significantly reduced serum triglyceride values and boosted high-density lipoprotein (HDL) and lowered total cholesterol in the test subjects (Torres-Duran et al., 2007).

8.4.2 INSULIN RESISTANCE

In a randomized study with Korean subjects, spirulina was shown to reduce serum triglycerides. It also reduced inflammatory response by reducing inflammatory cytokines IL-6 and TNF-α and increased adiponectin (Lee et al., 2008). Similar reduction in inflammatory cytokines (IL-6 and TNF-α) was observed in a randomized double-blind, placebo-controlled study in elderly Korean subjects. This study also showed an increase in anti-inflammatory cytokine IL-2 (Park et al., 2008).

Spirulina is a primitive organism originating some 3.5 billion years ago that has established the ability to utilize carbon dioxide dissolved in sea water as a nutrient source for their reproduction. Spirulina is a photosynthesizing cyanophyte (blue-green algae) that grows vigorously in strong sunshine under high temperatures and highly alkaline conditions (Habib et al., 2008).

Spirulina blue-green algae are fast emerging as a whole answer for varied demands of diabetics. From the study by Layam and Reddy (2010), it can be concluded that spirulina has hypoglycemic and hypocholesterolemic effect which helps the diabetics to have control on blood glucose levels and also to deter the complications.

Layam and Reddy (2010) studied the effect of supplementation of spirulina on fasting blood glucose, fasting HbA1C, and lipid profile of male non-insulin dependent diabetes mellitus (NIDDM) volunteers belonging to 45–60 years of age. A total of 160 diabetics were selected ($n = 160$) for the study. The total sample was divided into four groups depending upon their diet, drugs, and insulin dosage (groups–I, II, III, and IV). Group I was control group, and groups II, III, and IV were experimental groups. Group II experimental groups were only on dietary regime, group III with diet and drugs and group IV were on diet, drugs, and insulin. Spirulina was supplemented for 12 weeks (90 days) of 2 capsules (each capsule 500 mg glascobrand) per day. There was a statistically significant reduction ($P < 0.001$) from pre to post levels of fasting blood glucose, HbA1C, and lipid-profile levels of the diabetics. There was an increase in HDL cholesterol levels from pre to post.

Gupta et al. (2010) studied the protective effect of *Spirulina fusiformis* extract against rosiglitazone-induced osteoporosis and pharmacodynamic effects of rosiglitazone with spirulina in treating hyperglycemia and hyperlipidemia of insulin-resistant rat, by using the 30 Wistar albino rats, equally divided into five groups as control (C), diabetes mellitus (DM), diabetes mellitus + rosiglitazone (DM + R), diabetes mellitus + spirulina (DM + S), and diabetes mellitus + rosiglitazone + spirulina (DM + R + S). Serum glucose, triglyceride, HDL, low-density lipoprotein (LDL), and insulin concentrations were estimated by routine standard methods in blood samples collected on day 21. A significant decrease in total bone mineral density was observed in group DM + rats ($P < 0.05$). The number and depth of resorptive pits on surface of the bone in rosiglitazone-treated rats improved clearly with spirulina administration. The intactness and integrity of the bone surface as well as the bone strength improved due to the high content of calcium and phosphorous in spirulina. Besides, chromium and GLA in spirulina helped to decrease the fasting serum glucose, HDL, LDL, and triglycerides levels in insulin-resistant rats.

The drugs prescribed to treat diabetes often lead to many side effects. Common issues such as nausea, headache, weight gain, bloating, constipation, diarrhea may arise or serious problems such as liver damage, heart complication, pancreatitis, tumor, bone loss, erectile dysfunction, psychosis, and muscle spasm are encountered (Patel & Goyal, 2013). The protective effects of *S. fusiformis* extract against rosiglitazone (a standard type II diabetes drug)-induced osteoporosis was assessed in insulin-resistant rats. After 45 days, the integrity of the bone surface as well as the bone strength improved. The bone restoration was assumed to be due to the high content of calcium and phosphorous in spirulina. The chromium and α-linoleic acid content was held responsible for decline in the fasting serum glucose, HDL, LDL, and triglycerides levels. These findings suggested that synergistic therapy of rosiglitazone and spirulina can be recommended for attenuating the risk of osteoporosis (Gupta et al., 2010). The intake of *S. maxima* extracts, 2 weeks prior to and 4 weeks during streptozotocin administration was reported to reverse the detrimental effects of the drug on male reproductive organ. The extract significantly increased the body and testis weight, metabolic parameters, normal seminiferous tubules, Leydig cell number, testosterone levels, and mRNAs for steroidogenic enzymes (Nah et al., 2012). Clinical prospects of spirulina in alleviating other side effects are worth exploring.

These studies show that spirulina can be used as a functional food to manage type 2 diabetes.

8.4.3 HYPERTENSION

Dietary fructose was shown to increase vasoconstriction and induce hypertension in rats. Inclusion of spirulina in the fructose-rich diet prevented these effects (Paredes-Carbajal et al., 1998).

In a study where rats were fed on sucrose, the rats became overweight and had elevated blood pressure and hyperlipidemia. Ethanolic extract of spirulina reduced the vascular resistance. This effect may be due to the enhanced endothelial NO release after spirulina supplementation (Mascher et al., 2006). According to Reddy et al. (2000), C-PC, one of the constituents of spirulina, is a selective inhibitor of cyclooxygenase 2. This mechanism of action may inhibit the synthesis/release of cyclooxygenase-dependent vasoconstrictor metabolite of arachidonic acid and induce vasodialation.

In humans, spirulina reduced blood pressure in hypertensive Mexican subjects when given 4.5 g spirulina for 6 weeks. Spirulina supplementation reduced both systolic and diastolic blood pressure in subjects (Torres-Duran et al., 2007).

Studies were conducted to see the effect of dietary spirulina on vasomotor reactivity of aortic rings from lean Wistar rats. Spirulina improved vasodialation in the experimental group. This effect was even more pronounced in fructose-fed obese rats, suggesting that dietary spirulina is able to prevent the effects of fructose-induced obesity and vasoconstriction (Juarez-Oropeza et al., 2009).

8.5 SPIRULINA AND CHOLESTEROL-LOWERING

Cardiovascular disease remains the number one cause of death in developed countries, despite increased awareness, and high cholesterol is one of the most important risk factors in atherosclerosis. Nakaya et al. (1988), in the first human study, gave 4.2 g/day of spirulina to 15 male volunteers and, although there was no significant increase in HDL levels, they observed a significant reduction of LDL cholesterol after 8 weeks of treatment. The atherogenic effect also declined significantly in the above group (Nakaya et al., 1988). The addition of spirufina to normal diets reduced the cholesterol level in experimental animals (Gonzalez de Rivera et al., 1993; Hayashi et al., 1993).

Ramamoorthy and Premakumari (1996) in a more recent study administered spirulina supplements to ischemic heart disease patients and found a significant reduction in blood cholesterol, triglycerides, and LDL cholesterol and an increase in HDL cholesterol. More research is needed before spirulina can be recommended to lower cholesterol levels but its role as a natural food supplement in combating hyperlipidemia, in combination with other therapeutic options, should not be overlooked. Finally, Mani et al. (2000) in a clinical study, found a significant reduction in LDL:HDL ratio in 15 diabetic patients who were given spirulina. However, this study was small and better studies are needed before spirulina can be recommended in diabetes. They also tested spirulina on patients with high cholesterol and ischemic heart disease, and concluded that it plays a key role in reducing blood cholesterol levels and improving lipid profiles (Ramamoorthy et al., 1996).

Nagoaka et al. (2005) have presumed that C-phycocyanin content of spirulina inhibits pancreatic lipase activity and leads to decrease in jejunal cholesterol absorption and hence blood pressure.

Awareness of the need to lower cholesterol levels in order to lower the risks of heart diseases has been well established. Besides dietary improvements, the research is underway to identify natural foods having a cholesterol-reducing effect. Spirulina is one of such identified foods.

Spirulina, now named arthrospira, is a photosynthetic, filamentous, spiral-shaped, multicellular blue-green microalga that grows naturally in the alkaline water of lakes in warm regions. Spirulina means "little spiral". It measures about 0.1 mm across, generally takes the form of tiny green filaments coiled in spiral of varying tightness and number, depending on the strain. *S. platensis* and *S. maxima* are the two most commonly grown species (Smitha, 2006).

8.6 SPIRULINA AND ANEMIA

According to the WHO, there are 2 billion people affected with anemia; in Egypt, anemia prevalence among preschool children is 25.2% and among mothers of surveyed children to be 14.8%. In the meantime, the food consumption study showed that 14% of mothers and 52.5% of the children used to get less than 75% of the Rccommended Dietary Allowances (RDA). Moreover, iron deficiency anemia was found to be prevalent in different parts of world, particularly in rural areas.

Anemia is a common case throughout the world. Its main cause is the iron deficiency which is the most common known form of nutritional deficiency affecting more than 700 million persons all over the world (WHO, 1997).

In a study conducted by Stolzfus (2003), anemia is characterized by a low level of hemoglobin in the blood. Hemoglobin is necessary for transporting oxygen from the lungs to other tissues and organs of the body. Anemia usually results from a nutritional deficiency of iron, folate, vitamin B_{12}, or some other nutrients. This type of anemia is commonly referred to as iron-deficiency anemia. Iron deficiency is the most widespread form of malnutrition in the world, affecting more than two billion people.

Iron deficiency anemia is a major nutritional problem in India, Egypt, and in many other developing countries. A range of 20–40% of natural deaths are due to anemia during pregnancy (Park, 2005).

Anemia itself is not considered a disease. It is considered a therapeutic case resulted from different nutritive disorders, for example, bleeding and bone marrow disorders. Anemia is effected by decrease in red blood cell (RBC) count or decrease in the amount of hemoglobin.

Iron is an extremely important dietary mineral, as it takes part in many body functions. Iron is needed to carry oxygen to the brain and to many tissue cells, and is also involved in the immune mechanisms. The recommended daily intake of iron is 10 mg/day for men and 15 mg/day for women.

It has been estimated by the WHO that nearly 3.7 billion people were iron-deficient and the problem was severe enough to cause anemia in two billion people.

Anemia continues to be a major public health problem worldwide, particularly among growing children, females of reproductive age, and elderly people, especially in the developing countries. WHO estimates that anemia affects approximately 1.62 billion people worldwide, corresponding to 24.8% of the human population (Jeniferpricillia, 2010).

The causes of anemia are diverse and multifactorial, but among the leading etiologies in the developing countries are: nutritional deficiencies (especially of iron, folate, and vitamin B_{12}), chronic or acute blood loss, inherited genetic defects (e.g., thalassemia), chronic diseases and/ or inflammatory disorders, malaria, parasitic infestations (e.g., hookworm), hemolytic disorders, drug-induced hemolysis or marrow suppression or it may be unexplained (Jeniferpricillia, 2010).

According to the WHO reference criteria, anemia is said to be present in an adult, if the blood hemoglobin concentration falls below 13.0 g/dL in men or below 12.0 g/dL in non-pregnant women or below11.0 g/dL in pregnant women. However, reference range for normal blood hemoglobin levels may vary in individuals depending on the age, gender, race, geographical location, and food habits of the study Population (WHO, 2014).

Anemia is one of the most commonly recognized disorders. Its prevalence in women and preschool children is high; therefore, it needs more attention. Evidence from studies showed that adolescents are at an increased risk of developing anemia due to increasing iron demand during puberty, menstrual losses, limited dietary iron intake, and faulty dietary habits.

A study conducted by Seshadri (1993) showed that the spirulina feeding among rural 5000 preschool children with 1 g/day reduced the prevalence of Bitot's spot and prevented the occurrence of severe form of vitamin A deficiency. Giving 1 g/day increased the serum retinol levels. Thus, the prevalence of B-complex deficiency was reduced in preschool children with spirulina supplementation.

A study was conducted by Kauser and Parveen (1999) among 20 malnourished children in the age group of 6 years and was divided into two groups, 10 as experimental and 10 as control. The children of experimental group were given nutritional supplement spirulina (1 g/day) for a period of 3 months and the control group was given placebo for the same period. Diet pattern was same for all the 20 subjects. Results showed that there was an increase in the serum hemoglobin level and serum protein level in the experimental group after the supplementation with spirulina. There was also a definite change in the academic performance and intelligence level of the children after the supplementation. There was no difference in the serum levels of the control group in the study period of 3 months. The intelligence test showed a difference in the control group, but it was not significant. It was concluded that the increase in serum levels and change in intelligence and academic performance in the experimental group as compared to control group was definitely due to the effect of spirulina.

A comparative study was conducted by Vilasini (2001) to determine the efficacy of spirulina and iron tablets on the hemoglobin levels of anemic adult women. Twenty-one subjects with blood hemoglobin levels below 13 g/dL were selected for the study and were divided into three groups. Subjects in Group I were supplemented with spirulina capsules at 2 g/day, subjects in Group II were supplemented with iron tablets and those in group III were treated as controls. The study was carried out for a period of 90 days. The study concluded that both synthetic supplements like iron tablets and natural supplements like spirulina were effective in increasing the blood hemoglobin levels of anemic adult women and the effect produced by both the supplements was more or less alike. The results of the study also showed that both iron and spirulina supplements were effective in the treatment of (marginal) iron deficiency anemia without causing any side effects.

Juvekar (2001), in his study, evaluated the effect of spirulina against cold stress-induced changes in blood glucose levels, blood ascorbic acid levels, adrenal gland weight, adrenal gland ascorbic acid content, and histopathology of adrenal glands. Study revealed that spirulina at doses 100, 200, and 500 mg/kg (oral) prevented adrenal gland hyperactivity, which was indicated by inhibition of weight variation in adrenal gland, recovery of fall in adrenal gland ascorbic acid content, reduction of rise in blood glucose and blood ascorbic acid levels and normalization of morphology in histology of adrenal glands. Thus, the study revealed that spirulina at the doses 200 and 500 mg/kg (oral) was found to be effective against chronic restraint stress-induced increase in blood glucose levels and restraint stress-induced immunosuppression.

A study conducted by Thirumani and Uma (2005) assessed the extent of prevalence of anemia among the adolescent girls and observed the impact of Spirulina supplementation. A quantity of 4 g spirulina was supplemented to the experimental groups of 20 subjects for a period of 90 days. Conclusions drawn were that spirulina improved the biochemical profile of anemic subjects especially the hemoglobin and serum iron content.

The impact of an alimentary integrator composed of spirulina on the nutritional status of undernourished HIV-infected and HIV-negative children was assessed by Jacques et al. (2005). Eighty-four children were HIV infected and 86 were HIV negative, and the duration of the study was 8 weeks. Anthropometric and hematological parameters allowed us to appreciate both the nutritional and biological effect of spirulina supplement to traditional meals. Rehabilitation with spirulina showed on average a weight gain of 15 and 25 g/day in HIV-infected and HIV-negative children, respectively. The level of anemia decreased during the study in all children, but recuperation was less efficient among HIV-infected children. In fact 81.8% of HIV-negative undernourished children recuperated as opposed to 63.6% of HIV-infected children. The results confirmed that spirulina is a good supplement for undernourished children. In particular, rehabilitation with spirulina also seems to correct anemia and weight loss in HIV-infected children, and even more quickly in HIV-negative undernourished children.

S. platensis, a blue-green alga (Oscillotoreaceae), is rich in proteins, lipids, carbohydrates, and some minerals such as selenium, magnesium, manganese, and vitamins including alpha tochopherol, alpha lipoic acid, and riboflavin (Upasani & Balaraman, 2003). Spirulina is known for its wide-ranging biological activities: antioxidative, anti-inflammatory, antimutagenic, antiviral, cardioprotective, anticancer properties, and immune enhancing (Simsek et al., 2007). Spirulina increases the hemoglobin concentration, RBC and WBC counts, and erythropoiesis during chemotherapy. The effect of *S. platensis* in alleviating toxic impacts of heavy metal-adulterated diet was investigated. The results suggested that the algal supplementation may be useful in treatment of leukemia and anemia caused by lead and cadmium. Based on experimental outcome, it was inferred that a 12-week supplementation of spirulina may ameliorate anemia in senior citizens. Steady increase in the corpuscular hemoglobin content in the blood samples of the subjects was recorded (Selmi et al., 2011).

8.7 DAILY INTAKE OF SPIRULINA

Increase in serum beta-carotene level from deficient level to a range close to normal individual level was seen in a study among 30 preschool children with the supplementation of spirulina at 2 g/day for a period of 30 days (Sindhu, 1999).

For children, who are not able to swallow the capsules, the powder can be taken out of the capsules and can be mixed with honey and then can be given. The powder can also be mixed with fruit juice, butter milk, salads, and convenient soups. The amount of spirulina needed depends on metabolism degree to physical exertion, lifestyle, and an individual child's unique body needs. By starting with a small amount—1 capsule/tablet or 1/4 teaspoon—and gradually increasing until the optimal daily amount is found, children can enjoy the benefits of this super nutritious food from babyhood throughout their lives. (Children of all ages, including infants can be given 2 g of spirulina per day.) Spirulina is not a drug, but a natural food supplement, and is not habit forming. Its effects can be sustained by taking it regularly at approx imately 2 g/day. To see any benefits of spirulina, it should be taken at least for 6–8 weeks (Parry, 2014).

8.8 ANTIMICROBIAL EFFECTS OF SPIRULINA

Spirulina contains a whole spectrum of naturally mixed carotene and xanthophyll phytopigments which, together with PC, seem to be related to its antioxidant activity (Pineiro Estrada et al., 2001). *S. platensis* was also reported to present antimicrobial activity as well as to inhibit the replication of several viruses, such as herpes simplex and HIV-1 (Hernandez-Corona et al., 2002).

Cyanobacteria may contain significant quantities of lipids with the composition similar to those of vegetable oils (Singh et al., 2002). The lipids of some cyanobacterial species are rich in essential fatty acids such as linoleic 18:2n6 and α-linolenic 18:3n3 acids and their C20 derivatives, eicosapentaenoic acids 20:5n3, and arachidonic acids 20:4n6. Where microalgae can be cultured, PUFA in algae have profound benefits and functions in dietetics and therapeutic uses. They are believed to have positive effects for the treatment of hypertension, premenstrual tension, various atopic disorders, diabetes, and a number of other cases. Cyanobacteria have a glycerolipid composition very similar to that of the chloroplasts of leaves, the major lipids being monoglactosyl diacylglycerols, digalactosyl diacylglycerols, sulfoquinovosyl diacylglycerols, and, to a minor degree, phosphotidylglycerol. Glycolipids molecular species and their fatty acid composition in *S. platensis* were recently reported (Xue et al., 2002).

Recent studies reported that *S. platensis* could be used as a matrix for the production of selenium-containing compounds and proved to be successful in transforming inorganic selenium to organic selenium *in vivo* when cultivated in selenium-rich medium (Li et al., 2003).

The antimicrobial activity of *S. platensis* was studied against various Gram-positive, Gram-negative bacteria, and fungal species. The methanol extract showed maximum antimicrobial potency. GC–MS analysis identified the volatile components of *S. platensis* to be heptadecane and tetradecane (Ozdemir et al., 2004).

The supercritical fluid extraction and ethanol fractionation of *S. platensis* demonstrated some degree of activity toward *Staphylococcus aureus*, *Escherichia coli*, *Candida albicans*, and *Aspergillus niger* (Mendiola et al., 2007). The antioxidant activities of selenium-containing PC and its different aggregates (monomer, trimer, and hexamer) against free radicals of superoxide, hydrogen peroxide, and 2,2-diphenyl-1-picrylhydrazyl (DPPH) were found to be variable (Huang et al., 2007), and more recently, Mendiola et al. (2007) applied supercritical fluid extraction to obtain functional extracts with antioxidant and/or antimicrobial activities from *S. platensis*.

S. platensis lipids exhibited a strong radical scavenging activity toward stable DPPH free radicals, whereas 27% of DPPH radicals were quenched after 2 h incubation. TL and lipid classes inhibited the growth of different microorganisms except Gram-negative bacteria. At high concentrations, the tested lipids appeared more effective against *A. niger* (28.3 ± 1.53 mm) (Ramadan et al., 2008).

S. platensis was tested for its probiotic efficacy and inhibitory effect against several pathogens (Bhowmik et al., 2009). The doses of 5 and 10 mg/ml promoted growth of *Lactobacillus acidophilus* up to 171.67% and 185.84%, respectively. Maximum inhibition was reported against *Proteus vulgaris*, the pathogen notorious for urinary tract and wound infections. The water extract of *S. platensis* demonstrated significant antimicrobial activity against *Klebsiella pneumoniae* and *P. vulgaris* (NCIM2027), whereas the acetone extract shows pronounced biological activity against

K. pneumoniae followed by *Salmonella typhi*, *Pseudomonas aeruginosa*, *E. coli*, and *S. aureus* (Mala et al., 2009). The effect of spirulina supplement in combating HIV patients was assessed in a 6-month follow-up. The patients administered with spirulina at a dose of 10 g/day showed significant improvement in weight, arm girth, number of infectious episodes, $CD4^+$ count, and protidemia (protein level in blood) (Yamani et al. 2009). HIV-infected patients develop abnormalities of glucose metabolism due to the virus and antiretroviral drugs. The normalizing effect of *S. platensis* was assessed in HIV-infected patients. The results of the 2-months-long study suggested that, the insulin sensitivity in HIV patients improves significantly (about 225%) when spirulina supplement (19 g/day) is taken. Further study is needed to evaluate efficacy of spirulina in combating HIV symptoms (Marcel et al., 2011).

8.9 USE OF SPIRULINA AS FOODS

S. platensis, a blue-green microalga, has been used since ancient times as a source of food because of its high nutritional value (Dillon et al., 1995).

Spirulina is a super food, full of nutritional wonders, truly an amazing food. Spirulina is a free-floating filamentous microalga growing in alkaline water bodies. With its high nutritional value, spirulina has been consumed as food for centuries in Central Africa.

Fradique et al. (2010) observed that the incorporation of *S. platensis* increased raw pasta firmness and imparted it a stable color. Sensory analysis also showed better acceptance scores. The effect of *S. platensis* enrichment in semolina was observed. Addition of 2 g spirulina in 100 g semolina resulted in higher swelling index, lower cooking loss, and increase in pasta firmness (Zouari et al., 2011). It was discovered that date and spirulina powder-based food tablets are suitable for consumption by those having dysphagia (difficulty in swallowing food), to fulfill the nutritional requirement. Also, these tablets are expected to act as natural and cheap drug delivery carriers (Adiba et al., 2011). It was observed that at lower heating or cooling rates, *S. maxima* gel exerted viscoelastic functions akin to that of pea protein, κ-carrageenan, and starch. This finding may lead to the use of spirulina as thickener in food industry like the above hydrocolloids (Batista et al., 2011). It was observed that *S. platensis* stimulates proliferation of probiotic lactic acid bacteria. The addition of dry algal biomass at 10 mg/mL promoted growth of *Lactobacillus acidophilus* to 186%, suggesting the prebiotic potential of the microalga (Bhowmik et al., 2009). A protective medium with spirulina as an ingredient was optimized for enhancing viability of *Lactobacillus rhamnosus* during lyophilization. It was observed that the algal additive at 1.3%, along with lactulose and sucrose promotes viability of the microbe (Kordowska-wiater et al., 2011). The dermoprotective potential of raw spirulina and its lactic acid bacteria-fermented product was compared. The results showed that though both forms exert skin ameliorative functions; the fermented product performed better in terms of radical scavenging, anti-inflammation, and UV protection. It was inferred that the fermentation process released unidentified polyphenols and converted PC to phycocyanobilin. Based on the results, it was suggested that fermented spirulina can be a potent supplement for skin health (Liu et al., 2011). Spirulina is an incredibly rich source of proteins that could efficiently fight against food deficiency in developing countries (Ravelonandro et al., 2011). The safety profile of spirulina was investigated and its microcystin toxin-free status was suggested. So, it is clear that the long-term dietary supplementation of this alga does not pose any health risks if consumed in moderation (Yang et al., 2011).

In 1967, spirulina was established as a "wonderful future food source" in the International Association of Applied Microbiology (Sasson, 1997; Layam & Reddy, 2010). While no organism fulfilled its promise of cheap protein to combat malnutrition in the underdeveloped and developing countries, spirulina continued to give rise to research and increasing production, reflecting its perceived nutritional assets (Falquet, 2000).

A study conducted by Mahalakshmi (2000) proved the effect of spirulina supplementation on the blood hemoglobin. The participants were supplemented with spirulina 2 g/day for a period of 45 days. The results indicated that there was a significant elevation of the blood hemoglobin level after spirulina supplementation.

Spirulina is a microscopic blue-green aquatic plant and it is the nature's richest and most complete source of organic nutrition. The concentrated nutritional profile of spirulina occurs naturally, so it is ideal for those preferring a whole food supplement to artificial nutrient sources. Spirulina, the blue-green alga, has a unique blend of nutrients that no single source can provide. It contains a wide spectrum of nutrients that include B-complex vitamins, minerals, good-quality proteins, GLA, and the super antioxidants, beta-carotene, vitamin E, and trace elements.

Spirulina is fast emerging as a whole answer to the varied demands due to its impressive nutrient composition which can be used for therapeutic uses (Venkataraman, 1998; Layam and Reddy, 2006).

8.10 FOOD FORTIFICATION WITH SPIRULINA AS A RESPONSE TO CHRONIC MALNUTRITION

In many countries, supplementation policies may have limited effects because populations suffer from several deficiencies. They may also be inefficient if provided out of the "time window" of early brain development. Supplementation policies bringing several nutrients may have a better impact on health. Furthermore, given the impact of infectious diseases on mental and cognitive development of children, it is likely that micronutrient supplementation would further enable children's cognition through a reinforcement of immunity. Finally, the study by Walker et al. (2005) illustrates the need for strengthening physical and psychosocial stimulation of infants. This parameter has even been recently included into recommendations by the WHO (WHO, 2006). From the literature, the combination of enhanced stimulation and appropriate nutrition is expected to have a strong and long-lasting effect on children's cognition and behavior.

In poor settings, complementation of traditional meals with the microalga spirulina, which contains high levels of essential micronutrients such as iron, vitamin A, B_1, and B_2, as well as macronutrients such as essential fatty acids (EFA) and proteins, is a promising source for food fortification. Vitamin A and iron, associated together, may efficiently reduce the incidence of a large number of mental diseases—or disabilities—attributable to nutrient deficiencies. The presence of EFA in spirulina is interesting, although n-3 fatty acids are lacking; other dietary sources of iodine (sea fish, sea algae) or n-3 fatty acids (fish, vegetable oils such as soybean, sesame, canola) need to be made available. Animal sources of nutrients (milk or meat) also contain large amounts of essential nutrients (iron, zinc, proteins) (Ardiet & Von Der Weid, 2006).

The complementation of meals with spirulina could be a solid and cost-effective option to provide to the most vulnerable populations a solid basis of physical and mental health. In one of the projects started by *Antenna Technologies* in R. D. Congo, a mixture of 3 cereals mixed with spirulina, water, and sugar is currently provided to 2500 children suffering from mild or severe malnutrition. This meal, named SOSPISOMA (100 g sorghum, 15 g spirulina, 100 g soya, 200 g corn, sugar, and 1 l of water), is more efficient to rehabilitate children than the meal provided without spirulina complementation (Zaccharie Kasongo, personal communication). In Burkina Faso, Simpore et al. (2006) have shown that daily supplementation with spirulina, added for 8 weeks to MISOLA, another nutritional complement for children widely used in Western Africa, was efficient in rehabilitating undernourished children. Those meals, already without spirulina, have a high nutritional value and are distributed to malnourished children or convalescent people; the combination with spirulina powder is more efficient for rehabilitation. Finally, the experience from our partners on the field is worth to report: S Valerie Kingombe from a dispensary in Goma/Himbi (RDC) reported that "People living with HIV/AIDS are the first beneficiaries of Spirulina because they recovered physical strength due to an enhanced appetite," "several people suffering from diabetes recovered their strength," "a young patient suffering from tuberculosis became more healthy." These are some of the multiple evidences that we regularly collect, and which we aim to further validate in a scientific way.

8.11 USE OF SPIRULINA IN FOOD PROCESSING

The spirulina industry in China is developing rapidly as a national strategic program. Currently, there are more than 80 production factories, with a total annual production of more than 350 t dry powder and total production area of over 106 m^2. Spirulina products are being used as food, forage, and medicine. Besides spirulina pills and capsules, there are also pastries, blocks, and spirulina-containing chocolate bars, marketed as health food. Other spirulina products are formulated for weight loss and as an aid for quitting drug addiction. Cosmetics containing spirulina extracts are also available on the market (Li & Qi, 1997).

Sixteen food formulas were prepared by Sharoba (2014) as complementary food for babies (1–3 years age) by using spirulina at 0%, 2.5%, 0.5%, and 7.5% for the production of two types of baby food one of them is ready to eat by using some fruits and vegetables. Papaya with good nutritional values and cheap price as an essential ingredient of 30% in the four formulas and banana, rich in potassium in four formulas addition to potatoes purée and carrot purée by adding 10% for each and apple purée, guava puree, and mango juice by adding 15% for each been mobilized mixes in jars glass and thermal treatment was carried out at 100°C for 40 min. The other type of baby food-based formulas were produced by using cereals, legumes, and some dried green vegetables, where it was manufactured through eight dried formulas: four of them by 30% wheat, flour 72%; and four others by 30% milled rice in addition to the 30% crushed pearl barley and dryer lentils and dried spinach, dried cauliflower by adding 10% for each formulas. After production, formulas was done were packaged in bottles court lock. Then, evaluation of all formulas microbiologically to study their safety before sensory evaluation was done and was found to be microbiologically safe. Sensory evaluation of produced formulas was acceptable significantly. After that, chosen four formulas containing 5% spirulina based on the results of sensory evaluation, and conducted analysis chemotherapy were selected. The chemical composition indicated that these formulas were suitable as a food supplement for children aged 1–3 years.

A novel pasta product was successfully produced by adding microalgae biomass to semolina flour by Fradique et al. (2010). Pastas prepared with *Chlorella vulgaris* and *S. maxima* presented a chemical composition richer than the control pastas, namely, in protein, total fat, and ash. The cooking quality of pastas was not affected by including microalgae in the fresh pastas.

Some physical properties (hardness, friability, disintegration time, and erosion) of food tablets containing various food powders obtained from dates (*Phoenix dactylifera* L.), spirulina (*Spirulina* sp.), and oranges (juice and zest) were investigated by Adiba et al. (2011). Also, experimental data were related to the release kinetic of PC (antioxidant substance of spirulina). So, the date and spirulina powder-based food tablets could be of various uses: (1) consumption as such by all categories of consumers, (2) feeding of patients for whom it is difficult to chew or swallow food, knowing that these tablets can be either sucked or swallowed, and (3) as natural and cheap drug delivery carriers.

The cassava cake was developed enriching it with a biomass of *S. platensis* and a type of bran made out of its own starch by Navacchi et al. (2012). This biomass, apart from being rich in protein, also contains vitamins, essential fatty acids, and minerals. The produced food was free of gluten and was given to celiac people. In this complex, a solid by-product is generated, which is rich in starch and fibers. Developed energetic food based on cassava lacks protein, but this can be supplied by adding the biomass of *S. platensis*. Different formulations of this cassava cake were developed with varying the concentration of *S. platensis* and cassava bran. The formulation that presented the best features received chocolate before being submitted to sensory tests by children in the public education system. The results showed an excellent acceptance which made viable the development of this product because of aspects like nutrition, technology, and sensorial.

Sharma and Dunkwal (2012) were assessing the nutritional composition of spirulina powder and the development of spirulina-based value-added products and their nutritional composition and shelf life. Value-added biscuits were prepared by using refined wheat flour, sugar powder, ghee, milk, ammonia, baking powder, custard powder, milk powder, vanilla, and pineapple essence and 10% level of spirulina powder. Mean score for overall acceptability of value-added biscuit was 7.5 against the control sample, 7.9 on nine-point hedonic ranking scale. The developed value-added biscuit contained 2.95% moisture, 19.6 g protein, 26.71 g fat, 2.08 g crude fiber, 1.83 g ash, 46.83 g carbohydrate, and 506.11 kcal energy per 100 g on dry weight basis. Spirulina-supplemented biscuits had beta-carotene, vitamin C, iron, and potassium contents in the range of 349.75 g/100 g, 2.75 mg/100 g, 17.62 mg/100 g, and 292 mg/100 g, respectively. Fat acidity revealed satisfactory quality of the value-added biscuit at the end of 3 months of storage period. Thus, spirulina-based value-added products may be beneficial for vulnerable population due its high nutritive value. These would also be advantageous for those who are suffering from degenerative diseases because of its therapeutic properties.

In recent years, several products prepared from spirulina microalgae biomass have been developed, rich in carotenoid and polyunsaturated fatty acids, namely, emulsions colored with natural pigments and microalgae biomass (Fradique et al., 2010), gelled desserts (Gouveia et al., 2008), and biscuits colored with microalgae biomass and enriched with polyunsaturated fatty acids (Gouveia et al., 2008).

Spirulina has found wide applications in the areas of agriculture, food, pharmaceuticals, perfumeries, medicine, and science. Nowadays, this organism is used as a food supplement and is marketed in the form of pills, capsules, and powder or incorporated into various types of food-like cakes, biscuits, noodles, and health drinks, and others. Various countries are developing strategic programs for the production and use of spirulina.

In the world, the spirulina industry is developing rapidly with several factories producing hundreds of tons of spirulina in the form of dried powder that is used as food, fodder, and medicine (Desai & Sivakami, 2004).

As functional food, additive and prebiotic, Spirulina has earned scientific validation regarding its role as healthy food component. This nutritious and easily digestible food can be consumed in several forms. It was demonstrated that the whole spirulina or its PC-rich fraction could be a suitable functional ingredient in soy milk, fruit juices, and whole fruits (McCarty, 2007). It was suggested that the ingestion of cocoa and spirulina powder mix can promote antioxidant status and vascular health (McCarty et al., 2010). The flavonol-rich cocoa and PC-rich spirulina are assumed to work in synergy to increase the endothelial production of nitric oxide and act as a potent inhibitor of NADPH oxidase. Spirulina is expected to enhance the nutritional content of conventional foods when incorporated as colorant, texturizing agent, gelling agent, and prebiotic. The pigments PC and allophycocyanin are used in the food and beverage industry as a natural colorant. This blue colorant finds use in ice cream, sweets, chewing gum, candy, jelly, cake decorations as well as soft drinks, alcoholic drinks (Liu et al., 2012; Li and Qi, 1997).

At present, there are two categories of spirulina food. The first is the pills and capsules made from dry spirulina. Food containing spirulina and other ingredients belongs to the other category. Examples are instant noodles, stylish noodles, nutritious blocks, beverages and cookies. The first three food items are recommended luncheon food for middle-grade school students by the State Commission of Education Committee.

8.12 BIOLOGICAL EVALUATION OF SPIRULINA

In a study by Mitchell et al. (1990), *S. maxima* adversely affected the utilization of vitamin E (reduced plasma and liver levels of α-tocoferol) when fed to male rats at a level as low as 2.7%; plasma levels of retinol were also decreased, whereas liver retinoid levels were increased.

8.13 BIOCHEMICAL PARAMETERS EVALUATION OF SPIRULINA

Vijayalakshmi and Rema (1994) determined the biochemical profile of 50 normal adult men and women volunteers (20–39 years) in Coimbatore city. Results revealed that, the mean levels of hemoglobin (men: 14.8 g%, women: 12.2 g%), blood glucose (men: 98.5 mg/dL, women: 95.0 mg/dL), serum cholesterol (men: 176.7 mg/dL, women: 170.6 mg/dL), serum iron (men: 117 mcg/dL, women 107.8 mcg/dL), serum protein (men: 6.95 g/dl, women: 6.78 g/dL), serum retinol (men: 40–50 mcg/dL, women: 38.28 mcg/dL), and serum calcium (men: 10.36 mg/dL, women: 10.69 mg/dL) were comparable with the western standards.

Parikh et al. (2001) studied the effect of spirulina supplementation at 2 g/day dose for 2 months on blood glucose levels, HbA1C, and lipid profile of 25 diabetic type I subjects. They found a reduction of fasting, postprandial blood glucose, and HbA1C level. Change in lipid profile observed demonstrated a reduction of total cholesterol.

Anuradha and Vidhya (2001) studied the impact of administration of spirulina (4 g/day) for 60 days on blood glucose levels of selected diabetic patients. The result proved that spirulina has hypoglycemic effect on non-insulin-dependent diabetic patients. Significant decrease in body weight was also observed.

During the 2 weeks of spirulina intake in the study by Milasius et al. (2009), the physical development of the sportsmen showed no statistically significant changes. Immediately following the 14-day period of spirulina administration, the sportsmen's blood erythrocyte count and hemoglobin concentration had a tendency to increase, accompanied by a decrease of the mean erythrocyte volume and unchanged blood hematocrit. Leukocyte count also showed an increasing tendency, and the percent ratio of agranulocytes and granulocytes was leveling off. Spirulina intake caused no violation of the recommended standard levels of blood biochemical indices. The levels of cholesterol, triglycerides, and bilirubin in the blood showed a decreasing and urea and uric acid an increasing tendency. Most of the positively changed morphological and biochemical indices of the sportsmen's blood composition retained similar levels for another 2 weeks following the withdrawal of spirulina from their diet. The spirulina food supplement, based on microalgae, exerts a positive effect on the quantitative indices of immune response.

An experimental study was conducted by Ramesh et al. (2013) to study the effect of spirulina on anthropometric parameters and the biochemical parameters before and after its use as nutritional supplement in school children. The study was conducted in a residential girl's school in the age group between 11 and 13 years. Three capsules of spirulina were given during dinner time for 3 months. At the initial survey and at the end of 6th month the anthropometric and biochemical findings were recorded. Blood samples were taken to analyze hemoglobin (Hb), serum ferrtin, serum zinc, serum protein, and serum albumin levels at 0, 3, and 6 months. There was a significant increase in anthropometric measurements and hemoglobin, serum ferrtin, serum zinc, serum protein, and serum albumin levels in the study sample after 6 months. Finally, the anthropometric and biochemical parameters improved after the use of spirulina.

8.14 HISTOLOGY EVALUATION OF SPIRULINA

The study by Upasani and Balaraman (2003) investigated the protective effect of spirulina on lead-induced changes in the levels of lipid peroxidation and endogenous antioxidants in liver, lung, heart, kidney, and brain of rats. Levels of elemental lead were also measured in the organs of rats in all experimental groups. In the liver, lung, heart, and kidney of lead-exposed animals, there was a significant ($p < 0.001$) increase in the lipid peroxidation and a decrease in the levels of endogenous antioxidants. Although spirulina did not affect the deposition of lead in organs apart from the brain, simultaneous administration of spirulina to lead-exposed animals significantly ($p < 0.001$) inhibited lipid peroxidation and restored the levels of endogenous antioxidants to normal. To conclude, spirulina had a significant effect on scavenging free radicals, thereby protecting the organs from damage caused by the exposure to lead. Furthermore, spirulina showed a significant ($p < 0.05$) decrease in the deposition of lead in the brain.

KEYWORDS

- ***Spirulina***
- **amino acids**
- **white blood cells**
- **fatty acid**
- **diabetes mellitus**

REFERENCES

Adiba, B. D.; Salem, B.; Nabil, S.; Abdelhakim, M. Preliminary Characterization of Food Tablets from Date (*Phoenix dactylifera* L.) and Spirulina (*Spirulina* sp.) Powders. *Powder Technol.* **2011,** 208, 725–730.

Anuradha, V.; Vidhya, D. Impact of Administration of Spirulina on the Blood Glucose Levels of Selected Diabetic Patients. *Indian J. Nutr. Diet.* **2001,** *38*, 40–44.

Ardiet D.-L.; Von Der Weid D. *Spirulina as a Food Complement to Improve Health and Cognitive Development*. Nutrition and Cognitive Development, Antenna Technologies: Geneva, Switzerland, 2006. www.antenna.ch.

Babu, M. Evaluation of Chemoprevention of Oral Cancer with *Spirulina fusiformis*. *Nutr. Cancer* **1995,** *24*, 197–202.

Batista, A. P.; Nunes, M. C.; Fradinho, P.; Gouveia, L.; Sousa, I.; Raymundo, A.; Franco, J. M. Novel Foods with Microalgal Ingredients—Effect of Gel Setting Conditions on the Linear Viscoelasticity of *Spirulina* and *Haematococcus* Gels. *J. Food Eng.* **2011,** *110*, 182–189.

Bhowmik, D.; Dubey, J.; Mehra, S. Probiotic Efficiency of *Spirulina platensis*—Stimulating Growth of Lactic Acid Bacteria. *World J. Dairy Food Sci.* **2009,** *4*, 160–163.

Bucaille, P. *Effectiveness of Spirulina Algae as Food for Children with Protein–Energy Malnutrition in a Tropical Environment*. University Paul Sabatier: Toulouse, France, 1990.

Desai, K.; Sivakami S. Spirulina: The Wonder Food of the 21st Century. *APBN* **2004,** *8*(23), 1298–1302. www.asiabiotech.com Clinical Application. www.biol.tsukuba.ac.jp/~inouge/ino/cy/spirulina.gif.

Dillon, J. C.; Phuc, A. P.; Dubach, J. P. Nutritional Value of the Alga Spirulina. *World Rev. Nutr. Dietetics* **1995,** *77*, 32–46.

Dolly W. D. *Biotechnological Potentials and Role of Cyanobacteria in Agriculture and Industry*. Division of Microbiology, Indian Agricultural Research Institute: New Delhi, India, 2014.

Falquet, J. *A Sustainable Response to Malnutrition in Hot Regions: The Local Production of Spirulina*. Antenna Technologies: Geneva, 2000.

Ferreira-Hermosillo, A.; Torres-Duran, P. V.; Juarez-Oropeza, M. A. Hepatoprotective Effects of *Spirulina maxima* in Patients with Non-alcoholic Fatty Liver Disease: A Case Series. *J Med Case Rep.* **2010,** *4*, 103.

Fradique, M.; Batista, A. P.; Nunes, M. C.; Gouveia, L.; Bandarra, N. M.; Raymundo, A. Incorporation of *Chlorella vulgaris* and *Spirulina maxima* biomass in Pasta Products. Part 1: Preparation and Evaluation. *J. Sci. Food Agric.* **2010,** *90*, 1656–1664.

Gonzalez de Rivera, C.; Miranda-Zamora, R.; Diaz-Zagoya, J. C.; Juarez-Oropeza, M. A. Preventive Effect of *Spirulina maxima* on the Fatty Liver Induced by a Fructose-Rich Diet in the Rat, A Preliminary Report. *Life Sci.* **1993,** *53*(1) 57–61.

Gouveia, L.; Batista, A. P.; Raymundo, A.; Bandarra, N. M. *Spirulina maxima* and *Diacronema vlkianum* Microalgae in Vegetable Gelled Desserts. *Nutr. Food Sci.* **2008,** *38*, 492–501.

Gupta, S.; Hrishikeshvan, H. J.; Sehajpal, P. K. Spirulina Protects against Rosiglitazone Induced Osteoporosis in Insulin Resistance Rats. *Diabetes Res. Clin. Pract.* **2010,** *87*, 38–43.

Hayashi, H.; Hayashi, T.; Morita, N. An Extract from *Spirulina platensis* is a Selective Inhibitor of Herpes Simplex Virus type 1 Penetration into HeLa Cells. *Phytother. Res.* **1993,** *7*, 76–80.

Hayashi, T.; Hayashi, K.; Maeda, M.; Kojima, I. Calcium Spirulan, an Inhibitor of Enveloped Virus Replication, from a Blue Green Algae *Spirulina platensis*. *J. Nat. Prod.* **1996,** *59*, 83–87.

Hernandez-Corona, A.; Nieves, I.; Meckes, M.; Chamorro, G.; Barron, B. L. Antiviral Activity of *Spirulina maxima* against Herpes Simplex Virus Type 2. *Antiviral Res.* **2002,** *56*(3), 279–285.

Huang, Z.; Guo, B. J.; Wong, R. N. S.; Jiang, Y. Characterization and Antioxidant Activity of Selenium Containing Phycocyanin Isolated from *Spirulina platensis*. *Food Chem.* **2007,** *100*, 1137–1143.

Jacques, S.; Frederic, Z.; Youssouf, O.; Fatoumata, K.; Deleli, D.; Augustin, B.; Jean-Baptiste, N.; Salvatore, M. Nutrition Rehabilitation of the HIV-infected and Negative Under Nourished Children Utilizing Spirulina. *Pak. J. Biol. Sci.* **2005,** *8*(4), 589–595.

Jeniferpricillia, S. S. *A Study to Assess the Effectiveness of Spirulina Upon Stress and Anemia Among Late Adolescent Girls at E.T.C.M. College of Nursing, Kolar. M.Sc.* E.T.C.M. College of Nursing, Rajiv Gandhi University of Health Sciences: Bangalore, Karnataka, 2010.

Juarez-Oropeza, M. A.; Mascher, D.; Torres-Duran, P. V.; Farias, J. M.; Paredes-Carbajal, M. C. Effects of Dietary Spirulina on Vascular Reactivity. *J. Med. Food* **2009,** *12*(1), 15–20.

Juvekar, A. B. Effect of Spirulina against Cold Stress. *Nutr. Metab.* **2001,** *27*, 212–215.

Kamalpreet, K.; Rajbir, S.; Kiran, G. Effect of Supplementation of Spirulina on Blood Glucose and Lipid Profile of the Non-insulin Dependent Diabetic Male Subjects. *J. Dairying, Foods Home Sci.* **2008,** *27*, 3–4.

Kapoor, R.; Mehta, U. Effect of Supplementation of Blue Green Algae on Outcome of Pregnancy of Rats. *Plants Food Hum. Nutr.* **1993,** 43, 131–148.

Kauser, F.; Praveen, S. Effect of Spirulina as a Nutritional Supplement on Malnourished Children. *Indian J. Nutr. Dietetics*. **1999,** *38*, 269–271.

Khan, Z.; Bhadouria, P.; Bisen, P. S. Nutritional and Therapeutic Potential of Spirulina. *Curr. Pharm. Biotechnol.* **2005,** *6*, 373–379.

Kim, M. H.; Kim, W. Y. The Change of Lipid Metabolism and Immune Function Caused by Antioxidant Material in the Hypercholesterolemin Elderly Women in Korea. *Korean J. Nutr.* **2005,** *38*, 67–75.

Kordowska-wiater, M.; Wasko, A.; Polak-Berecka, M.; Kubik-Komar, A.; Targonski, Z. Spirulina Enhances the Viability of Lactobacillus Rhamnosus E/N after Freeze-Drying in a Protective Medium of Sucrose and Lactulose. *Lett. Appl. Microbiol.* **2011,** *53*, 79–83.

Kumari, D.; Jalaja, B. B.; Jaffar S. K.; Prasad M. G.; Ibrahim M. D.; Siddque A. K. M. D. Potential Health Benefits of *Spirulina platensis*. *Pharmanest* **2011,** *2*(5–6), 417–422.

Layam, A.; Reddy, C. L. K. Antidiabetic Property of Spirulina. *Diabet. Croat.* **2006,** *35*(2), 29–33.

Layam, A.; Reddy, C. L. K. Effect of Supplementation of Spirulina on Blood Glucose, Glycosylated Hemoglobin and Lipid Profile of Male Non-insulin Dependent Diabetics. *Asian J. Exp. Biol. Sci.* **2010,** *1*(1), 36–46.

Lee, E. H.; Park, J. E.; Choi, Y. J.; Huh, K. B.; Kim, W. Y. A Randomized Study to Establish the Effects of Spirulina in Type 2 Diabetes Mellitus Patients. *Nutr. Res. Pract.* **2008,** *2*(4), 295–300.

Li D.-M.; Qi Y.-Z. Spirulina Industry in China: Present Status and Future Prospects. *J. Appl. Phycol.* **1997,** *9*(1) 25–28.

Li, Z. Y.; Guo, S. Y.; Li, L. Bioeffects of Selenite on the Growth of *Spirulina platensis* and its Biotransformation. *Bioresour. Technol.* **2003,** *89*, 171–176.

Li-Kun, H.; Dong-Xia, L.; Xiao-Jie, G.; Yasumasa, K.; Isao, S.; Hiromichi, O. Isolation of Pancreatic Lipase Activity—Inhibitory Component of *Spirulina platensis* and It Reduce Postprandinal Triacylglycerolemia. *Yakugaku Zasshi* **2006**, *126*(1), 43–49.

Lisheng, L. Inhibitive Effective and Mechanism of Polysacharide of *Spirulina platensis* on Transplanted Tumor Cells in Mice. *Marine Sci. Qindao China* **1991,** *5*, 33–38.

Liu, J. -G.; Hou, C.-W.; Lee, S.-Y.; Chuang, Y.; Lin C.-C. Antioxidant Effects and UVB Protective Activity of Spirulina (*Arthrospira platensis*) Products Fermented with Lactic Acid Bacteria. *Process Biochem.* **2011,** *46*, 1405–1410.

Liu, Y.; Feng, Y.; Lun, J. Aqueous Two-Phase Countercurrent Distribution for the Separation of C-Phycocyanin and Allophycocyanin from *Spirulina platensis*. *Food Bioprod. Process.* **2012,** *90*, 111–117.

Mahalakshmi, B. Effect of Spirulina Supplementation on the Blood Hemoglobin. *J. Nutr.* **2000,** *168*(4), 312–315.

Mala, R.; Sarojini, M.; Saravanababu, S.; Umadevi, G. Screening for Antimicrobial Activity of Crude Extracts of *Spirulina platensis*. *J. Cell Tissue Res.* **2009,** *9*, 1951–1955.

Mani, U. V.; Desai, S.; Iyer, U. Studies on the Long-Term Effect of Spirulina Supplementation on Serum Lipid Profile and Glycated Proteins in Non-insulin Dependent Diabetics Patients. *J. Nutraceut., Funct. Med. Foods* **2000,** *2*(3), 25–32.

Marcel, A. K.; Ekali, L. G.; Eugene, S.; Arnold, O. E.; Sandrine, E. D.; von der Weid, D.; Gbaguidi, E.; Ngogang, J.; Mbanya, J. C. The Effect of *Spirulina platensis* versus Soybean on Insulin Resistance in HIV-infected Patients: A Randomized Pilot Study. *Nutrients* **2011,** *3*, 712–724.

Martelli, G.; Folli, C.; Visai, L.; Daglia, M.; Ferrari, D. Thermal Stability Improvement of Blue Colorant C-Phycocyanin from *Spirulina platensis* for Food Industry Applications. *Process Biochem.* **2014,** *49*, 154–159.

Mascher, D.; Paredes-Carbajal, M. C.; Torres-Durán, P. V.; Zamora-González, J.; Díaz-Zagoya, J. C.; Juárez-Oropeza, M. A. Ethanolic Extract of *Spirulina maxima* Alters the Vasomotor Reactivity of Aortic Rings from Obese Rats. *Arch. Med. Res.* **2006,** *37*(1) 50–57.

McCarty, M. F. Clinical Potential of Spirulina as a Source of Phycocyanobilin. *J. Med. Food* **2007,** *10*(4), 566–570.

McCarty, M. F.; Barroso-Aranda, J.; Contreras, F. Potential Complementarity of High-Flavanol Cocoa Powder and Spirulina for Health Protection. *Med. Hypotheses* **2010,** *74*, 370–373.

Mendiola, J. A.; Jaime, L.; Santoyo, S.; Reglero, G.; Cifuentes, A.; Ibanez, E.; Senorans, F. J. Screening of Functional Compounds in Supercritical Fluid Extracts from *Spirulina platensis*. *Food Chem.* **2007,** *102*, 1357–1367.

Milasius, K.; Malickaite, R.; Dadeliene, R. Effect of Spirulina Food Supplement on Blood Morphological Parameters, Biochemical Composition and on the Immune Function of Sportsmen. *Biol. Sport* **2009,** *26*(2), 157–172.

Mitchell, G. V.; Grundel, E.; Jenkins, M.; Blakely, S. R. Effects of Graded Dietary Levels of *Spirulina maxima* on Vitamins A and E in Male Rats. *J. Nutr.* **1990,** *120*, 1235–1240.

Nagaoka, S.; Shimizu, K.; Kaneko, H.; Shibayama, F.; Morikawa, K.; Kanamaru, Y.; Otsuka, A.; Hirahashi, T.; Kato, T. A Novel Protein C-phycocyanin Plays a Crucial Role in the Hypocholesterolemic Action of *Spirulina platensis* Concentrate in Rats. *J. Nutr.* **2005,** *135*(10), 2425–2430.

Nah, W. H.; Koh, I. K.; Ahn, H. S.; Kim, M. J.; Kang, H.-G.; Jun, J. H.; Gye, M. C. Effect of *Spirulina maxima* on Spermatogenesis and Steroidogenesis in Streptozotocin-Induced Type I Diabetic Male Rats. *Food Chem.* **2012,** *134*, 173–179.

Nakaya, N.; Homma, Y.; Goto, Y. Cholesterol Lowering Effect of Spirulina. *Nutr. Rep. Int.* **1988,** *37*(6), 1329–1337.

Navacchi, M. F. P.; Monteiro de Carvalho, J. C.; Takeuchi, K. P.; Danesi, E. D. G. Development of Cassava Cake Enriched with its Own Bran and *Spirulina platensis*. *Acta Sci. Technol. (Maringá)* **2012,** *34*(4), 465–472.

Ozdemir, G.; Karabay, N. U.; Dalay, M. C.; Pazarbasi, B. Antibacterial Activity of Volatile Component and Various Extracts of *Spirulina platensis*. *Phytother. Res.* **2004,** *18*, 754–757.

Paredes-Carbajal, M. C.; Torres-Duran, P. V.; Rivas-Arancibia, S.; Zamora-Gonzalez, J.; Mascher, D.; Juarez-Oropeza, M. A. Effects of Dietary *Spirulina maxima* on Vasomotor Responses of Aorta Rings from Rats Fed a Fructose—Rich Diet. *Nutr. Res.* **1998,** *18*(10), 1769–1782.

Parikh, P.; Mani, U.; Iyer, U. Role of Spirulina in the Control of Glycemia and Lipidemia in Type 2 Diabetes Mellitus. *J. Med. Food* **2001,** *4*(4), 193–199.

Park, H. J.; Lee, Y. J.; Ryu, H. K.; Kim, M. H.; Chung, H. W.; Kim, W. Y. A Randomized Double-Blind, Placebo-Controlled Study to Establish the Effects of Spirulina in Elderly Koreans. *Ann. Nutr. Metab.* **2008,** *52*(4), 322–328.

Park, J. Y.; Kim, W. Y. The Effect of Spirulina on Lipid Metabolism, Antioxidant Capacity and Immune Function in Korean Elderly. *Korean J. Nutr.* **2003**, 36, 287–297.

Park, K. *Preventive and Social Medicine*, 18th ed. Banarsidhas Bhanot Publishers: Jabalpur, 2005; p 715.

Parry, E. I. D. (India) Limited. *Spirulina for Children*. Parry Nutraceuticals Division: Dare House, Chennai, India, 2014. www.parrynutraceuticals.com.

Patel, S.; Goyal, A. Current and Prospective Insights on Food and Pharmaceutical Applications of Spirulina. *Curr. Trends Biotechnol. Pharm.* **2013,** *7*(2) 696–707, ISSN 0973-8916 (Print), 2230–7303 (Online).

Pineiro Estrada, J. E.; Bermejo Besco, S. P.; Villar del Fresno, A. M. Antioxidant Activity of Different Fractions of *Spirulina platensis* Protean Extract. *Farmaco* **2001**, *59*, 497–500.

Ponce-Canchihuaman, J. C.; Perez-Mendez, O.; Hernandez-Munoz, R.; Torres-Duran, P. V.; Juarez-Oropeza, M. A. Protective Effects of *Spirulina maxima* on Hyperlipidemia and Oxidative-Stress Induced by Lead Acetate in the Liver and Kidney. *Lipids Health Dis.* **2010,** *9*, 35.

Ramadan, M. F.; Asker, M. M. S.; Ibrahim, Z. K. Functional Bioactive Compounds and Biological Activities of *Spirulina platensis* Lipids. *Czech J. Food Sci.* **2008,** *26*(3), 211–222.

RamamoorthyRamamurthy, A.; Premakumari, S. Effect of Supplementation of Spirulina on Hypercholesterolemic Patients. *J. Food Sci. Technol.* **1996,** *33*(2), 124–128.

Ramesh, S.; Manivasgam, M.; Sethupathy, S.; Shantha, K. Effect of Spirulina on Anthropometry and Bio-Chemical Parameters in School Children. *IOSR J. Dental Med. Sci.* **2013,** *7*(5), 11–15. (IOSR-JDMS) eISSN: 2279-0853, p-ISSN: 2279–0861.

Ravelonandro, P. H.; Ratianarivo, D. H.; Joannis-Cassan, C.; Isambert, A.; Raherimandimby, M. Improvement of the Growth of *Arthrospira* (Spirulina) *platensis* from Toliara (Madagascar): Effect of Agitation, Salinity and CO_2 Addition. *Food Bioprod. Process.* 2011, 89, 209–216.

Reddy, C. M.; Bhat, V. B.; Kiranmai, G.; Reddy, M. N. Selective Inhibition of Cyclooxygenase-2 by C-phycocyanin, a Biliprotein from *Spirulina platensis*. *Biochem. Biophys. Res. Commun.* **2000,** *277*(3), 599–603.

Samuels, R.; Mani, U. V.; Iyer, U. M.; Nayak, U. S. Hypocholesterolemic Effect of Spirulina in Patients with Hyperlipidemic Nephrotic Syndrome. *J. Med. Food* **2002,** (5), 91–96.

Sasson, A. *Micro-Biotechnologies: Recent Developments and Prospects for Developing Countries*. United Nations Educational, Scientific and Cultural Organization (UNESCO), BIOTEC Publication 1/2542: Place de Fontenoy: Paris, France, 1997, pp 11–31.

Schwartz, J.; Shklar, G.; Reid, S.; Trickler, D. Prevention of Experimental Oral Cancer by Extracts of Spirulina—Duraliella Algae. *Nutr. Cancer* **1988,** *11*(2), 127–134.

Selmi, C.; Leung, P. S.; Fischer, L.; German, B.; Yang, C. Y.; Kenny, T. P.; Cysewski, G. R.; Gershwin, M. E. The Effects of Spirulina on Anemia and Immune Function in Senior Citizens. *Cell. Molecular Immunol.* **2011,** *8*, 248–254.

Seshadri, C. V. Large Scale Nutritional Supplementation with Spirulina Alga. All India Coordinated Project on Spirulina. Shri Amma Murugappa Chettiar Research Center (MCRC): Madras, India, 1993.

Sharma V.; Dunkwal V. Development of Spirulina Based "Biscuits": A Potential Method of Value Addition. *Ethno. Med.* **2012**, *6*(1), 31–34.

Sharoba, A. M. Nutritional Value of Spirulina and Its Use in the Preparation of Some Complementary Baby Food Formulas. *J. Agroalimentary Processes Technol.* **2014,** *20*(4), 330–350, Available online at http://journal-of-agroalimentary.ro.

Simpore, J.; Kabore, F.; Zongo, F.; Dansou, D.; Bere, A.; Pignatelli, S.; Biondi, D.; Ruberto, G.; Musumeci, S. Nutrition Rehabilitation of Undernourished Children Utilizing Spiruline and Misola. *Nutr. J.* **2006,** *5*(3), 1–7.

Simsek, N.; Karadeniz, A.; Karaca, T. Effects of the *Spirulina platensis* and *Panax ginseng* Oral Supplementation on Peripheral Blood Cells in Rats. *Rev. Med. Vet.* **2007,** *158*, 483–488.

Sindhu, V. Effect of Supplementation of Spirulina on Serum Betacarotene Levels of Preschool Children. M. Sc. Thesis, Nagarjuna University, Andhra Pradesh, 1999.

Singh, C. S.; Sinha, P. R.; Hader, P. D. Role of Lipids and Fatty Acids in Stress Tolerance in Cyanobacteria. *Acta Protozool.* **2002,** *41*, 297–308.

Smitha, M. K. Effect of Spirulina on Lipid Profile of Hyperlipidemics. Master of Home Science in Food Science and Nutrition. Department of Food Science and Nutrition, College of Rural Home Science, University of Agricultural Sciences, Dharwad, 2006.

Stoltzfus, R. J. Iron Deficiency: Global Prevalence and Consequences. *Food Nutr. Bull.* **2003,** *24*(4 suppl. 2), S99–S103.

Thirumani, D. A.; Uma, K. R. Effect of Supplementation of Spirulina on Anemic Adolescent Girls. *Indian J. Nutr.* **2005,** *42*, 534.

Thomas, S. S. *The Role of Parry Organic Spirulinain Health Management*. Parry Nutraceuticals, Division of EID Parry (India) Ltd: Chennai, India, 2010, www.parrynutraceuticals.com.

Torres-Duran, P. V.; Ferreira-Hermosillo, A.; Juarez-Oropeza, M. A. Antihyperlipemic and Antihypertensive Effects of *Spirulina maxima* in an Open Sample of Mexican Population: A Preliminary Report. *Lipids Health Dis.* **2007,** *6*(1), 33–41.

Upasani, C. D.; Balaraman, R. Protective Effect of Spirulina on Lead Induced Deleterious Changes in the Lipid Peroxidation and Endogenous Antioxidants in Rats. *Phytother. Res.* **2003,** *17*, 330–334.

Venkataraman, L. V. Spirulina: Global Reach of a Health Care product. Souvenir, IFCON 98, 4th International Food Convention, 1998, 175.

Vijayalakshmi, P.; Rema, S. R. Biochemical Profile of Selected Normal Healthy Adult Men and Women. *Indian J. Nutr. Dietetics* **1994,** *31*(10), 277–284.

Vilasini, B.. The Effect of Supplementation of Spirulina and Iron Tablets on Anemic Adult Working Women Belonging to Lower Income Group. *Indian J. Nutr. Dietetics* **2001,** *22*; 234.

Walker, S. P.; Chang, S. M.; Powell, C. A.; Grantham-McGregor, S. M. Effects of Early Childhood Psychosocial Stimulation and Nutritional Supplementation on Cognition and Education in Growth-Stunted Jamaican Children: A Prospective Study. *Lancet* **2005,** *366*, 1804–1807.

Wang, Y.; Chang, C. F.; Chou, J.; Chen, H. L.; Deng, X.; Harvey, B. K.; Cadet, J. L.; Bickford, P. C. Dietary Supplementation with Blueberries, Spinach or Spirulina Reduces Ischemic Brain Damage. *J. Exp. Neurol.* **2005,** *193*(1), 75–84.

WHO. *Iron/Multi-micronutrient Supplements for Young Children. Summary of UNICEF/ WHO Consultation, Copenhagen, August, 1996*. ILSI Press: Washington, DC, USA. 1997.

WHO. *Mental and Severe Food Shortage Situations: Psychological Considerations*. 2006. Available online at www.who.int.

WHO. www.who.com, 2014.

Xue, C.; Hu, Y.; Saito, H.; Zhang, Z.; Li, Z.; Cai, Y.; Ou, C.; Lin, H.; Imbs, A. B. Molecular Species Composition of Glycolipids from *Sprirulina platensis*. *Food Chem.* **2002,** *77*, 9–13.

Yamani, E.; Kaba-Mebri, J.; Mouala, C.; Gresenguet, G.; Rey, J. L. Use of Spirulina Supplement for Nutritional Management of HIV-infected Patients: Study in Bangui, Central African Republic. *Med. Trop. (Mars.)* **2009,** *69*, 66–70.

Yang, Y.; Park, Y.; Cassada, D. A.; Snow, D. D.; Rogers, D. G.; Lee, J. In Vitro and In Vivo Safety Assessment of Edible Blue-Green Algae, *Nostoc commune* var. *Spharoides kutzing* and *Spirulina platensis*. *Food Chem. Toxicol.* **2011,** *49*, 1560–1564.

Zouari, N.; Abid, M.; Fakhfakh, N.; Ayadi, M. A.; Zorgui, L.; Ayadi, M.; Attia, H. Blue-Green Algae (*Arthrospira platensis*) as an Ingredient in Pasta: Free Radical Scavenging Activity, Sensory and Cooking Characteristics Evaluation. *Int. J. Food Sci. Nutr.* **2011,** *62*, 811–813.

INDEX

B

I

N

Q

R

S

T

For Product Safety Concerns and Information please contact our EU representative GPSR@taylorandfrancis.com Taylor & Francis Verlag GmbH, Kaufingerstraße 24, 80331 München, Germany